Estuarine and Marine Bivalve Mollusk Culture

Editor
Winston Menzel, Ph.D.
(Deceased)
Service Professor
Department of Oceanography
Florida State University
Tallahassee, Florida

CRC Press, Inc.
Boca Raton Ann Arbor Boston

Library of Congress Cataloging-in-Publication Data

Estuarine and marine bivalve mollusk culture / editor, Winston Menzel.
p. cm.
Includes bibliographical references.
ISBN 0-8493-4936-2
1. Shellfish culture. I. Menzel, Winston, 1920-1989.
SH370.E88 1990
639.4—dc20

89-48813
CIP

This book represents information obtained from authentic and highly regarded sources. Reprinted material is quoted with permission, and sources are indicated. A wide variety of references are listed. Every reasonable effort has been made to give reliable data and information, but the author and the publisher cannot assume responsibility for the validity of all materials or for the consequences of their use.

Direct all inquiries to CRC Press, Inc., 2000 Corporate Blvd., N.W., Boca Raton, Florida 33431.

International Standard Book Number 0-8493-4936-2
Library of Congress Card Number 89-48813
Printed in the United States

IN MEMORIAM

Robert Winston Menzel, Sr.
1920—1989

With the passing of Dr. Robert Winston Menzel, Sr. on June 11, 1989, those of us who had the pleasure of knowing him lost a treasured friend, one who left an indelible impression on us all. In estuarine and marine biology, higher education lost one of its most productive professors as evidenced by the 40 Masters and 11 Ph.D. degree recipients who graduated under his guidance. In addition, fisheries science lost one of its major researchers as evidenced by the 54 papers he authored during his career.

Born in James City County, Virginia on January 29, 1920, Dr. Menzel grew up in a family which commercially fished the Chickahominy River near Toano, Virginia. There he developed both his interest in aquatic natural resources and his appreciation for the needs of commercial fishermen.

In 1940 he received a B.S. degree in Botany at the College of William and Mary. He followed that accomplishment in 1943 with an M.A. degree in Aquatic Biology from the College as part of its cooperative program with the Virginia Fisheries Laboratory, later to become the School of Marine Science/Virginia Institute of Marine Science of the College. This was the first degree awarded from the Institute; his thesis was entitled, "Notes on the Biology and Commercial Fishery of the Catfish of the James River, Virginia." Fittingly, he made contributions to knowledge of fish which his family harvested commercially.

After a short time in graduate studies at the University of Virginia, Dr. Menzel went to College Station, Texas where, in 1954, he obtained his Ph.D. in Biology at Texas A&M University with a dissertation entitled, "Some Phases of the Biology of *Ostrea equestris* Say, and a Comparison with *Crassostrea virginica* (Gmelin)." From that time on, his research interests were strongly oriented toward bivalve molluscs and, in particular, oysters and clams.

Although he is associated with a specialty in which he made significant contributions, he is remembered also for being an excellent generalist in marine biology. Those of us who were privileged to be his students were impressed with his breadth of knowledge about organisms and ecosystems of the coastal and estuarine waters. As part of my Ph.D. degree program, I worked on a disease of oysters under his guidance and had the advantage of acquiring knowledge from him about the host and parasite as related to basic organismic biology and ecology. When we students needed advice from other biologists, he was able to introduce us personally to researchers in many disciplines, from fishery science to mycology to protozoology. All of these contributions as a major professor were made in an unpretentious and even humble manner. His students were never made to feel intimidated. Because of his influence I was led to believe that a student of nature should never become pompous and overly impressed with his knowledge. The complexities of the natural world and life sciences do not permit pomposity in the true scientist.

Dr. Menzel will be remembered by those of us who knew him as a gentleman and scholar. We truly miss him and feel regrets for those who will not have the chance to know him.

Dr. Frank Perkins
October, 1990

PREFACE

Bivalve mollusks have long been used by mankind for food and as food for domesticated animals. The shells are used for construction, as food supplements for certain domesticated animals, for certain chemical manufacturing, and for conditioning the pH of farm lands. In addition, ornaments and jewelry are made from shells and, of course, pearls.

Bivalves are filter feeders and efficiently collect diffuse plankton (mainly phytoplankton) and convert these into nutritious and palatable animal meat. Since they are mainly primary consumers, the loss of energy, when several food levels occur, is avoided.

An attempt is made to include all worldwide important species of bivalve mollusks occurring in marine and estuarine habitats into one volume. Unfortunately, several areas are not included. Scallops and clams are cultured in Europe and there is some culture in South Africa (e.g., *Crassostrea gigas* Thunberg). The northwest coast of the U.S. and British Columbia, Canada are not included. The pressure of other duties prevented the contributors from these areas from furnishing manuscripts. This is an important region, producing considerable tonnage of oysters as well as clams and scallops. It is in this region that commercial applications of new experimental techniques have been undertaken in recent years. These include the extensive use of hatcheries to supply seed, especially for the exotic Pacific or Japanese oyster. Also, the use of induced triploid oysters is beginning to be exploited. Triploid oysters are not as fertile as the normal diploid, and thus the energy used in production of gametes is channeled into continued growth. In addition, they maintain their meat weight during the spawning season when spawned-out oysters may decrease as much as one-half. Hatcheries have now developed techniques of shipping reared, ready-to-attach larvae to buyers (growers), that avoids the transportation and handling costs of seed attached to cultch material or even the less bulky cultchless seed.

All the contributors have followed a more or less general outline. The inclusion or omission of data is the responsibility of each contributor. The use of the metric system for weights and measures is used throughout the manuscripts. In some instances, the editor has changed the classification of genera, especially for oysters, to conform to more recent classification. Avoidance of duplication was not always possible because some species, e.g., the Japanese oyster, have been introduced to many areas throughout the world and the biology and culture practices are similar. Also some species, e.g., the blue mussel, occur naturally over wide areas and the same basic techniques for culture are used.

The arrangement of sequence of species and chapters has no satisfactory solution. It was decided to group all species cultured within a region together, as much as possible.

Except for several species, all the discussions, including references cited, are brief because of the need for inclusion in one volume. The exceptions are those relatively new techniques developed for species that have been only recently cultured; therefore, fuller treatises seemed warranted.

The editor thanks the contributors, all recognized authorities for the species discussed, who have taken the time from their busy schedules to compile the data. The editor thanks Dr. Anne Thistle and Ms. Terri Olson for editing and proofreading. Special thanks are given to Ms. Patricia Klein for the laborious task of typing, retyping, editing, and proofreading. Dr. Wu Yung, Zhejiang Mariculture Institute, is thanked for facilitating contacts in China. The preparation has taken almost three years and CRC Press has shown forbearance and patience, as have some of the early contributors. For some species, culture techniques are developing rapidly, and some of the newer and more promising ones are not included.

Winston Menzel, Editor
Service Professor
Department of Oceanography
Florida State University
Tallahassee, Florida

THE EDITOR

Winston Menzel, Ph.D., Professor in the Department of Oceanography at Florida State University, Tallahassee, Florida, (when he completed preparation for this volume), died of illness shortly before its publication.

Dr. Menzel received his B.S. degree In Botany and subsequent M.A. in Aquatic Biology from the College of William and Mary in 1940 and 1943, respectively, and his Ph.D. in Biology from Texas A & M University in 1954. That same year he began his career of teaching and research at Florida State University, where he remained until his death in 1989.

Dr. Menzel was an active member of the American Fisheries Society and the Estuarine Research Federation. He served as president of the Gulf Estuarine Research Society and also served as president of the National Shellfisheries Association. In 1985, he became a member of the executive board of the Florida Aquaculture Association. The National Shellfisheries Association presented him with their Meritorious Award and in 1981 awarded him honorary membership for contributions to molluscan biology.

From 1978 to 1988 he served on the Special Scientific and Statistical Committee on Mollusks, for NMFS, NOAA, Gulf of Mexico Fishery Management Plan. Since 1982 he served on the Manuscript Review Board of the American Malacological Bulletin and since 1980, as official reviewer for Marine Fisheries Review.

Dr. Menzel's research ranged from projects in selective breeding and hybridization in clams and oysters, to clam mariculture. In 1972 Dr. Menzel, working with the Florida State Board of Natural Resources, developed a management program for the oyster resource in Apalachicola Bay, Florida.

From 1955 through 1987 he served as major professor for fifty-one students who were awarded masters or doctoral degrees at Florida State University.

During his career he authored 54 articles, numerous abstracts, book reviews, and reports, as well as studies and evaluations for federal and state agencies. He served as consultant to laboratories and private industry, and made numerous presentations as invited lecturer at workshops on mariculture and estuarine ecology.

CONTRIBUTORS

Jay D. Andrews, Ph.D.
Senior Marine Scientist (Retired)
Department of Malacology
Virginia Institute of Marine Science
Gloucester Point, Virginia

Erik Baqueiro C., M.Sc.
Subdirector
Centro Regional de Investigaciones Pesqueras
Campeche, Mexico

M. F. Bull, Ph.D.
Fisheries Management Division
Ministry of Agriculture and Fisheries
Nelson, New Zealand

Keashne Chalermwat
Institute of Marine and Coastal Sciences
New Jersey Agricultural Experimental Station
Rutgers University
New Brunswick, New Jersey

Paul Chanley, M.A.
President and Secretary-Treasurer
Indian River Mariculture, Inc.
Grant, Florida

Matoira H. Chanley
Indian River Mariculture, Inc.
Grant, Florida

J. M. Deslous-Paoli
Institut Francais Recherche Exploitation Mer
La Tremblade, France

Parameswar Dinamani, Ph.D.
Fisheries Scientist
Fisheries Research Division
Ministry of Agriculture and Fisheries
Wellington, New Zealand

Trevor G. Dix, Ph.D.
Aquaculture Development Services
Tasmania, Australia

Antonio J. Figueras, Ph.D.
Colaborador Cientifico
Consejo Superior Investigaciones Cientificas
Institute Investigaciones Marinas
Vigo, Spain

William K. Fitt, Ph.D.
Assistant Professor
Department of Zoology
University of Georgia
Athens, Georgia

Osamu Fukuhara
Senior Scientist
Nansei Regional Fisheries Research Laboratory
Ohno, Saeki, Hiroshima, Japan

Richard G. Gustafson
Institute of Marine and Coastal Sciences
New Jersey Agricultural Experimental Station
Rutgers University
New Brunswick, New Jersey

Maurice Héral
Chef de Laboratoire
Institut Francais Recherche Exploitation Mer
La Tremblade, France

Robert Hickman, M.Sc.
Ministry of Agriculture and Fisheries
Fisheries Research Division
Wellington, New Zealand

Herbert Hidu, Ph.D.
Department of Zoology
Rutgers University
New Brunswick, New Jersey

Susumu Ito
Director of Aquarium
Aomori Enterprise Public Service Corporation
Aomori Prefectural, Japan

Yutaka Kusuki
Chief
Division of Fisheries Resources and Environment Research
Hiroshima Prefectural Fisheries Experimental Station
Aki-gun, Hiroshima, Prefectural, Japan

Rene E. Lavoie, Ph.D.
Assistant Director
Biological Sciences Branch
Fisheries and Oceans of Canada
Dartmouth, Nova Scotia, Canada

David L. Leighton, Ph.D.
President
Marine Bioculture, Inc.
Leucadia, California

Shieh Lianchen, M. A.
Professor
Department of Aquaculture
Qingdao University of Oceanology
Qingdao, Shandon, China

Richard A. Lutz
Institute of Marine and Coastal Sciences
New Jersey Agricultural Experimental Station
Rutgers University
New Brunswick, New Jersey

U. H. Mane
Reader
Zoology Department
Marathwada University
Aurnagbad, India

Winston Menzel, Ph.D.
Service Professor (deceased)
Department of Oceanography
Florida State University
Tallahassee, Florida

R. Nagabhushanam, Ph.D.
Professor and Head
Zoology Department
Marathwada University
Aurangbad, India

Iracema Andrade Nascimento, D.Sc.
Professor
Department of Biology
Federal University of Bahia
Salvador-Bahia, Brazil

Carter R. Newell, M.S.
Biologist
Great Eastern Mussel Farms, Inc.
Tenants Harbor, Maine
President
Pemaquid Oyster Company
Darmariscotta, Maine
Board of Directors
Maine Aquaculture Association
Darmariscotta, Maine

Zong-Oing Nie
Research Associate Professor
Fuzhou Fisheries Research Institute
Fuzhou, Fujian Province, China

Kazuhiko Nogami
Coastal Fisheries Development Division
Japan Sea Regional Fisheries Research Laboratory
Suido-cho, Niigata, Japan

Neon C. Rosell, Ph.D.
Professor
Institute of Biology
University of Philippines
Quezon City, Philippines

Wang Rucai, M.M.C.
Associate Professor
Department of Aquaculture
Qingdao University of Oceanology
Qingdao, Shandong, China

Anibal Velez, M.Sc.
Professor
Department of Biologia Pesquera
Universidad de Oriente
Cumana-Sucre, Venezuela

Katsuhiko T. Wada, Dr. of Agriculture
Chief of Genetics Section
Division of Genetics and Breeding
National Research Institute of
Aquaculture
Nansei, Mie, Japan

Peter H. Wolf, B.Sc.
Oyster Pathologist (retired)
Department of State Fisheries
Bondi Juction, New South Wales,
Australia

ACKNOWLEDGMENTS

We, the children of Dr. R. Winston Menzel, wish to express our gratitude to those whose efforts have made the completion of this publication possible. Without their efforts it would have been difficult.

Dr. Frank O. Perkins and Dr. Roger Mann of the School of Marine Science, Virginia Institute of Marine Science, The College of William and Mary, for their assistance in the editing and proofing of the book.

Patricia A. Klein of the Department of Oceanography, Florida State University, who has been the coordinator for us in this effort.

Robert W. Menzell, Jr.
Gary P. Menzel
Mary Menzel Johnson

PUBLISHER'S ACKNOWLEDGMENT

Due to the untimely death of Dr. Winston Menzel during the editing process of this book, the publishers had to obtain the services of other professionals in the field for the final editing process. We would like to thank Dr. James P. McVey, of the National Sea Grant College Program and Dr. Roger Mann, of the Virginia Institute of Marine Science, who brought this volume together in its final form.

TABLE OF CONTENTS

Chapter 1

THE SOFT-SHELL CLAM *MYA ARENARIA* (LINNAEUS) IN NORTH AMERICA

Carter R. Newell

TABLE OF CONTENTS

I. BIOLOGY

A. MORPHOLOGY

The soft-shell clam *Mya arenaria* (Linnaeus) (family Myidae; common names: soft-shell clam, long-necked clam, nannynose, sand gaper, steamer clam) has an elongate, elliptical shell (Figure 1), with a large siphonal gape at the elongate posterior end and a small gape anteriorly. The left valve has a distinct chondrophore, which encloses the hinge ligament attached to the right valve. Distinct growth lines are apparent on the shell surface and in cross-section. In living position, a long contractile siphon extends from the posterior end of the clam to the sediment surface. The incurrent siphon has a ring of tentacles. The foot is small and muscular and extends from an opening in the anterior end of the shell. The mantle lobes are fused except at the siphon and pedal gape.

B. RANGE

The species occurs on the east and west coasts of the North Atlantic and has been successfully introduced into Pacific waters from California to Alaska. It is intertidal and subtidal to a depth of 200 m along the Atlantic coast from Labrador to South Carolina and extends, in lesser abundance, south to Florida;[1] the fishery extends from Chesapeake Bay northward. It is most abundant intertidally along New England and Canadian Maritime coasts and subtidally in Chesapeake and Delaware bays.

C. ENVIRONMENTAL TOLERANCES

Larvae are more sensitive to salinity than adults. The salinity stress point is 5‰ for Chesapeake Bay clams, 15‰ for Massachusetts clams, and 22 to 24‰ for Medomak River, ME, clams. Mortalities were as high as 90% in Chesapeake Bay when salinities dropped to 2‰ after hurricane Agnes.[2] Clams located in runoff streams can suffer salinity stress in some intertidal environments. Clam survival at low salinity is greater at lower temperatures;[3] however, salinity acclimation, which is regulated by amino acids in the hemolymph, occurs more rapidly in warmer waters.[4]

Oxygen utilization is governed by body size, water temperature, rate of metabolism, and oxygen concentrations in the water. Oxygen intake is dependent on oxygen concentration down to 2.8 mg/l.[5] Clams living in the intertidal zone can switch to anaerobic metabolic pathways at low tide, functioning as facultative anaerobes. When the clams are reimmersed, they oxidize the end products of anaerobic respiration by pumping rapidly.

At the northern end of its range, the soft-shell clam is limited by temperatures too low (12 to 15°C) for spawning, and in the south, its range is limited by high temperatures.[6] Temperature influences the timing of spawning and the length of larval life. There is a single spawning north of Cape Cod with two spawns prevalent to the south. In Maryland, a spawning takes place when water temperatures reach 10°C in April and another when later summer temperatures fall to 25°C.[7] Clams in Chesapeake Bay rarely survive temperature above 28°C.[8] Optimal temperature for larval development is about 20°C.[9] The pumping rates of soft-shell clams increase up to 16°C.[10] Temperature also influences the rate of burrowing; rates are highest at 18°C in Chesapeake Bay and lower at temperatures above 21°C or below 9°C.[11] Clams tend to grow faster in warmer waters, and greater shell growth occurs during the warmer temperatures in clams from Maine.[12] Oxygen intake is greatest at about 20°C.[13] Appeldoorn[14] found that temperature accounted for 68% of the growth rate differences in clams collected along the east coast of North America. At low temperatures, clams continue to filter, but food assimilation rates may be low.[15]

D. SEXUALITY

Sexes are separate; the sex ratio is 50:50 and there is very low incidence of hermaphrodites in most populations. Clams longer than 20 mm in shell length are usually capable of

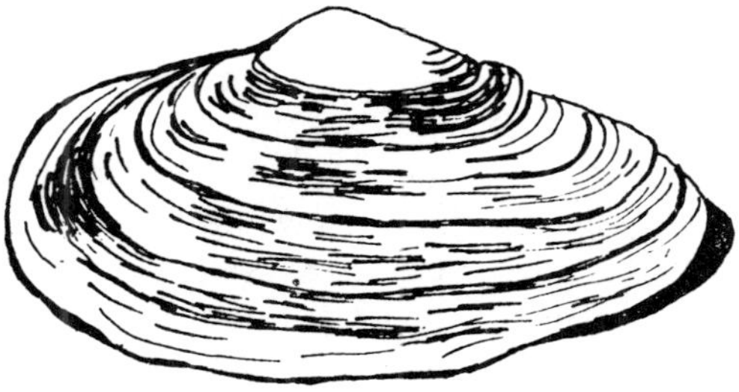

FIGURE 1. Soft-shelled clam shell.

spawning.[16] Clams from cold Maine waters spawn at a smaller size than those from Massachusetts or Long Island Sound. In the North Atlantic, gametogenesis begins in late winter and early spring, and spawning peaks from June to September, depending on location.[17] In Maryland, clams begin to spawn in April when water temperature reaches 10°C, and larval abundance peaks when water temperatures are 20°C.[18] Several studies have elucidated regional differences in timing of reproduction. In Ipswich, MA, clams had an early spawn in March and April, followed by a large spawn in June and July. Clams from Long Island Sound (Connecticut) spawned twice annually in one location and just once a year in two other locations.[16] On a Maine mudflat, spawning was observed in June when water temperatures fluctuated between 13°C at high tide and a seasonal maximum of 21°C on a sunny day in shallow water at low tide.[19] There is an increase in fecundity with size; a female clam 60 to 70 mm long produces as many as 3 million eggs a year.[20] However, fecundity also varies with location and food availability. In the middle intertidal zone, females only produced 120,000 eggs a year at Cape Ann, MA.[19]

E. LARVAL AND JUVENILE BIOLOGY

Fertilization is external, and the spherical egg (66 μm in diameter) is white and gelatinous.[21] The embryo forms a top-shaped trochophore larvae after about 12 h. At about 80 μm in size, a shell is formed, and the larva becomes a veliger, which passes through the prodissoconch I (straight hinge) and prodissoconch II (shell with umbo) stages. Locomotion is provided by a ciliated velum. Clams reared in a hatchery until settlement have been measured and photographed,[22] but identification of clam larvae from plankton samples is still difficult because the shape of the *Mya* shell is easily confused with that of *Hiatella* sp. However, the larvae can be identified on the basis of hinge structure.[23]

The larval period lasts from about 14 to 21 d, depending upon water temperature, and is longer at temperatures below 20°C. In culture, larval temperature optima were between 17 and 23°C and salinity optima between 16 and 32‰; highest salinities and lowest temperatures depressed the Maryland stocks.[24] Along the New Hampshire coast from June to October, veliger abundance peaked in late summer with numbers as high as 1000/m^3. Numbers were higher in inshore waters than in offshore waters, and higher at depths from 5 to 9 m than near the surface or at 13 m.[25]

At settlement, the velum is lost, and the clam attaches to suitable substrate with a byssal thread. Larvae are thought to attach first to filamentous algae, eelgrass and its associated epiphytes, and other projections in the subtidal zone.[26] Addition of adult pallial fluid stimulates mature larvae (256 μm) to attach the byssal thread, suggesting a gregarious component to clam settlement.[27] Initial byssal attachment may be ephemeral as the clam searches for a suitable substrate. After settlement, the spat (0.25 to 1 mm long) undergoes a swimming-crawling stage that lasts for 2 to 5 weeks.[20] During this stage the foot and byssal thread are active in the migration, which is thought to be partially mediated by resuspension by wind-induced waves and tidal currents.[28] It is unclear whether recently set spat have a separate drifting thread such as is found in the blue mussel *Mytilus edulis* L. When the clam burrows into the sediment and reaches 5 mm in size, it is referred to as a clam seed until it reaches market size (50 mm).

Juvenile clams have a complex behavior in the first year; substantial repositioning occurs until they reach about 15 mm in shell length. In New England, several authors have attributed a shoreward migration of seed clams to physical factors. Clam flat sediments are often dynamic, with erosion and deposition of up to 10 cm occurring over the course of a year,[29] indicating that shallow-burrowing seed clams could well be moved along with the sediments. Clams at a Massachusetts beach that set in the summer and grew to over 5 mm the following spring were moved shoreward along with coarse sediment particles during spring storms.[30] In Maine, movement of first-year juveniles peaked in the fall during early growth and again in May after growth resumed in the spring.[31] Clams migrated up to several hundred yards toward the shore.

Clam distribution is aggregated in intertidal and shallow subtidal areas,[32] possibly because of gregarious settlement, predation effects, or concentration by hydrodynamic factors. In some environments, clams were concentrated during changes in shoreline profile.[30] Large numbers of seed clams may be found in small areas. The highest densities are usually in eddies, along the mouths of rivers or streams emptying into shallow water, or in slack water adjacent to rapid currents. Densities of 538,000 clams per square meter have been recorded.[20] There are several reports of good clam sets after sand has been added to a mudflat, presumably because it provides an improved surface for byssal attachment. In Virginia in the subtidal zone, density of clam spat dropped from 6000/m^2 to 0 in 1 month as a result of predation.[28] However, clams that burrow as deep as 10 cm can avoid most predators in Chesapeake Bay.[33]

F. GROWTH

Computer-generated von Bertalanffy growth curves of soft-shell clams from different geographic areas demonstrate widespread differences.[34] In a review of the published growth rates of clams to a market size of 51 mm, there was a general trend of increasing growth rates with decreasing latitude.[35] Clams in southern New England grew to 51 mm in 1 to 3 years, but took 6 to 7 years in more northerly locations. Clam growth is generally positively correlated with increasing temperature up to about 20°C. Clam growth is affected by sediment type,[36] current and food availability,[20] and intertidal height. Constant immersion in the subtidal zone up to 10% tidal exposure gives the highest growth rates for a given area.[37] Clams will grow faster in silty or sandy sediments than in gravel or out of sediment (Figure 2).[36] Clams in the upper intertidal zone in coarse sediments generally have the lowest growth rates. Slow-growing clams have higher percent shell by weight and larger shell length-depth regression slopes (shell inflation) than faster-growing clams.

Seasonal growth rates of 15-mm clam seed stained with alizarin and planted in a Maine mudflat showed the most rapid growth during the months of April, May, June, and July.[12] Instantaneous growth rates of meat were the highest during the spring, whereas the shell grew the fastest during the late spring and summer. Glycogen content of the meats in planted juvenile and adult clams showed a distinct seasonal cycle, with highest levels in late spring and early summer, intermediate levels in later summer and fall, and lowest levels in winter (Figure 3).

II. FISHERY

A. COMMERCIAL FISHERY

Annual U.S. commercial landings from 1977 to 1981 averaged 4.2 million kg, worth $15 million; the 1984 landed value was $19.8 million. In New England, the resource has

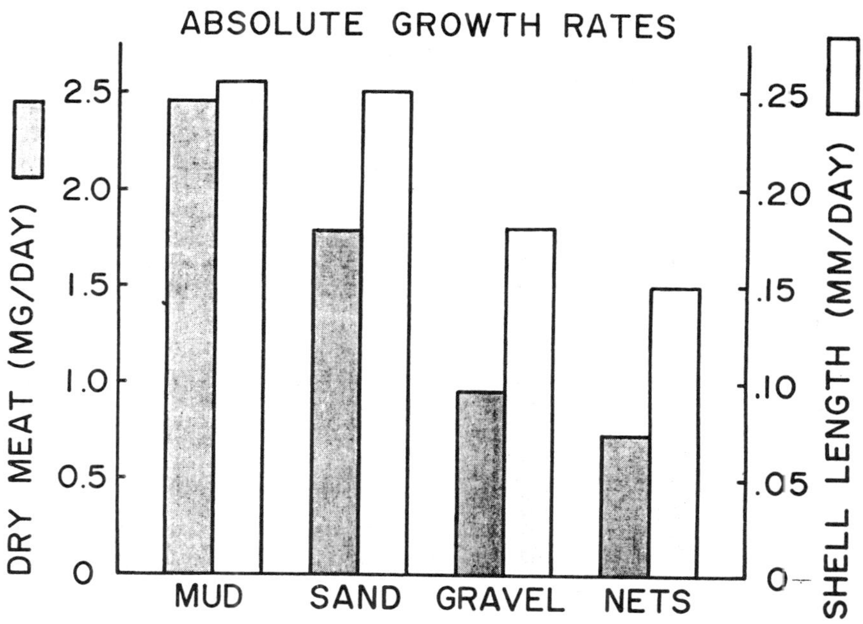

FIGURE 2. Effects of sediment type on growth of clams.

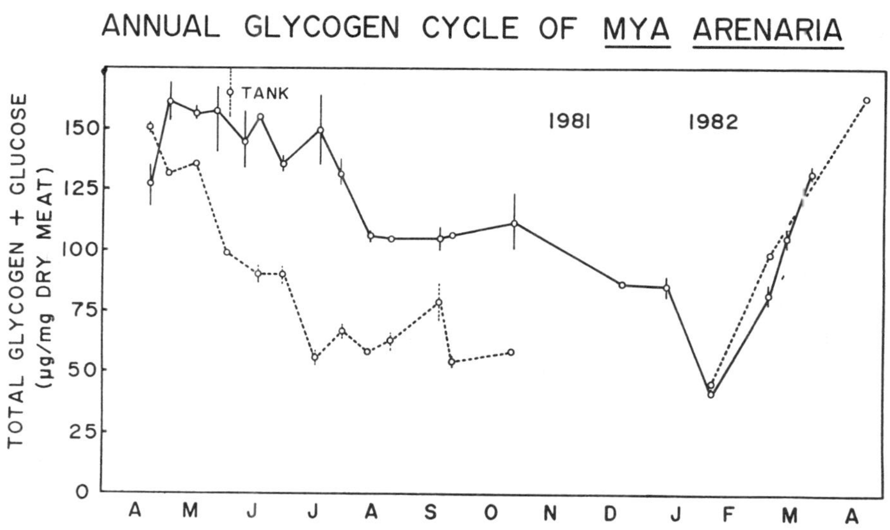

FIGURE 3. Annual glycogen cycle of adult and juvenile clams in Maine.

been important since colonial times, where a hand-digging public fishery still persists, and major landings are from Maine and Massachusetts. A second important commercial fishery has developed since the 1950s in Maryland, where the resource is subtidal and the clams are harvested with a hydraulic escalator dredge. Where the resource is intertidal, harvesting is limited to hand implements such as a clam hoe. Price per kilogram has tripled in the last decade. The rapid increase in value of this resource in states like Maine (Figure 4) makes it a likely candidate for cultivation.

Fluctuations in landings are due to predation, variable recruitment, and digging pressure. Predation by the green crab during periods of warm temperatures in Maine caused drastic

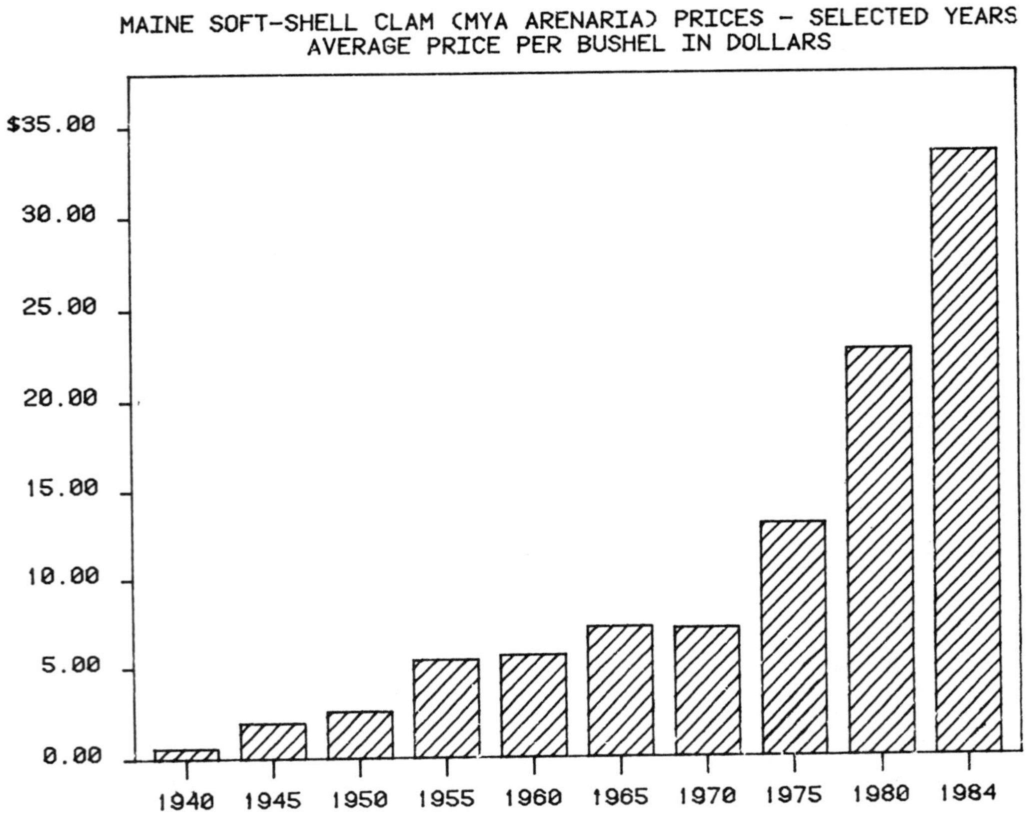

FIGURE 4. Maine soft-shell clam landings from 1940 to 1944 in 5-year intervals; average price per bushel.

reductions in landings during the 1950s and 1960s.[38] Predator fencing and protective mesh were put out to protect good sets of clams in areas where crab predation was heavy. Other important predators are mummichogs, sandworms, blue crabs, horseshoe crabs, and moon snails. Seals in eastern Maine have been reported to eat clams after digging a funnel-shaped hole. Mortality rates are highest in summer, when predators are most abundant and active.

Sometimes good sets of clams can result in a strong year class, which can grow to yield a considerable biomass; for example, 1 ha of mudflat contained clams about 62 mm long at a density of 269 clams per square meter and 9834 kg of meats. In New Hampshire, high larval densities in 1975 (530/m^3) were followed by high spat densities (8200/m^2) in 1976, followed by a strong fishery from 1978 to 1980.[39]

B. MANAGEMENT

Because most of the clam resource occurs in the public fishery, state and municipal management of the fishery is a major method of increasing production. Management includes the opening and closing of flats on the basis of period population surveys, limiting harvesting pressure, predator control, wild seed transplantation, and public hatchery, nursery, and grow-out programs. Management by population surveys and periodic closures is based on the finding that, when market-sized clams are dug, high mortality rates of the seed clams result from burial and breakage.[40] If harvesting is restricted until most of the clams are market size, losses due to harvesting pressure are minimized. In Maine, towns with a clam management program averaged yields 15% higher than those of unmanaged towns.

Management by size limits is based on increased clam survival with increasing size in addition to increased fecundity with size. Once the clams reach a predator-refuge size of

FIGURE 5. Hydraulic seed-clam harvester used in subtidal collection of seed clams for transplantation.

about 35 mm, mortality rates are low, and landings can be significantly increased if the population is simply permitted to grow to a larger size. Shell length-to-volume relationships are such that a small increase in shell length yields a large increase in volume. Clam growth rates, based on the presence of annual rings in the external shell layer or the more accurate acetate peel technique, can be used to plot shell length against age and, thus, to determine when a given cohort should reach market size for each site. A flat with a high percentage of 35-mm clams should be closed to harvesting for about 1 year for those clams to grow to market size.

Because seed clams may be concentrated by hydrographic factors and dense sets result in intense competition for food and space, the transplantation of natural seed is perhaps the most cost-effective method of increasing clam production. After the seed is located from a clam survey, it can be dug by hand or moved with an intertidal or subtidal hydraulic seed-clam dredge such as those developed by the University of Maine Agricultural Engineering Department (Figure 5). The advantages of enhanced growth in the lower intertidal and subtidal zones due to increased immersion times must be weighed against increased predation along the lower shore.

III. AQUACULTURE

Since the early 1900s, clam management by periodic surveys and closures, predator control, and seed transplantation has occurred in New England and can be considered a form of extensive public aquaculture. Intensive soft-shell clam aquaculture has not been feasible up to the present because the relatively low unit value of less than 5 ¢ a clam was no higher than the hatchery and nursery costs of rearing that clam to a predator-resistant planting size. Hatchery culture of *Mya* was demonstrated in 1964 by Stickney[24] and was used in several clam-hatchery projects in Maine through the Ira C. Darling Center of the University of Maine. Clams are successfully induced to spawn when ripe broodstock are placed in seawater heated to 24°C and cultured algae are added. The addition of striped gametes stimulates spawning, and larvae can be cultured with standard culture techniques. At metamorphosis, larvae attach by a byssal thread to the rearing vessel, where they can be grown to 1 mm in size.

The major bottleneck in obtaining large numbers of clam seed is the cost of nursery culture from 1 mm to a planting size of over 25 mm. Upweller pipes and floating trays or raceways have been successfully used in the nursery culture of clams. A new approach to clam nursery culture was developed in Washington state for geoducks.[41] Clams are set in sand beds in a raceway in which water is pumped over the surface of the sand and up through the sand through lateral slotted pipes. This system has great potential for the nursery culture of *Mya*.

The development of fast-growing strains is another potential of hatchery-reared clams. Induced triploidy has been achieved in soft-shelled clams[42] and results in clams that grow faster after they reach sexual maturity, probably because of the blockage of gametogenesis and reallocation of energy to somatic growth. The selection of superior strains through the classic selection of fast-growing, disease-resistant individuals also has potential.

IV. CONSTRAINTS

Pollution of clam beds continues to be a significant factor in limiting the growth of the fishery, especially in areas along the coast where there is increasing population pressure. For example, in Massachusetts, more than 1500 ha are closed to shellfishing because of bacterial contamination, and 500 ha more are borderline. In addition, ponds and embayments are feeling the effects of eutrophication caused by the increasing levels of sewage in coastal areas. Purification of clams in depuration plants allows for the elimination of bacterial contamination of clams under controlled conditions and provides for the use of otherwise polluted stocks.

Clam hematopoietic neoplasm, a leukemia-like disease that was rare in clam populations until recent years, now affects a large number of clam beds from Maryland to Maine.[43,44] The etiology of the disease and its effects on the clam are not known at this time.

REFERENCES

1. **Theoroux, R. B. and Wigley, R. L.,** Distribution and abundance of east coast bivalve mollusks based on specimens in the National Marine Fisheries Service Woods Hole Collection, U.S. Dept. Commer. NOAA, Natl. Mar. Fish. Serv., Tech. Rep. 768, 1983, 172.
2. **Shaw, W. N. and Hamons, F.,** The present status of the soft-shell clam in Maryland, U.S. Fish Wildl. Serv. Spec. Sci. Rep. Fish. No. 508, 1974, 5.
3. **Allen, J. A. and Garrett, M. R.,** The excretion of ammonia and urea by *Mya arenaria* L. (Mollusca: Bivalvia), *Comp. Biochem. Physiol.*, 39A, 633, 1971.
4. **Creaser, E. P. and Clifford, D. A.,** Salinity acclimation in the soft-shell clam, *Mya arenaria, U.S. Natl. Mar. Fish. Serv. Fish. Bull.*, 75, 225, 1977.
5. **van Dam, L.,** On the utilization of oxygen by *Mya arenaria, J. Exp. Biol.*, 12, 86, 1975.
6. **Lawson, D,** The genus *Mya* in the Arctic region, *Malacologia*, 3, 399, 1966.
7. **Pfitzenmeyer, H. T.,** Periods of spawning and setting of the softshelled clam, *Mya arenaria*, at Solomons, Maryland, *Chesapeake Sci.*, 3, 114, 1962.
8. **Pfitzenmeyer, H. T.,** Tentative outline for inventory of molluscs: *Mya arenaria* (soft-shelled clam), *Chesapeake Sci. Suppl.*, 13, 182, 1972.
9. **Chapman, S.,** personal communications, 1985.

10. **Harrigan, R. E.,** The Effect of Temperature on the Pumping Rate of the Soft-Shelled Clam, *Mya arenaria,* M.S. thesis, Columbian College, George Washington University, Washington, D.C., 1956, 54.
11. **Pfitzenmeyer, H. T. and Drobeck, K. G.,** Some factors influencing reburrowing activity of the soft-shell clam, *Mya arenaria, Chesapeake Sci.,* 8, 193, 1967.
12. **Newell, C. R.,** The Soft-Shell Clam *Mya arenaria* L.: Growth Rates, Growth Allometry and Annual Growth Line Formation, M.S. thesis, University of Maine, Orono, 1982, 139.
13. **Kennedy, B. S. and Mihursky, J. A.,** Effects of temperature in the respiratory metabolism of three Chesapeake bivalves, *Chesapeake Sci.,* 13, 1, 1972.
14. **Appeldoorn, R. S.,** Variation in the growth of *Mya arenaria* and its relation to the environment as analyzed through principal components analysis and the W parameter of the von Bertalanffy equation, *U.S. Natl. Mar. Fish. Serv. Fish. Bull.,* 81, 75, 1982.
15. **Gilfillan, E. S., Mayo, D., Hanson, S., Donovan, D., and Jiang, L. C.,** Reduction in carbon flux in *Mya arenaria* caused by a spill of No. 6 fuel oil, *Mar. Biol.,* 37, 115, 1976.
16. **Brousseau, D. J.,** A comparative study of the reproductive cycle of the soft-shell clam, *Mya arenaria* in Long Island Sound, *J. Shellfish Res.,* 6, 7, 1987.
17. **Ropes, J. and Stickney, A. P.,** Reproductive cycle of *Mya arenaria* in New England, *Biol. Bull. (Woods Hole, Mass.),* 128, 315, 1965.
18. **Pfitzenmeyer, H. T.,** Annual cycle of gametogenesis of the soft-shelled clam, *Mya arenaria,* at Solomons, Maryland, *Chesapeake Sci.,* 6, 52, 1965.
19. **Brousseau, D. J.,** Spawning cycle, fecundity, and recruitment in a population of softshell clam, *Mya arenaria,* from Cape Ann, Massachusetts, *U.S. Natl. Mar. Fish. Serv. Fish. Bull.,* 76, 155, 1978.
20. **Belding, D. L.,** The soft-shelled clam fishery of Massachusetts, Mar. Fish. Ser. 1, Div. Fish Game, Department of Conservation, Commonwealth of Massachusetts, 1930, 65.
21. **Loosanoff, V. L. and Davis, H. C.,** Rearing of bivalve mollusks, in *Advances in Marine Biology,* Russell, F. S., Ed., Academic Press, London, 1963, 1.
22. **Loosanoff, V. L., Davis, H. C., and Chanley, P. E.,** Dimensions and shapes of larvae of some bivalve mollusks, *Malacologia,* 4, 351, 1966.
23. **Lutz, R. A., Goodsee, J., Castagna, M., Chapman, S., Newell, C., Hidu, H., Mann, R., Jablonski, D., Kennedy, V., Siddall, S., Goldberg, R., Beattie, J., Falmagne, C., Chestnut, A., and Partridge, A.,** Preliminary observations on the usefulness of hinge structures for identification of bivalve larvae, *J. Shellfish Res.,* 2, 65, 1982.
24. **Stickney, A. P.,** Feeding and growth of juvenile soft-shell clams, *Mya arenaria, U.S. Fish. Wildl. Serv. Fish. Bull.,* 63, 635, 1964.
25. **Normandeau Associates, Inc.,** *Seabrook Ecological Studies,* Tech. Rep., Public Service Company, Nashua, NH, 1978.
26. **Kellogg, J. L.,** Observations on the life history of the common clam, *Mya arenaria, U.S. Fish. Wildl. Serv. Fish. Bull.,* 19, 193, 1900.
27. **Hidu, H.,** personal communication, 1984.
28. **Lucy, J. A.,** The Reproductive Cycle of *Mya arenaria* L. and Distribution of Juvenile Clams in the Upper Portion of the Nearshore Zone of the York River, Virginia, M.S. thesis, College of William and Mary, Williamsburg, VA, 1976, 131.
29. **Anderson, F. E., Black, L., Watling, L. E., Mook, W., and Mayer, L. E.,** A temporal and spatial study of mudflat erosion and deposition, *J. Sediment. Petrol.,* 51, 729, 1981.
30. **Matthiessen, G. C.,** Intertidal zonation in populations of *Mya arenaria, Limnol. Oceanogr.,* 5, 381, 1960.
31. **Dow, R. L. and Wallace, D. E.,** The soft-shell clam industry of Maine, U.S. Fish. Wildl. Serv. Circ. 110, 1961, 36.
32. **Saila, S. B. and Gaucher, T. A.,** Estimation of the sampling distribution and numerical abundance of some mollusks in a Rhode Island salt pond, *Proc. Natl. Shellfish. Assoc.,* 56, 73, 1966.
33. **Blundon, J. A. and Kennedy, V. S.,** Refuges for infaunal bivalves from the blue crab *Callinectes sapidus* (Rathbun), *J. Exp. Mar. Biol. Ecol.,* 65, 67, 1982.
34. **Brousseau, D. J.,** Analysis of growth rate in *Mya arenaria* using the von Bertalanffy equation, *Mar. Biol.,* 51, 221, 1979.
35. **Brousseau, D. J. and Baglivo, J. A.,** A comparative study of age and growth in *Mya arenaria* (soft-shell clam) from three populations in Long Island Sound, *J. Shellfish Res.,* 6, 17, 1987.
36. **Newell, C. R. and Hidu, H.,** The effects of sediment type on growth rate and shell allometry in the soft-shelled clam *Mya arenaria* L., *J. Exp. Mar. Biol. Ecol.,* 65, 285, 1982.
37. **Gillmore, R. B.,** Assessment of intertidal growth and capacity adaptations in suspension-feeding bivalves, *Mar. Biol.,* 68, 277, 1982.
38. **Welch, W.,** Changes in the abundance of the green crab, *Carcinus maenas* (L.) in relation to recent temperature changes, *Fish. Bull.,* 67, 337, 1969.

39. **Savage, N.,** Soft shell clam studies on the northern Massachusetts, New Hampshire and southern Maine coast by Normandeau Associates, Inc., in *8th Annu. Conf. Clam Res.,* Maine Dept. Mar. Res., Boothbay Harbor, ME, 1981, 40.
40. **Glude, J. B.,** Survival of soft-shell clams, *Mya arenaria,* buried at various depths, *Maine Dept. Sea Shore Fish. Res. Bull.,* 22, 1954, 26.
41. **Beattie, H. and Zebal, R.,** Geoduck nursery culture: a new approach, *Aquaculture Dig.,* December, 10, 1986.
42. **Allen, S. K., Jr., Gagnon, P. S., and Hidu, H.,** Cytogenic and allozymic confirmation of induced polyploidy in clams and oysters, *J. Shellfish Res.,* 2, 87, 1982.
43. **Sherbourne, S. W. and Bean, L. L.,** Hematopoietic neoplasms, accessory siphons and gross tumors in the soft-shelled clams *Mya arenaria* from Dennysville, Maine, *Res. Ref. Doc. 83/18,* Maine Dept. Mar. Res., Boothbay Harbor, ME, 1983.
44. **Gruson, L.,** Clams and oysters in the east hit hard by mystery diseases, *New York Times,* p. A12, January 23, 1986.

Chapter 2

THE OYSTER FISHERY OF EASTERN NORTH AMERICA BASED ON *CRASSOSTREA VIRGINICA*

J. D. Andrews

TABLE OF CONTENTS

I. BIOLOGY

A. DISTRIBUTION AND TAXONOMY

The American oyster *Crassostrea virginica* (Gmelin) (family Ostreidae) is confined to the Atlantic coast of North America from the Gulf of St. Lawrence to the Gulf of Mexico.[1,2] Oysters of this species transplanted to the west coasts of Europe and North America failed to reproduce in the cool oceanic temperature regimes of those regions.[3] *Crassostrea virginica* is the only commercial oyster grown along the coast of Canada and the U.S. except for tray culture of *Ostrea edulis* Linnaeus in Maine and occasional imports of this European flat oyster to Canadian provinces.

Oysters of the genus *Crassostrea* are characterized by cupped left valves and large promyal cavities on the right side of the body to facilitate handling of silty estuarine waters.[1-3] Oysters of the genus *Crassostrea* are oviparous and do not incubate eggs. Eggs are small (50 mm) and are extruded through small openings in gills (ostia) into the branchial chamber for release into open waters. Shells of *Crassostrea* species have limited taxonomic use because shape varies widely with habitat and conditions, including substrate, intertidal exposure, salinity levels, crowding, and growth rates.

B. HABITAT AND CLIMATE

Oysters are adapted to the nutrient-rich, shallow, coastal waters of continents.[4] The eastern coast of North America has many large rivers fed by rain and snow from drainage of mountains and wide coastal plains. The climate is continental along the coast, providing wide seasonal variations in temperature and salinity regimes in the tidal estuaries.[5] The temperature in Chesapeake Bay near the center of this coastline varies from nearly 0°C in winter to 30°C in summer, and salinities at most sites change regularly 10 to 5‰ from spring lows to fall highs. For a comparison of hydroclimographs of the west and east coasts of North America see Reference 5.

These annual fluctuations in the two most important environmental factors provide advantages and disadvantages for oysters. High river flows and resultant low salinities in spring provide generous nutrient replenishment and effective control of predators and diseases, most of which require mesohaline salinities to survive. Low temperature causes a period of dormancy for 3 to 6 months from Chesapeake Bay northward; low temperature also limits intertidal survival of oysters north of Chesapeake and Delaware bays. Culture of oysters tends to be increasingly intertidal southward along the Atlantic coast. The disadvantages of wide fluctuations in salinity and temperature are fresh-water kills in spring and increases in diseases during summer and fall. The period of culture of oysters must be short in the Gulf of Mexico because of diseases, and it tends to be increasingly longer with distance north along the coast, where summers are cooler. In Chesapeake and Delaware bays, periods of dry or wet years cause extension and contraction of disease activity.

When growing naturally, *C. virginica* tends to form reefs in which oysters set on each other to build shelly, raised beds above the bottom where they feed in tidal currents.

C. REPRODUCTION

Temperature regimes control reproduction of *C. virginica* along the Atlantic coast.[6-9] The species requires temperature above 20°C to spawn. Several races of oysters are adapted to their respective geographic climates in respect to spawning temperature and timing of reproduction. From Long Island Sound northward, oysters often encounter only one or a few periods when the temperature is adequate for spawning.[7,8,10-13] Therefore, they fatten intensively while waiting and spawn profusely and easily when the temperature is adequate. During the 19th century, when many Chesapeake oysters were imported to New England for sale or planting, it was found that they could be held on beds all summer in plump

FIGURE 1. Planting of fossil shells dredged from buried deposits by water-hosing the shells overboard.

condition without spawning. In Chesapeake Bay, oysters begin spawning in mid-June and continue sporadically until late September. They usually wait until a temperature of 25°C is reached; then only partial spawning occurs at intervals, and more spawn is produced. In the Gulf of Mexico, a long period of intermittent spawning is possible.

Dispersal and predation are the principal factors controlling survival of larvae.[6] Knowledge of natural food species eaten by larvae is quite limited although hatcheries utilize well-known species. In temperatures above 25°C, which are prevalent all summer in Chesapeake Bay, larvae mature and set in 10 d in hatcheries; however, the period may be extended in nature.

Setting of oysters tends to be intensive in coastal bays such as those on the seaside of Virginia, South Carolina, and Georgia, where little fresh-water input and extensive marsh systems allow retention of larvae. River estuaries tend to have lower spatfalls, which are more irregular by seasons and years. Extensive setting records exist for Maryland,[14,15] Virginia,[7] and Long Island Sound,[10,11] which permit designation of seed, self-sustaining, and growing areas for efficient cycles of culture. Two patterns of setting occur in Chesapeake Bay (see Figure 1). The large river estuaries with extensive drainage areas, such as the James River, exhibit decreasing intensity of setting with distance upriver. Small coastal plain estuaries with small drainage areas; shallow flats; and narrow, winding channels usually obtain river-wide sets with highest intensities at the upper ends of the oyster-growing regions. Growth is slow in these small, low-salinity estuaries, which adapts them to production of thick-shelled seed oysters.

Sex ratios and sex changes are not important factors in production of seed stocks of *C. virginica*.[8] All along the coast, males predominate among young oysters, whereas most market-sized oysters are females. Sexual maturity is often achieved by spat less than a year old in the south and progressively later northward. Often oysters from low-salinity seed areas are of poor quality and contain little glycogen because of inadequate food supply; therefore, less spawn is produced.

D. GROWTH

Growth of oysters is poorly understood because food studies are inadequate.[11,16] Some generalizations are possible, but they are hard to verify. Low-salinity areas generally foster poor growth, presumably because of lack of adequate food. During drought years oysters from these same beds exhibit much faster growth. In seed areas, slow growth can be useful— it allows thickening of shells, which increases survival of spat when planted in predator-infested areas. In contrast, in some high-salinity areas such as seaside of the Eastern Shore of Virginia, Maryland, and South Carolina, where predation and diseases cause high mortality, thin-shelled oysters grow fast and reach commercial size early (18 months). These rapidly maturing races of oysters exhibit other distinctive genetic traits with respect to shape, thickness of shell, and resistance to diseases.[8]

Because meat yields increase rapidly with size, it has long been a tendency to hold oysters too long on growing beds where losses from mortality exceeded growth gains.

A unique method for studying factors regulating or interfering with growth is underwater weighing for shell growth.[17] A healthy oyster of 15 g weight will grow 1 g/week in a suitable habitat. Failure of individuals to grow in consecutive weeks is associated usually with disease, whereas, if all oysters show little gain, water quality is suspected. By this method it has been demonstrated over many years that dinoflagellate blooms occurring in early May and more commonly during July and August caused cessation of growth in oysters. Nutrient pollution may be a common cause of poor oyster growth because it favors undesirable food species.

II. THE FISHERY

A. HISTORY

Nearly one hundred years ago, surveys of natural beds were made to designate "public beds" for management by state governments.[18,19] Regulation and conservation consisted of placing limits on season, quantity, or method of catching oysters. Replenishment of public beds never matched harvesting; therefore, depletion was inevitable. From Chesapeake Bay northward, high-salinity areas were depleted early. The decline of public beds has been caused by interrelated factors of overharvesting, failure of reproduction, and salinity levels that permitted predators and diseases to destroy surviving oysters.[14,18,19] Oystermen soon began leasing adjacent grounds upon which to plant private beds of small seed oysters. If bottoms were not firm enough to sustain oysters, shells were planted as cultch and foundation. This private planting led to the designation of public seed areas. Along the Atlantic coast, seed areas tend to be located in sectors of estuaries where predators and diseases are excluded by annual low salinities in spring. Seed oysters with thickened shells and ages of 1 to 3 years are transplanted to high-salinity private beds for rapid growth and fattening.

B. RECRUITMENT AND METHODS OF CULTURE

Most seed areas, in both Delaware and Chesapeake bays, are located in salinity zones between 10 and 20‰ in late summer.[14,16] Setting intensity has always been higher in Delaware Bay than in Chesapeake Bay, where setting rates are low in the bay proper and the larger estuaries. Retention of larvae within the Delaware estuary appears to be much higher. In the low salinity of the Maryland sector of Chesapeake Bay, seed areas are located in small rivers and creeks with little runoff and low discharge rates.[14] These are shallow, winding estuaries that tend to retain larvae by trapping them over shallow flats.[6]

Most estuaries in Chesapeake Bay that serve as seed areas are quite shallow (<3 m over beds); therefore, little density stratification of water occurs. Oyster larvae are rather uniformly distributed by depth, but are spotty in horizontal distribution.[6] Setting occurs at all vertical levels, but advanced larvae increase in abundance with depth; however, there is no commercial suspension of cultch along the coast. Setting is confined to natural cultch on the bottom, although on the seaside of the Eastern Shore of Virginia and in South Carolina and Georgia, intertidal seed beds are commonly used. Much dredged fossil shell is planted on public beds in Chesapeake Bay. Generally there are a scarcity of seed oysters from Chesapeake Bay northward and an oversupply of high-density seed oysters southward. Experience has shown that local races of seed oysters are most successful in each region although much trial transplanting has been done between regions.[8] Genetic stocks selected for disease resistance and fast growth have been used experimentally in hatcheries,[12,20] but not commercially except where natural selection occurred as for *Haplosporidium nelsoni* (MSX) resistance in Delaware Bay. In general, brood stocks are in low supply from Chesapeake Bay northward, with the possible exception of Delaware Bay. High cost of hatchery seed oysters and lack of suitable nursery areas have been impediments to production of genetically superior seed oysters.

Oyster shell is routinely used as cultch for spatfall all along the Atlantic and Gulf coasts (see Figure 1).[14,16] Deep deposits of buried shell have accumulated as the ocean level slowly has risen over the last 10,000 years. These shells are mined with hydraulic harvesters in Maryland to supply millions of bushels of cultch planted directly on the bottom. A recent management trend is for shells to be planted in areas where setting occurs regularly to produce seed oysters, but political manipulations have often wasted large quantities of shell to please local oystermen.

C. CONSTRAINTS

The first major constraint on culture of *Crassostrea virginica* is disease.[21-24] The growth period after planting must often be limited to 1 or 2 years to avoid excessive losses to diseases. In Delaware Bay, seed oysters are held on growing beds for only one growing season, and in Chesapeake Bay, the salty lower half of the oyster growing area has been abandoned altogether for 25 years. The disease that has enforced these changes of habitat use is caused by the haplosporidan (sporozoan pathogen) *Haplosporidium nelsoni* (Haskin, Stauber, and Mackin) (called MSX), which causes high annual mortality (50%).[16,21] Destruction of brood stocks in the lower James River by MSX has caused a decrease of setting rates in the seed area from one fifth to one tenth the level for the pre-MSX period of the 1940s and 1950s.[16]

The Delaware Bay seed area is located in the upper low-salinity sector, which unlike those in Chesapeake Bay is not a sanctuary from diseases and predators. Consequently, most oysters in Delaware Bay have been exposed to, and selected by, MSX since 1957; the result is a rather high level of resistance to the pathogen in all oyster populations. With setting restored to pre-MSX levels, oystermen began planting in the high-salinity area where MSX is endemic. In Chesapeake Bay, most surviving populations of oysters are in low-salinity sanctuaries where little or no selection has occurred. As a result, resistance to MSX in native oysters is limited to one river—the York River-Mobjack Bay system, which has no low-salinity sector. Recovery from the ravages of MSX has been slow, and oyster planting in high salinities has been unsuccessful in Virginia. In Delaware Bay, private beds continue to be planted in high salinities in the lower bay, and setting in the seed area has recovered to pre-MSX levels most years.[21]

The second major constraint on culture of *C. virginica* is fouling. In the silty waters of coastal estuaries, a problem is to keep oysters clean of deposits from wind-stirring of shallow beds.[7,8,14,16] Fouling organisms such as barnacles, sea squirts, mussels, bryozoans, and sponges also interfere with setting and survival of oyster spat.[8] Because all oysters are grown on the bottom, predators such as oyster drills, starfish, mud crabs, blue crabs, and flatworms have easy access to spat. Avoiding predators by means of growing oysters in low-salinity waters is not possible in most regions along the coast except in upper Chesapeake Bay, where large areas of natural oyster beds are available. Private planters use thick-shelled oysters from low-salinity seed areas, which are more resistant to predation. Young oysters are most vulnerable to predators, whereas older and larger oysters become victims of diseases; to avoid mortalities from both sources requires use of low-salinity waters. In the south early harvesting is necessary to limit losses. Control of epizootic diseases in high-salinity water requires breeding and selection of races of oysters resistant to the diseases or manipulation of beds to avoid infections and remove all infective sources.[4,22,25] Each local area must be judged for the optimum duration of growth by a variety of market, growth, fatness, and weather conditions and for the best time to harvest oysters.[25,26] Nearly all oysters harvested in New England are sold on the half shell. Culture on seaside of the Eastern Shore of Virginia also produces tasty, salty oysters, and there is a tendency to increase the proportion of half-shell oysters with increased demand. Most oysters from Chesapeake Bay are sold as fresh-shucked oysters, for which demand is more seasonal and greatest during holiday seasons. All Chesapeake Bay oysters are now grown in moderate- to low-salinity waters because of diseases.

D. RECENT TRENDS

The oyster industry in Chesapeake Bay has undergone drastic changes in recent years.[16,27-32] A cycle of winter-spring droughts (1980, 1985, 1986) has increased losses of oysters in areas considered safe (salinity <20‰) from MSX over the long term. To avoid the risks of planting seed oysters on private beds, oystermen have found an alternative supply—that of buying oysters from other regions to supply their extensive marketing systems, which were developed over many decades. Soon after MSX devastated oyster beds in lower Chesapeake Bay, the supply of marketable oysters declined to one sixth or less of the 1950s production. Processing houses (of fresh-shucked oysters) began buying market oysters from the Gulf of Mexico. These are brought by truck in the shell and shucked in Virginia, a practice that increases the risks of pollution and introduction of diseases and parasites. Importations of oysters from the Gulf of Mexico have increased steadily, resulting in failure of some local seafood wholesale businesses.

In only 2 years, the recent practice of buying shucked oysters brought by air from the West Coast for repackaging as Virginia oysters has risen to a purported half million gallons (no separate statistics available). These oysters grown on the West Coast (mostly in Washington State) are the large Pacific or Japanese oyster, *C. gigas* (Thunberg). The end result of this marketing practice remains to be seen. It involves less capital risk for Virginia shucking houses than does planting oysters in local waters threatened by diseases, overpopulation of people, and ensuing pollution.

From 1960 to 1980, the supply of seed oysters in the James River was adequate for the reduced planting demand, which resulted in a low price of $2 to $3/bushel. Then in the fall of 1986, the seed area was opened for harvest of seed oysters for shucking, without size limitation. Even though the oysters were small (mostly under the usual 7.5-cm minimum size limit) and poor (yielding about half the usual meat weight of shucking oysters), demand increased the price to $12 to $14/bushel, making the small oysters too expensive for planting as seed oysters.

The Campbell Soup Co. operated special processing plants in Virginia to utilize small oysters; these were opened by steaming for use in soups and stews. The soup companies demand for small oysters led them to buy poor, half-grown James River seed oysters beginning about 1960. During the mid-1960s more oysters from the seed areas were bought for soups than for planting on private beds; this was a wasteful use of poor seed oysters which would have produced several times the quantity of meat if grown to regular shucking size. The discovery in 1975 of Kepone pollution in the James River stopped use of seed oysters for soups.

FIGURE 2. Marine inspector with tongers for seed oysters in the James River, Virginia.

Seed oysters were caught by hand tongs in Virginia (see Figures 2, 3, 4), (by small winch-operated dredges in Delaware Bay) and transported by truck or boat to planting areas. In the Rappahannock River, seed oysters were planted by a specially designed hopper barge, but elsewhere they were often shoveled or washed overboard with a water hose resulting in uneven distribution. Seed oysters already 2 or 3 years old were commonly left another 2 years on growing beds before harvest. Private beds are dredged with power equipment, but seed and market oysters are harvested from public beds by hand tongs or by patent tong rigs lifted by engine power. Most oysters are shucked by hand with a stabbing knife although numerous mechanical shuckers have been built and tested.[16]

As of 1986, the traditional methods of planting and harvesting oysters were still being practiced in the Chesapeake Bay. There had been over two decades of steady decline in planting of James River seed oysters on private beds and in harvesting of market oysters from public beds. The second long drought of the 1980s (1985 to 1988) accelerated the decline throughout Chesapeake Bay. The two diseases of oysters that are usually regulated by spring low salinities spread more widely and rapidly, and become more intensive in the upper beds of rivers and the bay which were usually sanctuaries for oyster culture. There were three or four winter-spring droughts with river discharges less than half normal; these were new phenomena in the bay. Salinities increased by as much as 10‰ over normal levels (20‰ at the Maryland Bay Bridge).

Perkinsus marinus became the dominant disease of oysters in the 1980s. It invaded the James River seed oyster areas and by 1989 had killed most oysters not harvested by tongers. No public seed stocks remained available except some disease-infested ones in coastal plain rivers such as the Great Wicomico River. Public beds became essentially barren throughout the low-salinity areas which were sanctuaries for oysters after MSX invaded the bay in 1959. Tongers flocked to the James River seed area which has the strongest spring river discharge to control the diseases. They harvested the survivors for market at $20 to $25/bushel.

FIGURE 3. Buyboat surrounded by tongers boats in James River, Virginia.

FIGURE 4. Transferring seed oysters from tongers boat to buyboat in the James River, Virginia. Tongers boats usually have two tongers and a culler for removing shells returned overboard.

Broodstocks are now at record low levels throughout the Chesapeake Bay, which puts recruitment by natural setting in serious jeopardy. The high price for oysters led to interest in hatcheries and tray-grown oysters, a much more expensive method of culture than plantings on natural beds. The future of the oyster industry in Chesapeake Bay is uncertain; when weather conditions return to normal and the diseases recede to their enzootic boundaries, nature and man may be able to restore this fabulous fishery.

III. FUTURE OF THE OYSTER INDUSTRY

The greatest natural beds on earth for growing oysters are found on the Atlantic coast of North America.[14,16,25,27,30] The combination of flooded river valleys in wide, low, coastal plains with moderate rainfall provides unmatched opportunities. This coastal plain extends from Long Island Sound to Texas. The most favored in total area of shallow beds and large sectors of low-salinity zones is Chesapeake Bay. The state of Maryland has thousands of hectares of natural oyster beds, most of which are now barren or produce at very low levels because setting has failed. These beds plus more in Virginia are usually free of predators and diseases and have few serious competitors, such as fouling organisms. Most natural beds are designated as public oyster grounds, which are far too extensive for either state to replenish with seed oysters. Shells are planted in potential seed areas, but low setting rates impede progress. There is little need to consider use of expensive, labor intensive methods of off-bottom culture such as those used in Europe, Asia, and Australia.

Overfishing of natural beds combined with diseases, pollution, and severe weather conditions (drought and flood) have steadily reduced harvestable oyster stocks throughout Chesapeake Bay. Apparently, levels of broodstocks are too low to sustain previous setting levels, and survival and transport of larvae are reduced by adverse physical conditions such as low oxygen in channels in summer. Self-replenishing beds are rare now in comparison to past decades.

Public management of natural seed areas for harvesting of market oysters results in invasion of diseases because growth to adequate size takes too long. In addition, meat of oysters grown in low salinities is usually poor in quality. Seed oysters should be harvested within 1 or 2 years after setting if they are transplanted to growing beds essentially free of predators. This practice allows new shell to be planted and the cycle of culture before harvest to be shortened.

For almost one hundred years, biologists have recommended that natural, public oyster beds in Chesapeake Bay be leased to private companies for development, but watermen have consistently opposed attempts to permit leasing of the areas.[14,16,27-29,31] With brood stocks depleted and setting undependable, private investment is discouraged as well. Problems of pollution and nutrient discharge are increasing with larger human populations. Industrialization and urbanization are increasing rapidly. It may be getting too late for restoration of the oyster industry. The primary seed area for Chesapeake Bay is the James River, and it is no longer capable of producing 2 to 3 million bushels of seed oysters as it did in the 1950s, without restoration of brood stocks and setting levels. Purposeful management seems to be obscured by temporary political and economic expedience.

For two decades numerous public and private interests have explored the use of hatcheries to produce seed oysters.[12,30] With rare exceptions, all have failed because of high costs and failure of nursery culture. There are wide variations in price of market oysters along the Atlantic coast. Canada and New England obtain the highest prices for oysters on the half shell. Seed oysters there are scarce and bring high prices.

In Chesapeake Bay, a considerable difference in cost of seed ($2 to $3/bushel) and market oysters ($12 to $15) may seem to offer prospects for economical production of hatchery seed. A major argument for hatchery seed is the ability to use selected brood stocks for rapid growth and disease resistance. The state of Virginia initiated a new hatchery program in 1985. Eyed larvae are to be provided to oyster planters for setting and nursery care. The attempt to produce seed oysters in a state hatchery was encouraged by oystermen who visited successful hatcheries on the west coast of North America. Natural seed-producing areas are quite limited on the west coast, and private culture of oysters is more aggressive than on the east coast. The large tidal range on the west coast provides intertidal areas for growing spat to planting size. In Virginia, low-salinity areas or suspension must be used for nursery systems. These methods are likely to produce seed too expensive to compete with the 2- and 3-year old James River seed oysters that are low in price and high in quality.

The hurricanes that have recently destroyed oysters on the Gulf coast and the droughts endangering low-salinity sanctuaries in Virginia and Maryland keep the oyster industry in constant flux.[31] With rising costs for labor, new machines, and insurance, the oyster industry has little time to resolve basic biological problems such as pollution and setting. Neither land nor sea farming can compete with the capitalization of industry and housing developments. With marketing and cultivation in a state of flux and growing beds idle, it will take rare purpose and leadership to save Chesapeake Bay oysters. The bounty of oysters harvested around 1880 was 14 million bushels in Maryland alone.[30,31] Most oyster beds were self-sustaining in those times, and spatfalls were frequently high enough to maintain reproducing populations. Now even in seed areas good sets are rare, and failures of setting are frequent. West Coast oystermen using many innovative techniques and an introduced oyster *(C. gigas)* are outperforming the east coast in culture methods. The East Coast is handicapped by traditional attitudes toward oyster farming. The most serious problem on the East Coast is the tradition of public ownership of natural beds, which ensures low production and misuse of beds as a result of politics. Private culture has long been predominant in New England, where high prices have sustained the industry.

The embayments north of Chesapeake Bay, in New Jersey, New York, and New England, are mostly areas of high salinity and are more difficult to manage because of fouling organisms, predators (starfish and green crabs), and disease problems. South of Chesapeake Bay along the Atlantic coast, all these problems of predators, diseases, boring sponges, and fouling are intensified so that intertidal culture is more promising. These problems extend to the Gulf of Mexico, where subtidal culture on shallow beds is important.

A. FUTURE OF THE OYSTER INDUSTRY IN CHESAPEAKE BAY

Management of the Chesapeake Bay oyster industry is based essentially on politics and economics. In Virginia there are large acreages of leased beds available for private planting and a limited supply of low-cost seed oysters. Most planters have ceased culture or plant limited acreage. Until economics become more favorable for the 2-year growing cycle required to mature seed oysters to market oysters, planting may continue to decline as imports of marketable oysters increase. Opening abandoned or barren public beds for leasing may not entice watermen to grow oysters in Virginia. In Maryland, most natural oyster grounds are designated as public grounds and managed by state agencies. Despite more favorable growing conditions due to low-salinity waters, failures in spatfalls even in designated seed areas limit the amount of state repletion possible by transplanting of seed oysters. Much effort is currently being expended in an attempt to save the habitats of Chesapeake fisheries from growing pollution and urbanization problems.

The time would seem right to consider some biological management procedures to restore broodstocks and setting levels. Public oyster grounds are a source of disease infection for all beds because regular, complete harvesting is not achieved. If private culture is to succeed, more intensive management than just simple planting and harvesting of oysters in crop cycles must be applied. The following recommendations are based on the assumption that private culture is more successful and efficient than that resulting from public management. Therefore, entrepreneurs must be given the freedom of use of leased natural beds and a political climate that favors their proper use.

This overview of *Crassostrea virginica* emphasizes the fishery and its management. A monograph by Galtsoff[1] provides an extensive description of the biology of this species, for which the literature is extensive. A book by Korringa entitled *Farming the Cupped Oysters of the Genus Crassostrea* describes operation of the fisheries in Long Island and Louisiana around 1972.[3] More recently, Kennedy and Breisch have written an intensive review of the biology of *C. virginica* and have compiled social and economic commentaries on the fishery in Maryland.[31] The present review emphasizes fishery conditions of the middle Atlantic states in Chesapeake and Delaware bays, where high production formerly dominated the industry and where cultural problems now are most severe. The potential for restoring productivity is great in terms of suitable habitats and knowledge of biology of the oyster—the primary obstacles are political and economic. In Chesapeake Bay, the states remain stubbornly attached to public management of all natural beds. Seed beds should, in fact, remain under public management to ensure conservation of adequate broodstocks for successful reproduction. New Jersey provides a model for seed bed management in Delaware Bay. The catching season is limited to a few weeks in May and June, and small dredges are allowed to harvest until a catch limit set by biologists to save adequate broodstock is reached.

Disease has been a limiting factor in both bays although intensive selection in Delaware Bay continues to provide genetic resistance to the pathogen called MSX *(Haplosporidium nelsoni)*. In Virginia, about half the leased grounds of the 1950s (around 55,000 ha) were abandoned after 1960, and fear of MSX has inhibited planting in usually safe areas with low salinities. Since the record hurricane flood of June 1972, which killed oysters and most soft shell clams (*Mya arenaria* Linnaeus) in Maryland, cow-nose rays (*Rhinoptera bonasus* [Le Sueur]) have destroyed eelgrass and some oyster beds while searching for their favorite clam food. Fear of sudden destruction of oyster beds by rays has been blamed for failure to plant in low-salinity areas where diseases and other predators are rarely problems. The destruction by schools of rays is so rapid and complete that losses are prohibitive for future planting.

Some potential seed areas in relatively low salinities are compromised by another disease, *Perkinsus marinus* (Mackin, Owen, and Coller) (''Dermo''), which has become permanently established on public seed beds in rivers such as the Piankatank, Great Wicomico, and St. Marys.[33,34] Eradication of the diseases is impossible, but short cycles of seed production in which shells are planted and small oysters are harvested early would minimize the risks. Holding slow-growing oysters to market size in these estuaries provides regular sources of disease infection.

Virginia now imports most of the oysters it sells for shucking or repacking from the Gulf of Mexico and the West Coast of the U.S. Domestic production is low. Seed oyster supply is limited by inadequate reproduction and by low number of tongers available to catch the oysters by hand. Overfishing of public market oyster beds has depleted the broodstocks, and failure of private planters to restock their grounds compounds the failure of setting in most rivers. In contrast, a few private planters in New England are successfully applying intensive cultural practices. Oystermen in both New England and the Gulf of Mexico seem to be thriving. Chesapeake Bay is the last major area around the earth with a major free, public oyster fishery. It is time for the states to reorganize management of culture systems.

B. BIOLOGICAL RECOMMENDATIONS FOR MANAGEMENT OF THE OYSTER INDUSTRY IN CHESAPEAKE BAY

A list of biological recommendations for management of the oyster industry in Chesapeake Bay follows:

1. Restore the setting capacity of the James River seed area to supply the whole bay, including Maryland.
2. Lease most public beds to private groups or individuals for oyster culture – a gradual change may be politically feasible.
3. Revise rental terms to ensure usage of grounds and higher fees for repletion and management of seed oyster beds.
4. Require all shells to be replanted by shucking houses where spatfall is adequate for seed beds or self-sustaining market oyster grounds.
5. Plant shells at proper times and only in good setting areas where survival of spat is favorable.
6. Maintain broodstocks in seed areas by limitations on harvesting — the Delaware Bay seed oyster area is an example of good management.
7. Permit use of efficient harvesting gear – dredges on leased grounds — as in Delaware Bay. Light dredges can be permitted on seed beds if total catch is controlled.
8. Close public beds when necessary to protect good sets and to ensure high yields and prime condition of oysters when they are marketed.
9. Encourage rotation of private beds and fallowing between crops to control diseases and predators.
10. Monitor supplementary seed areas (intermittent setting) for occasional good spatfalls, which should then be used as seed sources and restocked with shells.
11. Integrate the oyster fishery in Chesapeake Bay by removing state restrictions on availability of seed oysters and acquisition and use of grounds.
12. Monitor diseases to avoid transplanting infected oysters to disease-free areas.

REFERENCES

1. **Galtsoff, P. S.,** *The American Oyster, Crassostrea virginica (Gmelin), U.S. Fish Wildl. Serv. Fish. Bull.,* 64, 1964.
2. **Gunter, G.,** The generic status of living oysters and the scientific name of the common American species, *Am. Midl. Nat.,* 43, 438, 1950.
3. **Korringa, P.,** *Farming the Cupped Oyster of the Genus Crassostrea,* Elsevier Science, New York, 1976.
4. **Korringa, P.,** Recent advances in oyster biology, *Q. Rev. Biol.,* 27, 266 and 339, 1952.
5. **Andrews, J. D.,** Climatic and ecological settings for growing shellfish, *Proc. Conf. Artif. Propagation Commer. Valuable Shellfish.,* Coll. Mar. Stud., University of Delaware, 77, 1971.
6. **Andrews, J. D.,** Transport of bivalve larvae in James River, Virginia, *J. Shellfish Res.* 3, 29, 1983.
7. **Andrews, J. D.,** Setting of oysters in Virginia. *Proc. Natl. Shellfish. Assoc.,* 45, 38, 1955.
8. **Andrews, J. D.,** Pelecypoda: Ostreidae, in *Reproduction of Marine Invertebrates. V. Mollusca: Pelecypods and Lesser Classes,* Giese, A. C. and Pearse, J. S., Eds., Academic Press, New York, 1979, 293.
9. **Menzel, R. W.,** Some phases of the biology of *Ostrea equestris* Say, and a comparison with *Crassostrea virginica* (Gmelin), *Tex. Inst. Mar. Sci.,* 4, 70, 1955.
10. **Loosanoff, V. L.,** Time and intensity of setting of the oyster, *Crassostrea virginica* in Long Island Sound, *Biol. Bull.,* 130, 211, 1966.
11. **Loosanoff, V. L.,** *The American or Eastern Oyster,* Bur. Commer. Fish. and Wild. Serv. Cir. 205, 1955.
12. **Loosanoff, V. L. and Davis, H. C.,** Rearing of bivalve mollusks, in *Advances in Marine Biology,* Vol. 1, Russell, F. S., Ed., Academic Press, New York, 1963.
13. **Mackenzie, C. L.,** Oyster culture in Long Island Sound 1966-1969, *Commer. Fish. Rev.,* 32, 27, 1970.
14. **Beaven, G. F.,** Various aspects of oyster setting in Maryland, *Proc. Natl. Shellfish. Assoc.* (1954), 29, 1955.
15. **Merritt, D. W.,** *Oyster Spat Set on Natural Cultch in Maryland Portion of the Chesapeake Bay (1939-1975),* University of Maryland Center for Environmental and Estuarine Studies Special Rep. 7, 1977.
16. **Haven, D. S., Hargis, W. J., and Kendall, P. C.,** *The Oyster Industry of Virginia. Its Status, Problems and Promise,* Virginia Institute Marine Science Special Papers Marine Science, 4, 1978.
17. **Andrews, J. D.,** Measurement of shell growth in oysters by weighing in water, *Proc. Natl. Shellfish. Assoc.,* 52, 1, 1961.

18. **Krantz, G. E. and Merritt, D. W.,** An analysis of trends in oyster spat set in the Maryland portion of Chesapeake Bay, *Proc. Natl. Shellfish. Assoc.,* 67, 53, 1977.
19. **Winslow, F.,** Present condition and future prospects of the oyster industry, *Trans. Am. Fish. Soc.,* 13, 148, 1984.
20. **Newkirk, G. F.,** Review of the genetics and the potential for selective breeding of commercially important bivalves, *Aquaculture,* 19, 209, 1980.
21. **Haskin, H. H. and Ford, S. E.,** *Haplosporidium nelsoni* on Delaware Bay seed oyster beds: a host-parasite relationship along a salinity gradient, *J. Invertebr. Pathol.,* 40, 388, 1982.
22. **Andrews, J. D. and Hewatt, W. G.,** Oyster mortality studies in Virginia. II. The fungus disease caused by *Dermocystidium marinum* in oysters of Chesapeake Bay, *Ecol. Monogr.,* 27, 1, 1957.
23. **Andrews, J. D. and Wood, J. L.,** Oyster mortality studies in Virginia. VI. History and distribution of *Minchinia nelsoni,* a pathogen of oysters in Virginia, *Chesapeake Sci.,* 8, 1, 1967.
24. **Sindermann, C. J.,** *Disease Diagnosis and Control in North American Marine Aquaculture,* Elsevier, New York, 1977.
25. **Mathiessen, G. C.,** *A Review of Oyster Culture and the Oyster Industry in North America,* Woods Hole Oceanogr. Inst. Contrib. No. 2528, 1971.
26. **Hopkins, S. H., Mackin, J. G., and Menzel, R. W.,** The annual cycle of reproduction, growth, and fattening in Louisiana oysters, *Natl. Shellfish Assoc. Conv. Papers,* 1953, 39.
27. **Brooks, W. K.,** *The Oyster, a Popular Summary of a Scientific Study,* 2nd ed., Johns Hopkins Press, Baltimore, 1905.
28. **Engle, J. B.,** The molluscan shellfish industry: current status and trends, *Proc. Natl. Shellfish. Assoc.,* 56, 13, 1966.
29. **Quittmeyer, C. L.,** *A Report on the Chesapeake Bay Fisheries of Maryland,* Seafood Advisory Committees of Wye Institute, Centerville, MD, 1966.
30. **Wharton, J.,** *The Bounty of the Chesapeake, Fishing in Colonial Virginia.* University Press of Virginia, Charlottesville, VA, 1957.
31. **Kennedy, V. S. and Breisch, L. L.,** *Maryland's Oysters: II. Oyster Fishery of Maryland,* University of Maryland Sea Grant Program, 1981.
32. **Breisch, L. L. and Kennedy, V. S.,** *A Selected Bibliography of Worldwide Oyster Literature,* University of Maryland Sea Grant Program, 1980.
33. **Andrews, J. D.,** Management strategies to control the disease caused by *Perkinsus marinus, Trans. Am. Fish. Soc.,* in review.
34. **Andrews, J. D.,** Epizootiology of the disease caused by the oyster pathogen *Perkinsus marinus, Trans. Am. Fish. Soc.,* in review.

Chapter 3

CULTURAL PRACTICES AND PRODUCTION OF THE AMERICAN OYSTER, WITH EMPHASIS ON THE SOUTH ATLANTIC AND GULF OF MEXICO

Winston Menzel

TABLE OF CONTENTS

I. INTRODUCTION

The purpose of this chapter is to supplement the preceding chapter with some additional data for the American oyster *Crassostrea virginica* (Gmelin), with emphasis on the South Atlantic and Gulf of Mexico states from North Carolina to Texas. This species is cultivated to some extent in both Canada and Mexico, but these operations will not be discussed.

II. BIOLOGY

The biology of this species was discussed by Andrews in the preceding chapter. In most respects, the biology is the same throughout its range. Shell growth, however, proceeds year-round in the more southern part of the range, particularly in the Gulf of Mexico, where the water temperature usually does not go below about 10°C, except during sporadic cold periods, when it may drop to freezing. The spawning season is also longer in warmer regions, starting usually in April and lasting through September, and sometimes October. There is usually a high peak in the first part of the season followed by some spawning throughout the summer and another, lesser peak in the fall. As a consequence, the period of less than prime meat quality is longer than in more northern regions, where spawning seasons are shorter.

III. FISHERY

A. HISTORY

The native American Indians used this mollusk as a source of food and probably practiced some primitive culture. There is evidence they transplanted oysters from distant places to more convenient locations near their settlements.

The European settlers quickly learned to exploit this food resource. Ingersoll[1] discusses the practices of the oyster industry in the eighteenth and nineteenth centuries. Oysters soon became depleted in northern New England, and oysters were shipped from southern regions such as Delaware and Chesapeake Bays. It is possible that, because of the perishable nature of the cargo and a need for more rapid transport, the oyster trade led to the development of the fast "Yankee Clipper" ships.

Considerable harvesting also took place near some of the southern centers of population, e.g., Charleston, SC; Jacksonville and Pensacola, FL; Mobile, AL; Biloxi, MS; New Orleans, LA; and Galveston, TX.

In these earlier times, oysters were cheap food and had not become the luxury item they are today. Ingersoll[1] gives accounts of barter: oysters for a newspaper subscription in Virginia and a bushel of oysters for a bushel of corn in North Carolina.

Partly because of inadequate rapid transportation and refrigeration, a canning industry developed in the southern regions, even as far north as Baltimore, MD. These canneries gradually declined in numbers, and today there are none.

As oysters decreased in supply, most regions established laws and regulations for their conservation. Sometimes the laws were in conflict with those in adjacent areas. These regulations included minimum-size laws, prohibition of summer harvest, and restrictions against dredging oyster reefs to use the shells for industrial purposes.

B. IMPORTANCE

Oysters on the East Coast have always been an important fishery, but the overall yield has declined steadily since the beginning of this century. Figure 1 shows an outline of the eastern U.S. with the oyster-producing states in the several regions. Table 1 and Figure 2 give the average yearly production along the coast for each region for decade intervals since

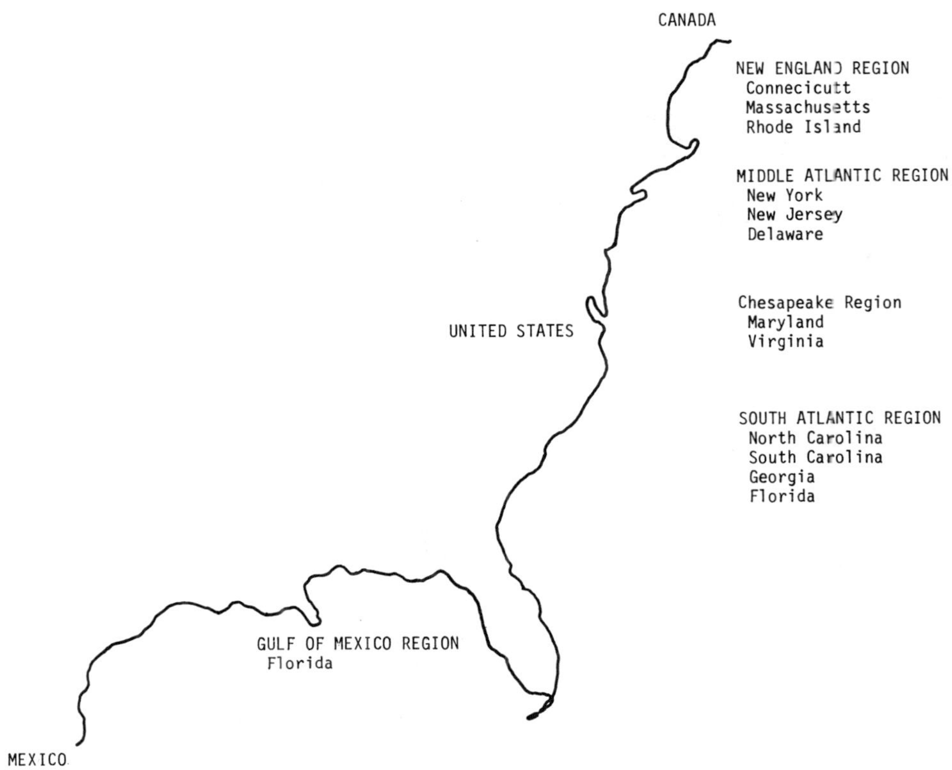

FIGURE 1. Outline map of the eastern U.S., giving the oyster producing states in the several regions.

TABLE 1
Average Yearly Harvest per Decade of Oyster Meat in Metric Tons and Percentage of Harvest from Each Region

	New England		Middle Atlantic		Chesapeake		South Atlantic		Gulf of Mexico		
Years	**t**	**%**	**t**	**%**	**t**	**%**	**t**	**%**	**t**	**%**	**Total t**
1880—89	3,833.2	5.7	14,425.5	21.5	45,376.1	67.5	1,386.1	2.1	2,176.7	3.2	67,197.6
1890—99	7,844.8	10.3	12,974.1	17.1	47,833.2	63.1	2,735.5	3.6	4,418.6	5.8	75,806.2
1900—09	6,635.1	9.8	13,148.8	19.2	31,571.7	46.2	8,063.0	11.8	8,932.7	13.0	68,351.3
1910—19	8,845.0	14.7	12,677.3	21.2	30,275.9	50.5	2,717.5	4.5	5,403.6	9.1	59,919.3
1920—29	2,735.0	6.0	12,206.0	27.0	20,266.2	44.8	3,261.6	7.2	6,771.8	14.9	45,246.6
1930—39	3,631.0	10.4	7,365.7	21.1	14,547.9	41.8	2,172.2	6.2	7,183.1	20.6	34,899.9
1940—49	1,762.4	5.6	6,496.9	20.8	15,014.6	48.0	1,218.9	3.8	6,768.0	21.6	31,260.8
1950—59	526.8	1.9	4,735.8	17.3	14,723.9	53.6	1,559.5	5.7	5,872.0	21.4	27,418.0
1960—69	135.6	0.6	612.2	2.8	10,362.5	46.9	1,636.9	7.4	9,331.3	42.2	22,078.5
1970—79	229.8	1.0	1,178.4	5.6	10,376.7	49.6	785.0	3.8	8,332.1	39.9	20,902.0
1980—85	344.1	1.7	897.1	4.5	7,433.6	37.5	867.0	4.4	10,252.2	51.7	19,794.0

1880, when the government[2] started collecting statistics. In the earlier years, the catch was not recorded every year for every state; therefore, the data are not very accurate but do show the downward trend.

Production from the New England region was considerable in the early years, but since 1950 has declined to less than 2% of the total. Production from the Middle Atlantic was roughly 29% of the total until the 1960s but has now fallen to 5%. The Chesapeake Bay

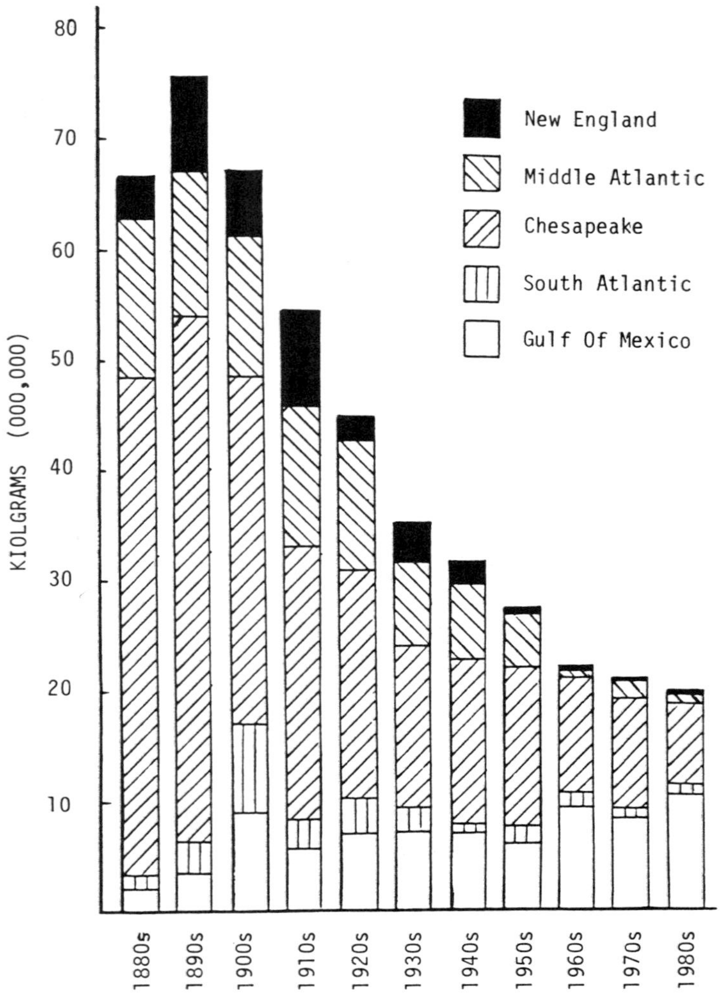

FIGURE 2. Average oyster production per year at decade intervals for the regions of the East Coast of the U.S.

area (Maryland and Virginia) formerly produced half of the oysters in the country but in the 1980s produced only a little over one third. The South Atlantic harvest was highest in the decade of 1900 to 1909, when it accounted for almost 12%, but has been only slightly higher than the New England region since the 1950s. Harvest from the Gulf of Mexico states, in contrast to that of the other regions, has shown a fairly steady increase, now accounting for over half of the harvest. Louisiana has been the leader, but other states have increased production also, especially Florida and Texas.

Table 2 gives weight, value, price per kilogram, and yield of meat per 0.1 m^3 in the shell for 1985, the latest year for which data are available.[2] The yield of meat per 0.1 m^3 in the shell has been calculated from the data given by the federal statistics for each state.

C. MANAGEMENT AND CULTURE

With some exceptions, private leasing is relatively unimportant in the southern states. The abundance of natural recruitment and fast growth, both on the public beds and on cultch planted by the states and individuals, lessens the need for the cultivation practices used in areas where oyster farming is more difficult. Several states do manage the public reefs to

TABLE 2
Harvest of Metric Tons of Oyster Meat, Value, Value per Kilogram, Yield in Kilograms per 0.1 m³ in Shell in 1985, and Peak Production Year since 1880

State	1985				Peak production	
	Ton	Value $000	$/Kg	Kg meat/0.1 m³	Year	Ton
Connecticut	372.3	3,870	10.39	9.7	1898	6,917.5
Massachusetts	20.9	240	11.48	8.2	1910	758.6
Rhode Island		Not available			1910	6,878.1
New York	135.9	1,203	8.86	9.5	1911	11,288.9
New Jersey	133.2	1,064	7.99	8.8	1887	10,692.5
Delaware	17.7	95	5.37	7.8	1954	1,972.7
Maryland	2,417.3	9,586	3.97	6.3	1880	32,667.3
Virginia	1,045.0	3,786	3.62	6.0	1880	20,698.5
North Carolina	247.7	1,037	4.19	6.0	1889	2,513.1
South Carolina	649.5	1,372	2.11	3.2	1929	1,424.3
Georgia	15.5	56	3.60	4.2	1908	3,368.3
Florida	1,975.0	4,960	2.52	5.4	1981	3,304.3
Alabama	580.5	1,581	2.72	6.5	1951	996.1
Mississippi	542.3	1,499	2.77	7.9	1937	5,860.9
Louisiana	6,180.5	22,163	3.59	6.0	1975	6,221.4
Texas	2,292.7	8,581	3.79	6.7	1983	3,609.4
Total	16,626.0	61,093	3.68			

varying extents, by planting cultch shell, enforcing closed seasons, restricting types of gear for harvest, and having minimum size laws (7.6 cm). States also sometimes plant seed, mainly from intertidal to subtidal bottoms, and sometimes relay oysters from polluted areas. A rigid monitoring system is maintained, and beds are closed when pollution threatens. A summary for each state is given below.

D. MANAGEMENT AND CULTURE IN THE SOUTH

In most states, public reefs are harvested by hand, either by picking from intertidal reefs or by hand tongs. In some instances dredges can be used on publicly controlled reefs. On public beds, most states require immediate culling and return of undersize oysters and shell to the bed from which they were taken.

As for the Chesapeake Bay, the majority of the oysters are hand shucked and sold fresh under refrigeration. Various shucking methods are used. In "stabbing", the blade of the shucking knife is forced between the two valves to sever the adductor muscle. Another method is to break the hinge ligament. The most common method is to break the posterior edge of the shell to allow easy entrance of the knife; some shucking houses use emery wheels to grind off the shell edge.

In some areas, as in the more northern areas, there is considerable shell trade. Oysters for the shell trade are culled "cleaner" and bring premium prices. A limited number are frozen, and some are shucked and breaded before freezing.

E. CONSTRAINTS

Some of the constraints are man-made, and others arise from natural causes. Probably the strongest constraint is man-made pollution. Thousands of hectares in all regions have been rendered unproductive by pollution, both domestic and industrial. All states constantly monitor for domestic pollution, and some oyster grounds are permanently closed because of the health threats. Others are closed temporarily, especially after runoff from excessive rainfall. Efforts have been made in the last several decades to curb pollution, but ever-increasing human population, factory sites, and boat traffic make it difficult. Dredging for maintenance of navigation channels has destroyed some productive beds in the past. In the past, indiscriminate removal of oysters along with the shells that acted as cultch for recruitment destroyed many productive areas.

Weather conditions such as storms and extended droughts cause much damage directly and indirectly. The South Atlantic and Gulf areas are subjected to many hurricanes, which directly destroy oysters by smothering or by moving to areas in which they cannot survive. Usually the storms are accompanied by heavy rainfall, resulting in freshets. Extended droughts cause damage indirectly: lessened nutrient runoff results in stress from decreased production of phytoplankton, the oysters' food. The water over productive outer beds increases in salinity, favoring invasion by destructive organisms, including some predators, diseases, and competitors that are less euryhaline than oysters.

Predators and diseases occur throughout the habitat of oysters. In the southern U.S., especially in the Gulf of Mexico, the two most destructive are the southern oyster drill *Thais haemastoma* Conrad, with planktonic larvae,[3] a serious predator, and the pathogenic sporozoan *Perkinsus* (= *Dermocystidium*) *marinus* (Mackin, Owen, and Colier).[4] Other enemies include shell boring sponges (*Cliona* sp.), flatworms (*Bucephalus cuculus* McGrady and *Stylocchus frontalis* Verrill), stone crabs (*Menippe mercenaria* [Say]), blue crabs (*Callinectes sapidus* Rathbun), and the ubiquitous xanthid mud crabs.[5] Myriad organisms attach to the shells, often fouling them so that oyster spat cannot attach. Several species of fish destroy oysters, including the black drum *Pogonias cromis* (Linnaeus) and several species of skates and rays.[6] The writer has observed sheepshead, *Archosargus probatocephalus* (Walbaum), and pin fish, *Lagodon rhomboides* (Linnaeus), grazing on attached spat.

F. CULTURAL PRACTICES AND COMMERCIAL IMPORTANCE IN SOUTHERN REGIONS

A brief description is given of production and value and of culture for each state in the South Atlantic and Gulf regions. The extent of public and leased oyster bottoms is given when known.

1. North Carolina

In 1985 North Carolina produced almost 250 t of oyster meats valued for the fishermen at over $1 million. The peak year was 1889, when over 2500 t were harvested (Table 2). The total area of the public bottom is not known, but in 1987 about 900 ha were leased, yielding an estimated 3% of the total production of the state. Management of the leased area consists of planting of cultch shell and some seed planting from intertidal reefs, as well as relaying of polluted oysters for depuration before harvest. The latter two practices are under state supervision and permit. The state manages the public bottom by planting of cultch shells, about 14,000 m^3 in 1987. There has been limited transplanting of intertidal oysters to subtidal areas. Harvesting is allowed on public beds from October 1 through March 31. Harvesting is mostly by hand, picking and tonging, but in the deeper waters of Pamlico Sound, restricted dredging is allowed.

There are some shucking establishments, but considerable numbers of oysters are sold in the shell by roadside vendors and to restaurants.

2. South Carolina

South Carolina had a yield of almost 650 t, valued at about $1.3 million, in 1985, considerably less than the peak year of 1889 when almost 1500 t were harvested (Table 2). In 1987, over 1200 ha were under lease, part of which were primarily for the harvest of quahog clam *Mercenaria mercenaria* (Linnaeus). The percentages of oysters harvested from private and public beds is unknown. In 1987 there were less than 150 ha of approved public intertidal oyster beds and a similar area of polluted beds. Oysters are harvested mainly by hand picking on the intertidal reefs, although dredging is used in some instances under permit.

South Carolina has an extensive recreational fishery, and the state manages over 20 reefs by planting cultch shell. In addition, property owners can harvest oysters adjacent to their property for their home use.

The state encourages private planting through education and research and by promoting the planting of cultch shell and the transplanting of intertidal oysters to subtidal beds.

3. Georgia

Georgia harvested only 15.5 t of oyster meats in 1985, valued at $56,000, but in 1908, the peak year, almost 3400 t were harvested (Table 2). A survey in 1977 found about 100 ha of live intertidal reefs and 15 ha of subtidal beds with at least 50% live oysters.[7]

Information on the amount of leased ground is not available. In 1978, only nine operators were licensed to harvest oysters, each employing several pickers and one shucking house. Formerly, up to six canneries were in operation.[7]

4. Florida

Florida has extensive coastlines on both the Atlantic and the Gulf of Mexico, but over 90% of its oyster harvest is from the Gulf side, and Apalachicola Bay, on the northwest coast, accounts for over 95%.[2] In 1985, production was almost 2000 t, valued at almost $5 million, down from 3300 t in 1981, the peak year (Table 2).

At present, nearly 900 ha of oyster bottom are leased, of which a little over 100 ha are in Apalachicola Bay. Leasing is permitted on any bottom that does not contain oysters. The yield from leased ground is not known.

There are about 4000 ha of public beds in Apalachicola Bay. All harvesting is by hand tongs. Although dredging is allowed by law on leased ground, after public hearings, the opposition by watermen is so intense that no dredging occurs in the state.

The state has an active program of management, including a closed season from April to September, except on some beds in Apalachicola Bay, which are usually opened for summer tonging. The bay is subject to periodic closing from pollution, especially after heavy rainfall. The state plants cultch shells on depleted beds and conducts some transfer of intertidal oysters to subtidal areas.

5. Alabama

Alabama harvested 580.5 t of oyster meat in 1985, valued at over $1.5 million, with a peak production in 1951, when 996 t were produced (Table 2). Private leasing is permitted, but the area of leased ground is insignificant at the present time. There is no closed season, and tongs are the only legal method of harvesting. Dredging was allowed in former years.

Upper Mobile Bay is permanently closed because of pollution, and other beds are closed from time to time if pollution becomes apparent. A survey in 1971 showed about 1200 ha of natural reefs.[8] The state maintains an active program of planting cultch shell, mainly of the brackish-water clam *Rangia cuneata* Gray.

6. Mississippi

In 1985, Mississippi harvested 542.3 t, valued at $1.5 million, down tenfold from 1937, the peak year, when 5860.9 t were harvested. At that time, up to 25 canneries were in operation whereas none operates today. In 1987, over 300 ha were privately leased, and there are about 3000 ha of public reefs. It is not known what percentage of the total production is produced from the private leases.

Hand tongs are used on the public reefs, except in some outer deeper-water areas, where restricted dredging is permitted. A closed season extends from May to September. About

400 ha are permanently closed because of pollution, but lease holders may move oysters for planting on their ground for depuration, under permit.

The state has a large planting program for cultch shell, especially on the outer reefs, which are destroyed periodically by hurricanes. Formerly, oysters shells were used, but recent plantings are mostly of *Rangia* shell secured from Louisiana. In 1987, the state planted over 16,000 m^3 of these.

7. Louisiana

The harvest from Louisiana in 1985 was over 6000 t, worth over $22 million, the largest for any state, only about 40 t lower than that in the peak year of 1975 (Table 2).

Louisiana is essentially an oyster-culture state, with over 100,000 ha leased in 1987. Some of the leased bottom is used for seed production, and it is the practice of some planters to lease unproductive land adjacent to leases to lessen poaching. Also, most planters do not plant all their leases every year. The planting of seed oysters allows an annual harvest. All bottom is open for leasing except that managed by the state.

The state retains control of and manages about 10,000 ha for seed oysters, of which 8900 ha are on the east side of the Mississippi River. There is no gear restriction, size limit, or daily limit. The state also manages one public bed (690 ha) west in Calcasieu Lake that is used for home use, has a daily catch limit, is restricted to tongs, and has a 7.6-cm minimum size limit. Some shell planting is done on the seed reefs, and the state relays polluted oysters from the upper Calcasieu Lake during the summer when harvest is restricted. Open season on public beds is from the first Wednesday after Labor Day to April 1.

A survey was made in 1897 to 1898,[6] and already many of the oyster beds were depleted. The reasons given were overfishing, periodic flooding from the Mississippi, which was not completely leveed then, and destruction by predators. Private leasing was permitted (about 5 ha), and some of the more enterprising oystermen used leased bottoms for seed production and for storage of oysters until they could be marketed or until enough were harvested to transport to market. Some areas were leased to produce superior shell stock oysters for sale in New Orleans.

The Mississippi delta affords much ground for the growing of oysters, but a new constraint has arisen besides the usual hurricanes, droughts, predators, and diseases. Since the Mississippi River levees were completed, the delta no longer has continual freshwater nourishment and settlement, resulting in inland encroachment of saline water. This process increases the suitable oyster growing regions to the north, but they are closer to the ever-increasing human population and the resulting pollution. More important, more natural enemies and diseases have intruded as formerly productive beds have become saltier.

8. Texas

The Galveston Bay area, south of Houston, produces more than 80% of the total oyster harvest in Texas. In 1985, almost 2300 t were landed, valued at over $8.5 million (Table 2), of which 220 t were from about 930 ha of leased bottom. The peak year was 1983, when 3610 t were harvested of which 418 t were from a little over 1000 ha of leased ground.[9] Leasing started in the 1800s and varies from year to year. Any area can be leased except natural oyster reefs. The maximum allowed for leasing is 400 ha. There are no minimum size limits, closed season, or restrictions on methods of harvest on leased ground. There are an estimated 2000 ha of public reefs in the Galveston Bay area.[10] The open season is generally from November 1 through April 30. From time to time the state closes certain reefs, mainly because of pollution; upper Galveston Bay is permanently closed because of pollution. Tongs are used, with restricted dredging in certain areas with daily catch limits.

IV. FUTURE DEVELOPMENTS

It is difficult to foresee the future development of the oyster industry on the East Coast. Much research and development has been done with the American oyster, but no effective practical methods have been developed for the control of some diseases, e.g., *Perkinsus marinus,* or for some of the predators. The devastations caused by storm action and droughts are uncontrollable. Oysters are perishable and are seasonal, partly because during the spawning season, the meats are inferior. Summer demand has been increasing in recent years, as a result of better refrigeration and transportation and the dispelling of the myth that oysters should not be eaten in any month spelled without an ''r''. Although prices have increased rapidly so that oysters are now a luxury food, the demand seems insatiable. In some states strong socioeconomic and political factors work against more efficient culture. The many hazards may result in complete failure in some years, and this threat discourages many new ventures.

Oyster culture is not as intensive on the East Coast of the U.S. as it is on the West Coast, where the exotic species *Crassostrea gigas* (Thunberg) is used. Hatcheries are not practical as yet on the East Coast, and the main constraint in the more northern areas is a bountiful supply of seed; mainly the largess of nature has been relied on. In the southern area, oysters are harvested mostly from controlled public beds. The planting of cultch shells and some seed is far from adequate.

The only southern state that has consistently increased production through the years is Louisiana, which has large private plantings coupled with an abundant seed supply managed by the state. Probably one of the reasons for the enlightened programs in Louisiana was the early depletion of the public beds. To echo the conclusion of Andrews in the preceding chapter, private, as opposed to public, culture seems the only way to achieve maximum production. Production probably will never reach that of around the turn of the century because of pollution, destruction of once productive beds for navigation purposes, and filling of many subtidal areas.

ACKNOWLEDGMENTS

The writer thanks the following people, who furnished data on culture and practices in their respective states: Michael Marshall and Hugh J. Porter (North Carolina), Victor Burrell (South Carolina), Mark Berrigan (Florida), Mark Van Hose (Alabama), William Demoran (Mississippi), Ronald Dugas (Louisiana), and C. E. Bryan (Texas).

REFERENCES

1. **Ingersoll, E.,** A Report on the Oyster, U.S. Fish Comm., Section X (Monogr. B), 1881, 1.
2. Fishery Statistics of the United States, U.S. Government Printing Office, Washington, D.C., numerous years.
3. **Butler, P. A.,** Synoptic Review of the Southern Oyster Drill *Thais haemastoma floridana,* NOAA Tech. Rep. NMFS 35, U.S. Department of Commerce, 1985, 1.
4. **Ray, S. M.,** Biological Studies of *Dermocystidium marium,* a Fungus Parasite of Oysters, Spec. Iss., Rice Institute, Houston, 1954, 1.
5. **Menzel, R., Hulings, N. C., and Hathaway, R. R.,** Oyster abundance in Apalachicola Bay, Florida in relation to biotic associations, influenced by salinity and other factors, *Gulf Res. Rep.,* 2(2), 73, 1966.
6. **Moore, H. F.,** Report of the oyster beds of Louisiana, *U.S. Commer. Fish Fish.,* 24, 1, 1899.
7. **Harris, C. D.,** Survey of the Intertidal and Subtidal Oyster Resources of the Georgia Coast Proj. No. 2-234-R, Georgia Department of Natural Resources, Brunswick 1980, 1.

8. **Anderson, W. F., Crance, J. H., and May, E. B.,** A Survey of the Oyster and Oyster Shell Resource of Alabama, Seafoods Div., Alabama Department of Conservation, Appendix to Mar. Res. Bull. 4, 1971, 1.
9. **Osburn, H. B., Saul, G. E., and Hamilton, C. L.,** Trends in Texas Commercial Fishery Landings, 1977—1985, Manag. Data Ser. 107, Coastal Fisheries Branch, Texas Parks and Wildlife Department, Austin, 1986, 1.
10. **Hofstetter, R. P.,** Trends in Population Levels of the American Oyster *Ostrea virginica* Gmelin in Galveston Bay, Texas, Tech. Ser. No 24, Texas Parks and Wildlife Department, Austin, 1977, 1.

Chapter 4

THE EUROPEAN OYSTER, *OSTREA EDULIS* L., IN MAINE AND EASTERN CANADA

Herbert Hidu and Rene E. Lavoie

TABLE OF CONTENTS

I. BIOLOGY

The European flat oyster *Ostrea edulis* L., family Ostreidae, has a shape typical of a reclining species. The valves are subcircular to circular; the left valve is deeply concave and the flat right valve acts as a lid. The species can get very large and old, as evidenced by the naturalized Maine specimens of Dutch stocks (Figure 1). European oysters are reported to exceed 12 years of age[1] and to reach sizes of 15 × 15.4 cm.[2] Both of the Maine oysters pictured exceed these limits and, from a rough estimate of age by growth bands, exceed 12 years of age.

The European oyster has a 2000-mi European range from Norway to Morocco, including the Mediterranean and Black seas. New World introductions include Connecticut, Maine, California, and Washington in the U.S.; eastern Canada, especially the Atlantic coast of Nova Scotia (see below); and Japan.[3]

The European oyster is a protandric hermaphrodite, changing sexes generally twice during a single season.[4,5] Some function as males early in the spawning season and later change to females and vice versa, so obviously both sexes are present at one time.[6] Oysters can also be ambisexual to some extent, with male and female gametes present in adjacent follicles.[7]

The species is thought to exist as a series of physiological spawning races.[8] Ahmed[3] reviewed literature indicating that the lowest-temperature race occurs in Spain and requires 12 to 13°C for spawning, whereas in Norway the spawning temperature is 25°C in the sheltered and warmed fjord waters.[8] The spawning temperature of Dutch stocks is thought to be 15 to 16°C.[9]

Earlier knowledge of larval biology and ecology has been reviewed in great detail by Korringa,[9] who helped develop the most efficient management techniques for the Dutch Oosterschelde. Briefly, the very telelolechethal eggs pass into the branchial or mantle chamber, where they are fertilized by externally released sperm. There they are incubated for a period of 6 to 18 d depending on ambient temperature from 13 to 23°C until they are released to the environment in a process called "swarming". Clouds of larvae are released by contraction of the adductor muscle, the entire act accomplished in a few hours. In the Oosterschelde neither swarming nor spawning can be correlated to near-term temperature peaks, but rather each is a product of cumulative temperature experience. Larvae then spend 8 to 10 d in the plankton growing from a shell length of 165 to 185 μm to 260 to 300 μm, feeding on "naked flagellates" not larger than 8 to 10 μm in size. The larvae are thought to be distributed passively by the current in the well-mixed Oosterschelde and have not been observed to undergo environmentally induced vertical migration, which might lead to differential transport. (See Andrews[10] for an excellent review of this subject.)

European oyster larvae are quite stenoholine. Korringa[9] found that the lower limit of good larval growth and setting is about 22.5‰, whereas Walne[11] obtained spatfall in experimental salinities only as low as 25.9‰. Davis and Ansell,[12] using Dutch stocks, obtained settlement in salinities as low as 20‰. Further, if larvae were reared to maturity at and above 22.5‰, they could set and survive at salinities as low as 15‰. Thus, it would be possible for larvae to colonize low-salinity areas if carried there after they had achieved maturity in higher salinities.

At setting, this species is definitely "gregarious".[13] Originally, the British believed that the spat (and adults) released a waterborne chemical to stimulate settlement in the adjacent area. In more recent times, they have come to believe that the gregariousness in settlement is mediated by surface chemistry; the setting response in benthically distributed mature larvae is released only after contact with specific compounds on the setting substrate, i.e., periostracum of oysters.[14] Our experiments indicate that the first viewpoint should not be dismissed.[15] Pallial fluid administered prior to and during exposure to cultch shells significantly

A

B

FIGURE 1. Maine-grown European oysters. The large oyster (A) in profile measures 17 × 14 × 7.5 cm. The other oyster (B) measures 16 × 15 × 6 cm.

increased settlement within a 1-h period. The mechanism of gregarious setting should be further resolved because it greatly affects our understanding of how oyster communities are formed and maintained.

Loosanoff[7] studied gametogenesis in the original Dutch implant at Boothbay Harbor, ME. He found the oysters in an early stage of gametogenesis in late May at 10 to 16°C. By mid-June the gonadal layer was still comparatively thin, and only in exceptional cases did follicles extend all the way to the digestive gland. Spawning began in the second week in July and continued until the end of August, and in September temperature conditions produced a resorption of gonad. Loosanoff believed that at prevailing Maine temperatures these oysters were able to complete only a single sex phase in a season.

Recent evidence and observation indicate that the unusually high summer water temperatures and their long duration may have played a major role in the recent population expansion in the Cundy's Harbor region of Maine. The unusually high temperatures in midsummer 1979 (up to 21°C) appear to be associated with the first large quantities of oysters 3 years later in Quahog Bay (Figure 2).[16] Similarly, in 1985, after a relatively warm summer, we observed abundant oyster settlement in this region, whereas the relatively cool summer of 1986 produced only sporadic settlement very late in the season.

II. FISHERY

The European oyster fishery in Maine and eastern Canada originated from intentional introductions in the 1940s and 1950s. The Maine introductions have stimulated a small but growing intensive aquaculture industry based on hatchery-produced seed, which is grown at leased sites. Now, 35 years later, new natural beds of European oysters are providing the beginnings of a new natural fishery in south central Maine. In eastern Canada, the species is confined to the Atlantic coast off Nova Scotia. Several off-bottom culture operations are still dependent on hatchery-reared seed. There is no known self-sustaining natural population in Canada.

A. MAINE FISHERY

European oysters came to Maine in 1949, when Dutch Oosterschelde stock was introduced at three coastal locations in cooperation with Korringa of the Netherlands and Loosanoff of the old Federal Bureau of Commercial Fisheries at Milford, CT.[17,18]

Original planting sites include Basin Cove, Harpswell, Boothbay Harbor, and the Taunton River in Franklin. Later (1954 to 1961) there were five more imports of Dutch stocks both directly from Holland and from progeny of Dutch stocks produced at the Milford laboratory. Planting locations included Boothbay Harbor (again), Small Point, Phippsburg, Peters Island, Bristol, and Mere Point Bay, in Harpswell (Figure 2). By 1962, small progeny populations had become established in at least 14 sites in some proximity to the original sites of introduction.[19]

Later, in the 1970s, additional European oyster populations were "introduced" as part of aquaculture raft culture operations and may have functioned as broodstock populations (Figure 2). The potentially most important of these were 3000 to 4000 oysters introduced by Montescue Moree in 1974 to 1975 at Leavitt Island, Cundy's Harbor, and 100,000 introduced in 1977 by Intertide Corp. at Wilson Cove, Middle Bay, Harpswell Neck. Intertide Corp. continued a nursery/grow-out operation through 1983.[15] Also, several other commercial operations, particularly on the Damariscotta River (Figure 2), cultured several million European oysters from 1975 to the present using, first, seed produced in California by Pacific Mariculture, Inc. of Pigeon Pt., CA, and International Shellfisheries, Inc., of Moss Landing, CA, and later seed from five Maine hatcheries including the University of Maine hatchery

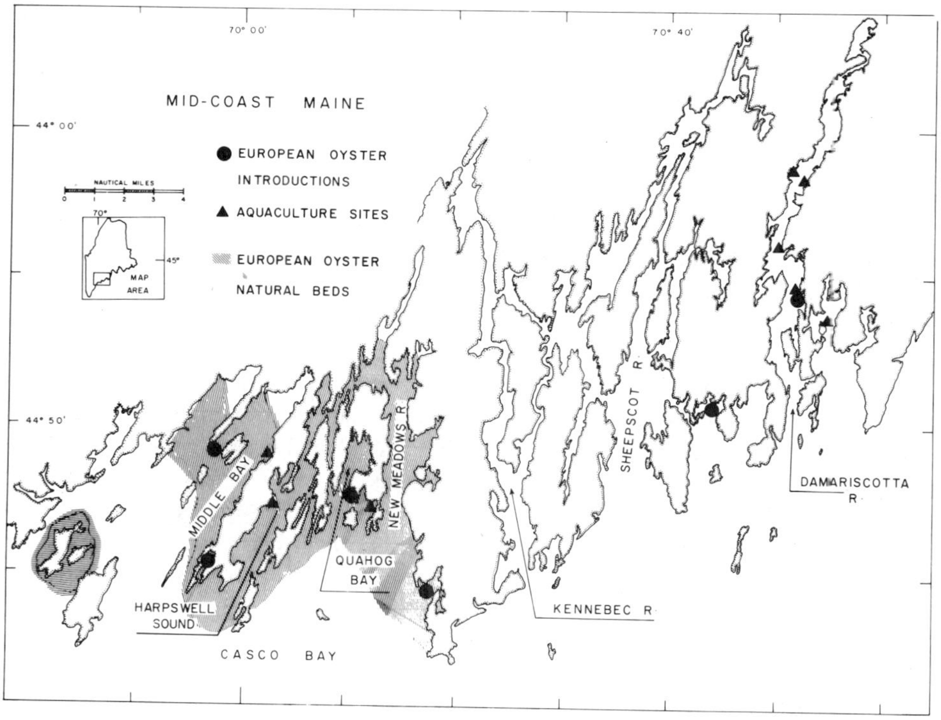

FIGURE 2. South central coastal Maine, sites of original introduction, aquaculture, and natural population establishment of the European oyster.

at Walpole, ME. Most or all of these stocks are thought to be of Dutch origin, although with the diversity of source we cannot be absolutely certain.

There is also a "French connection" to the introductions. Bowdoin College in 1978 introduced 500 Brittany, France, oysters at its Bethel Point Marine Laboratory as part of its *Amoco Cadiz* oil-spill project[16] (Figure 2). Loosanoff,[7] in studying gametogenesis of the original implant of Dutch oysters at Boothbay Harbor, noted widely disparate seasonal progression of gametogenesis and concluded that there may have been earlier intermixing of French stocks in Holland as part of the Dutch planting programs.

Through the 1960s and 1970s, small amounts of natural seed could be found around many of the sites. In 1983, significant commercial quantities of European oysters were found in the Quahog Bay region of Harpswell (Figure 2). With no initial state regulation, harvest levels rose from 551,000 oysters in 1983 to 615,000 in 1984 and 623,000 in 1985. Oysters were harvested mostly by Scuba divers and power vessels using sea-scallop dredges.

Recognizing the uniqueness and the possibly fragile nature of this new resource, the Maine Department of Marine Resources funded a biological survey of the resource as a basis for future regulation.[16] Although the precise distributional maps are kept confidential to protect the resource, certain generalizations have been made public to guide new regulations. These include the following:

1. The largest oyster populations are located in the Quahog Bay and Yarmouth Island areas, but oysters were found at nearly every dive site from Small Point to Cousin's Island with the exception of Harpswell Sound and Middle Bay (Figure 2).[16]
2. Seaward extension of the population appears limited by exposure to wave action and low temperatures, which decrease growth.

3. No oysters were found above −1 ft M.L.W., and this pattern appears to be related to winter freezing mortality.
4. The majority of oysters were found on firm mud or sand bottom with relic shell or small stones. These are associated with moderately strong currents associated with channels between land masses.

As a result of these studies a closed season, between June 15 and September 15, and a minimum size limit, 5 cm largest diameter, has been established in Maine to protect this resource. The closed season coincides with the period of gametogenesis and spawning, and the size limit protects oysters through their period of maximum fecundity as determined by Walne.[20]

It is important to speculate on the origin and future of the new natural European oyster fishery in Maine. It took 36 years from the original introduction for significant commercial populations to become apparent. However, it is entirely likely that the large aquacultured populations of recent times in this region were responsible for the colonization. Analysis of size-frequency distribution of this new population suggests this is the case. The modal size of oysters is in the 50- to 70-mm range (2 to 5 years old). Only occasionally are larger oysters (to 110 mm, 5 to 7 years of age) found. Because very extensive similar environment exists in south central Maine, it seems apparent that we may be seeing merely the first phase of a major new fishery. We must manage this resource judiciously and above all guard against importing the diseases that have decimated European populations of this species. More optimistically, should these oysters become really abundant, we must educate the American consumer to appreciate and purchase in quantity this true "gourmet's choice of fine oysters in the world".

B. MAINE AQUACULTURE

In the early 1970s the University of Maine, as part of its Federal Sea Grant program, began experimenting with the European oyster to develop it as a bona fide candidate species for intensive aquaculture. Environmental evaluation revealed that tray-reared European oysters grew well, up to 60 mm in the first year, along a wide estuarine gradient of the Damariscotta River that included a summer temperature range of 15 to 23°C (Figure 3). Winter temperatures below 0°C at the upper estuarine sites were seen, however, as a limiting factor to successful single-site culture.[21] As a follow-up, cultchless hatchery-reared seed was donated to over 35 private citizens along the Maine coast in an effort to stimulate a serious pilot commercial venture.

The Maine Coast Oyster Corp. (MCOC), Mark and Dean Richmond of Blue Hill, ME, was the first pilot venture, beginning in 1984. There, cultchless hatchery seed was reared to market size in two to three seasons on vertically stacked trays deployed on long-lines (Figure 4). The MCOC was very influential in stimulating additional commercial activity in the mid-1970s. In an innovative marketing move, the cultured European oyster was named "the Maine-grown Belon oyster", helping the product to become the highest priced half-shelled oyster ($0.85 each) at the time. The openness of MCOC in showing others their techniques stimulated dozens of others to begin their own businesses, especially on the Damariscotta River in south central Maine.

The mid- and late 1970s was a period of much trial and error, leading to interactive improvement in what now might be the beginnings of a more viable aquaculture industry.[22] Initially, the culturists mirrored the three-dimensional techniques initiated by MCOC at Blue Hill. European oyster cultchless seed was first reared in floating trays, then transferred to several mesh sizes of Japanese lantern nets or stacked trays. The winter of 1976 and 1977 was particularly severe, and many growers in upper-estuarine situations experienced severe

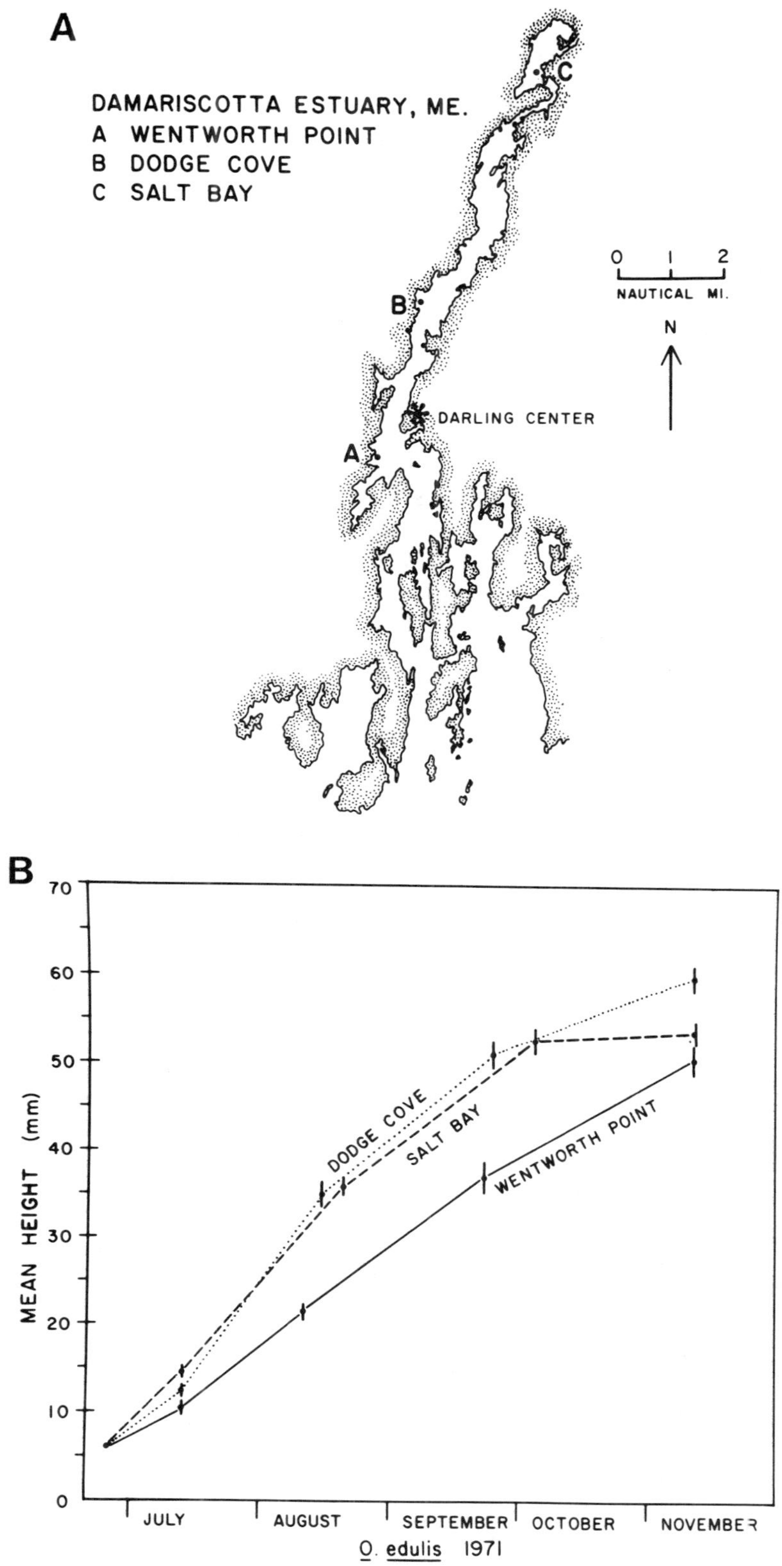

FIGURE 3. A comparison of European and American growth rates related to temperature at three sites on Damariscotta River estuary in south central Maine. Cultchless hatchery-reared seed suspended in cages 1 m deep.

losses. The danger of low temperatures necessitated a dual-site culture, in which oysters were moved to more marine sites for the winter, where higher water temperatures prevailed. All this mortality and manipulation squeezed the potential profits of the oyster culturists such that many were not economically viable.

The mid-1980s witnessed substantial changes in the price of oysters; European oysters dropped in price with the discovery and harvest of the new natural beds in Maine. Conversely, American oysters increased in price with the persistent scarcity of mid-Atlantic stocks. The result was that complete three-dimensional culture of European oysters was no longer considered viable, and new interest was sparked in on-bottom culture of both species at suitable leased sites. Foster[23] of the Maine Department of Marine Resources refined methods of on-bottom culture throughout this period.

The method of choice now is to culture oysters in their first season in floating tray fields until they are in the 35- to 50-mm range in the late fall. Fouling is controlled biologically by introduction of the browsing periwinkle *Littorina littorea*.[24] Oysters are then transferred to on-bottom leased sites for growth to market size. Problems now revolve around effective predator control methods for the major predators, which include eider ducks, the starfish *Asterias vulgaris*, and the rock crab *Cancer irroratus*.

C. EASTERN CANADA: INTRODUCTION AND AQUACULTURE

The European oyster was first introduced to the Maritime Provinces of Canada from Conwy, Wales, during the period 1957 to 1959. Introductions were made to locations near the St. Andrews Biological Station in New Brunswick and to the Ellerslie Biological Station in Prince Edward Island. The first trials were unsuccessful, as the oysters showed poor growth and suffered heavy mortalities. Downer and Medcoff[25] attributed the mortalities to low water temperatures and to an infestation of *Hexamita* sp.

The next attempt was made in 1961, when a strain of Dutch origin was imported from Boothbay Harbor, ME, to the Ellerslie Biological Station. The better survival of this stock led to a more important transfer, which took place in 1969 from Milford, CT.[26] In the fall of 1970, oysters of that stock were transplanted to Pleasant Point, Nova Scotia. They suffered rather severe mortalities during the following winter, but enough survived to encourage the Nova Scotia Department of Fisheries to attempt seed production at its Pleasant Point Hatchery.[27] The oysters from the 1969 introduction were used to refine hatchery procedures and to evaluate culture techniques and sites.

Genetic studies funded by the Nova Scotia Department of Fisheries were undertaken at Dalhousie University.[28] Two new strains of European oysters were imported from Maine and Wales through the Dalhousie quarantine. Eventually, Dalhousie started its own small-scale hatchery, and the program evolved into research on production technologies. The new strains obtained from the Dalhousie program are now utilized at both hatcheries. These produce more and larger larvae than the original strains; growth and survival rates of larvae from the new strains are also superior.

The European oyster is now grown at several locations along the Atlantic coast of Nova Scotia, where water temperature is too low for American oyster culture. The species grows well (3 to 4 years to marketable size) in off-bottom culture. All seed to date has been hatchery produced, because there is yet no wild stock in Maritime waters.

Natural spawning has, however, been observed for several years in Point Pleasant Pond, Spanish Ship Bay, East Dover, and Ketch Harbor. Although these observations indicate the successful spawning and setting of oysters held in nearby culture operations, no self-sustaining population has been located to date.

Production is still very modest. The 20,000 to 50,000 animals produced annually are of high quality and receive a good price in the market place ($0.36 each U.S.). European

A

FIGURE 4. Maine's first commercial oyster venture, the Maine Coast Venture Corp., on the Salt Pond, Blue Hill Bay, in the mid-1970s. Stacked wooden trays were deployed from a long-line accommodating oysters throughout their entire life cycle.

gourmets who have sampled the "Nova Scotia Belon" have generally rated it highly. In its development plan, the Ostrea Edulis Cooperative Association has set an ambitious annual production target of 6 million oysters per year by the year 2002. This annual production would be worth $2.2 million U.S. The most immediate constraints on development are low seed supply, the need for technical and business know-how, and the lack of venture capital.

At present, all the seed is produced by the two small-scale hatcheries. Dalhousie University in Halifax has expanded its experimental hatchery into a seed-production operation; the other is operated by the Nova Scotia Provincial Government at Point Pleasant. Together, these two establishments can produce approximately 3 million 3- to 5-mm seed oysters annually, which are sold to growers for $12.50 (U.S.)/1000 seed. This relatively modest seed production should prove sufficient to test several field culture sites, improve husbandry techniques, and eventually reach commercial scale. In the meantime, some of the seed is

FIGURE 4B

used in research. Research conducted on juvenile oyster diets with the objective of optimizing oyster growth rate per unit of culture effort is producing good results.[29-31] Eventually, these advances will produce increased profit margins through savings at the nursery stage and, potentially, better growth and survival to market size.

All but one of the present growers are part-time operators. For the most part, they came to the business with little knowledge of husbandry techniques and little familiarity with the ecological factors governing their culture sites. Through sometimes painful experiences, they have developed a body of useful knowledge that could now become part of an organized training program, which should also include small-enterprise business skills.

There are several reasons for optimism. As culture technology and genetic selection improve, production costs are dropping. Bottom culture techniques are under investigation in an attempt to reduce production costs even further. Producers are acutely aware of the value of maintaining high quality standards in order to retain their niche in the luxury seafood market. The market is eagerly buying all oysters produced and is asking for more. Several interested growers are lining up for seed to test new sites. There are solid indications that, once the seed-supply problem has been overcome, the European oyster will join the several other luxury seafood items already produced in Atlantic Canada.

III. FUTURE RESEARCH EMPHASIS

In view of the rapid changes taking place in culture of European oysters in the Northeast, several research and management areas should have priority.

1. First is protection and encouragement of the new natural populations in south central Maine. It would be valuable to learn the dynamics of population establishment so that the most efficient management methods can be employed. Important subjects for research include broodstock placement and concentration and, especially, efficient cultching, as the area is definitely substrate limited.

2. Second is monitoring the new natural populations for disease incidence so that baseline data will be available should mortality increase in the future.
3. Third is development of effective predator control methods for on-bottom aquaculture. Major predators include the starfish *Asterias vulgaris,* the rock crab *Cancer irroratus,* and the eider duck.
4. Last is development of superior strains of European oysters for mariculture, with higher growth rates, better shellform, and, especially, winter hardiness.

REFERENCES

1. **Altman, P. L. and Dittmer, D. S.,** *Biology Data Book,* Vol. 1, 2nd ed., Federation of American Societies for Experimental Biology, Bethesda, MD, 1972, 605.
2. **Nowak, W. S. W. and Stainsfield, J. D.,** The Worthington oyster, *Triton J. Br. Sub-aqua Club,* 16, 124, 1971.
3. **Ahmed, M.,** Speciation in living oysters, *Adv. Mar. Biol.,* 13, 357, 1975.
4. **Orton, J. H.,** Observations and experiments on sex-change in the European oyster *(O. edulis), J. Mar. Biol. Assoc. U.K.,* 14, 947, 1927.
5. **Orton, J. H.,** Observations and experiments on sex-change in the European oyster *(O. edulis).* III. On the fate of unspawned ova. IV. On the change from male to female, *J. Mar. Biol. Assoc. U.K.,* 19, 1, 1933.
6. **Cole, H. A.,** The fecundity of *Ostrea edulis, J. Mar. Biol. Assoc. U.K.,* 25, 243, 1941.
7. **Loosanoff, V. L.,** Gametogenesis and spawning of the European oyster *O. edulis* in waters of Maine, *Biol. Bull.,* 122, 86, 1962.
8. **Korringa, P.,** Water temperature and breeding throughout the geographical range of *Ostrea edulis, Ann. Biol.,* 33, 109, 1957.
9. **Korringa, P.,** Experiments and observations on swarming, pelagic life and setting in the European flat oyster, *Ostrea edulis* L., *Arch. Neerl. Zool.,* 5, 249, 1941.
10. **Andrews, J. D.,** Pelecypoda: Ostreidae, in *Reproduction of Marine Invertebrates.,* Vol. 5, *Molluscs: Pelecypoda and Lesser Classes,* Giese, A. C. and Pearse, J. S., Eds., Academic Press, New York, 1979, 293.
11. **Walne, P. R.,** Experimental rearing of the larvae of *Ostrea edulis* L. in the laboratory, *Min. Agric., Fish, Food, Fish. Invest. Series II,* 20, 23, 1956.
12. **Davis, H. C. and Ansell, A. O.,** Survival and growth of larvae of the European oyster, *O. edulis* at lowered salinities, *Biol. Bull.,* 22, 33, 1962.
13. **Cole, H. A. and Knight-Jones, E. W.,** The setting behavior of larvae on the European flat oyster, *O. edulis* L., *Min. Agric., Fish, Food, Fish Invest.,* 17, 1949.
14. **Crisp, D. J.,** Factors influencing the settlement of marine invertebrate larvae, in *Chemoreception in Marine Organisms,* Grant, P. T. and Mackie, A. M., Eds., Academic Press, London, 1974, 177.
15. **Hidu, H., Valleau, W. G., and Veitch, F. P.,** Gregarious setting in European and American Oysters—response to surface chemistry vs. waterborne pheromones, *Proc. Natl. Shellfish. Assoc.,* 68, 11, 1978.
16. **Heining, C. S. and Tarbox, B. P.,** A Range and Distribution Study of the Natural European Oyster, *Ostrea edulis,* population in Casco Bay, Maine, unpublished report to Maine Department of Marine Resources, Hallowell, 1984.
17. **Loosanoff, V. L.,** European oyster, *O. edulis,* in the waters of the United States, *Anat. Rec.,* 111, 126, 1951.
18. **Loosanoff, V. L.,** The European oyster in American waters, *Science,* 121, 119, 1955.
19. **Welsh, W. R.,** The European oyster, *Ostrea edulis* in Maine, *Proc. Natl. Shellfish. Assoc.,* 54, 7, 1966.
20. **Walne, P. R.,** Observations on the fertility of the oyster *(O. edulis), J. Mar. Biol. Assoc. U.K.,* 44, 293, 1964.
21. **Packie, R., Hidu, H., and Richmond, M. S.,** The Suitability of Maine Waters for Culturing American and European Oysters *Crassostrea virginica* and *Ostrea edulis* L., Maine Sea Grant Tech. Rep. TR-10-76, Ira C. Darling Center, Walpole, ME, 1976, 461.
22. **Clime, R. and Hamill, D.,** *Growing Oysters and Mussels in Maine,* Coastal Enterprises Inc., Middle St., Wiscasset, ME, 1979, 30.
23. **Foster, W.,** Evaluation of Bottom Culture of *Ostrea edulis,* unpublished mimeo report, Maine Department of Marine Resources, Hallowell, ME, 6, 1975.

24. **Enright, C. T., Krailo, D., Staples, L., Smith, M., Vaughn, C., Ward, D., Gaul, P., and Burgese, E.,** Biological control of fouling algae oyster, *Aquaculture J. Shellfish Res.*, 3, 1983, 41.
25. **Downer, P. J. and Medcoff, J. C.,** Trial Introduction of the European Oyster *Ostrea edulis* to the Canadian East Coast 1959, Manuscript Rep. No. 693, Fisheries Research Board of Canada, New Brunswick, 1959.
26. **Drinnan, R. E.,** The Introduction of the European Flat Oyster Ostrea edulis to Eastern Canada, Fisheries Research Board of Canada, 1970, 1.
27. **Muise, B., MacLeod, L., Henderson, K., and Truen, R.,** Cultivation of the European Flat Oyster *(Ostrea edulis)* in Nova Scotia: General Growth and Economic Model, Proj. Rep. No. 86-01, Nova Scotia Department of Fisheries, 1986, 183.
28. **Newkirk, G. F.,** Selecting oyster for faster growing, in *Proc. of the 8th Annu. Meet. World Mariculture Soc.*, Avault, J. W., Ed., Louisiana State University, 1977, 557.
29. **Enright, C. T.,** Determination of the Relative Value of Phytoplankton for Feeding the Juvenile Oyster, *Ostrea edulis,* Ph.D. thesis, Dalhousie University, Halifax, Nova Scotia, Canada, 1984.
30. **Enright, C. T., Newkirk, G. F., Craigie, J. S., and Castell, J. D.,** Evaluation of phytoplankton as diets for juvenile *Ostrea edulis* L., *J. Exp. Mar. Biol. Ecol.*, 96, 1986.
31. **Enright, C. T., Newkirk, G. F., Craigie, J. S., and Castell, J. D.,** Growth of juvenile *Ostrea edulis* L. fed *Chaetaceros gracilis* Schutt of various chemical composition, *J. Exp. Mar. Biol. Ecol.*, 96, 1986.

Chapter 5

QUAHOG CLAMS IN THE U.S.

Winston Menzel

TABLE OF CONTENTS

I. INTRODUCTION

Two species, the northern quahog *Mercenaria* (= *Venus*) *mercenaria* (Linnaeus) and the southern *M. campechiensis* (Gmelin), family Veneridae, are extant and native only on the east coast of North America. Northern quahogs have several common names—hard clams, round clams, and hard shell clams as well as the trade names of littlenecks, chowders, necks, nicks, sharps, and blunts. A recognized subspecies is the Texas quahog, *M. mercenaria texana* (Dall). A "form" *M. notata,* originally described as a subspecies, has zigzag mottlings on the exterior of the shell. The southern quahog is also called hard clam or Gulf quahog.[1] McHugh et al.[2] have published an extensive annotated bibliography on quahogs, containing over 2200 references.

II. BIOLOGY

A. MORPHOLOGY AND SPECIES DIFFERENCES

Typical northern and southern quahogs can be distinguished by the external shell morphology (see Figure 1). The northern quahog has a thick shell with heart-shaped lunule longer than wide; the exterior has concentric growth ridges that are more distinct near the beaks. The central exterior of the shell becomes "smoothish". The interior of the valves are whitish to gray, often with purple stains, especially near the posterior ends. The edges of the inner margins are crenulate, and there are three cardinals on each valve. The subspecies *M. mercenaria texana* differs from typical *M. mercenaria* in having external large coalescing flat-topped concentric ribs at anterior and posterior ends. The southern species has similar shells, but is best distinguished by persistent growth ridges completely around the shell. The shells become heavier than those in the northern species and seldom have interior purple coloring. Large southern quahogs have lunules as wide as long.[1] An important physiological difference is the shelf life. *M. mercenaria* will live for a week or more, whereas *M. campechiensis* will gape and die in several days when removed from the water.[3] The morphology of the soft parts are typically "clam-like".[3]

B. RANGE

The northern species occurs from Canada (Gulf of St. Lawrence) southward to the Gulf of Mexico, represented by the subspecies *M. texana* along the lower east coast of Florida and into the Gulf of Mexico, at least to Mexico.[3] The center of distribution is given as from Massachusetts to Virginia.[4] The southern species has been found as far north as New Jersey,[5] but is common only as far north as North Carolina.[3] The center of distribution seems to be the lower west coast of Florida.[3] Northern quahogs have been introduced to areas in California and Europe.[1]

C. SUBSTRATE AND DEPTH

Quahogs are infauna, occurring in a variety of substrates of sand, sand-mud, gravel, shells, and a combination of these.[3] They have relatively short siphons and are never buried very deep. The northern quahog may occur intertidally, especially in the more northern regions, up to depths of 15 m and more. The southern species has never been seen in the normal intertidal region and occurs to depths of 15 m and more.

D. SALINITY AND TEMPERATURE TOLERANCES

Northern quahogs are found in salinities of 20 to 30‰ in mainly estuarine areas although they occasionally occur in higher and lower salinities. Laboratory experiments have shown that "normal" burying is retarded at salinities below about 20‰.[6] High larval mortality occurs at salinities of 35‰ and below 20‰[7,8] although late larval stages will survive and

FIGURE 1. *Mercenaria mercenaria* (left) and *Mercenaria campechiensis* (right).

grow (slowly) at salinities as low as 17‰.[7] No controlled salinity tolerance experiments have been done with the southern quahog, but its occurrence in salinity only at 30‰ and above indicates it is less tolerant to low salinities than the northern species is.

Northern quahogs survive under an ice cover on the overlying water, but freezing temperature in Maine kills intertidal sets.[9] The northern species hibernates at a temperature of about 5 to 6°C,[10] with little feeding below 10°C or above 26 to 28°C.[11] Excessive mortality of hatchery-reared seed occurred in tropical St. Croix at continually high temperature.[12] On the basis of its distribution, the subspecies *M. texana* is tolerant of a higher temperature than is the typical northern quahog.[3]

Winter mortality occurred in hatchery-reared seed of southern quahogs when planted in shallow estuarine areas of Virginia and North Carolina.[13-14] Hatchery-reared seed of the southern species ceased shell growth at a low temperature of about 12 to 13°C, but continued growth at a temperature above 30°C.[11] Although the southern quahog has been found as far north as New Jersey, it occurs only in the more stable near shore oceanic waters.

E. GONADAL DEVELOPMENT AND SPAWNING

Gonadal development may occur in clams only several months old—in the autumn for clams that set in the spring—but these and even yearlings probably contribute little to the larval pool. Quahogs first develop as males, changing to an equal sex ratio after the first spawnings.[15]

Gonads are retained through the colder seasons, and gametes mature with rising spring temperature; spawning begins when the temperature rises above 20°C (22 to 25°C) throughout the latitudinal range for the northern quahog. The southern species spawns at about the same temperature.[3] The writer has found that both species usually have mature gametes throughout the winter in northwest Florida.[3] *M. mercenaria* has one peak of spawning in the northern latitudes (July through August), but two peaks (spring and fall) in the more southern regions, often with some spawning throughout the summer. The northern quahog is conditioned to spawn in the northern regions during the winter by a gradual increase in temperature.[16] Spawning is induced by temperature shock and the addition of sperm suspension.[16] Both species, when planted in northwest Florida, usually require no temperature conditioning to induce spawning.[3] The southern quahog follows the gonadal cycle of the northern species of the more southern latitudes.

Ova average 70 to 73 μm in diameter (range 60 to 85 μm) and have a clear gelatinous envelope giving a total diameter of 125 to 175 μm.[17] Fertilization and all larval development occur in the water column. After 8 to 15 d, the larvae (pediveliger) develop a foot and alternate between swimming and crawling. With the gradual loss of the velum, the benthic plantigrade or dissoconch stage attaches to an object on the bottom by a byssal thread, which may be broken and reattached, until it reaches a size of about 7 to 9 mm; then the young clam buries itself in the bottom.[17]

F. GROWTH

Growth studies have been made in nearly every area where quahogs occur. Growth rate is variable, differing from year to year within the same location, between locations, and even between individuals of a single male-female cross. Such factors as food availability, temperature regimes, current action, salinity, and genetic factors govern the growth rate.

Generally the northern quahog has faster shell growth in the more southern regions than farther north. The southern quahog has greater annual shell growth than the northern species, shown by experiments in which hatchery-reared seed of the two species were planted in the same area.[11-14] Cultured hatchery seed of the northern quahog takes from 1 $^1/_2$ to 3 years to reach the minimum market size (45 to 55 mm length). Both species may reach up to 15 cm or more in length, but the southern quahog usually becomes larger than the northern species does. Shell increases become progressively slower as the clams get older. Clams are estimated to live 25 years or more; there is one record of a northern quahog being at least 36 years old.[18]

G. PARASITES AND PREDATORS

No known catastrophic mortalities have been caused by microorganisms, as have been found in other commercial bivalves. Larval mortalities have occurred in hatcheries, and it is now standard practice to sterilize culture water with UV light and antibiotics.

Newly set and up to 10 to 15 mm clams are very vulnerable to a variety of predators. When first set, the small clams are near the surface of the substrate and are enticing prey for several decapod crustaceans and predacious gastropods, as well as certain fish species and water birds. Among the most serious predators, at least from New York southward, are the blue crab *Callinectes sapidus* Rathbun, and in the colder latitudes the green crab *Carcinus meanas* Linnaeus. Other decapods include the stone crab *Menippe mercenaria* (Say), from North Carolina southward, which can crush the shells of the largest clams. Several smaller species of xanthid mud crabs attack only the smaller clams. Several predacious gastropods prey on clams, including species of *Busycon, Eupleura, Polinicies,* and *Urosalpinx*. Both *Busycon* and *Polinicies* can attack larger buried clams, whereas these are usually safe from attack by *Eupleura* and *Urosalpinx*. Several fishes (rays, flounders, and puffers) eat clams. Gulls are known to seize clams, fly upward, and drop them to crack the shells. Diving ducks and geese feed on small clams.[3]

TABLE 1
Quahog Clam Production in the U.S.

Region	1951—1960 t	1951—1960 Value	1961—1970 t	1961—1970 Value	1971—1980 t	1971—1980 Value	1981—1985 t	1981—1985 Value
New England	2,695	2,156	1,515	2,027	1,084	3,146	2,628	17,730
Middle Atlantic	2,906	3,413	3,925	6,710	4,638	15,811	2,014	15,773
Chesapeake	555	640	949	1,271	633	1,260	756	3,175
South Atlantic	193	129	167	139	248	575	1,291	9,449
Gulf of Mexico	9	6	6	14	1	2	16	73
Total	6,358	6,344	6,562	10,161	6,604	20,794	6,705	46,200

Note: Metric tons (t) and value ($1,000) per year (mean of intervals), since 1951.

III. FISHERY

A. HISTORY

Quahogs have long been an important seafood in the U.S. The first European settlers found the native Indians harvesting them as they had been doing for a long time. The natives not only used them for food, but also ground "buttons" from the shells, especially those with the purple pigmentation, and used them for wampum, a form of currency. The meats were sometimes dried and used for trade with inland tribes. Harvesting was by hand in shallow water and intertidal areas or sometimes by diving.[19] The European settlers introduced rakes, tongs, and later dredges. The northern quahog has always supported the majority of the fishery although immense numbers of the southern species were harvested on the lower west coast of Florida in the period from about 1913 to 1947;[20,21] some are harvested currently.

B. COMMERCIAL FISHERY

Eastern Canada harvests a small but unknown quantity of quahogs. Harvest statistics have been collected in the U.S. since about 1880.[22] In the earlier years statistics were not always taken yearly nor from each area. Table 1 gives the landings of meat in metric tons (t) for the annual production for decade intervals since 1951 for each region along the eastern seaboard. Table 2 gives the annual production and value to the fishermen for each state in regions from 1981 to 1985. Ingersoll[19] estimated that about 5500 t were harvested annually in the latter half of the 19th century, mostly from New York. The national statistics show that nearly 3500 t were landed annually from 1928 to 1932 and 7500 t, from 1944 to 1948.[22] The average annual yield has been over 6000 t for the last two decades (see Table 1), with a high of almost 8600 t in 1981 (see Table 2). The yield was over 7000 t in 1985, the latest year for which statistics are available. The annual value to the clammers has steadily increased to over $49 million in 1985 (see Table 2). New York was formerly the leading producer, with often over half of the entire catch, until 1981 when Rhode Island became the leading state. In 1985 the harvest in Florida, which had an abundant set along the lower east coast, exceeded that of New York.

At the present time, harvest of the northern quahog is probably at the maximum from Chesapeake Bay northward and in some of the South Atlantic states. The northern quahog fishery has shown considerable variation in production within areas, but total annual production has been relatively constant overall. The changes in production from year to year within an area are due to erratic recruitment and to social and economical factors. The southern quahog is underexploited all along the south Atlantic and Gulf of Mexico coasts.

C. RECREATIONAL FISHERY

A considerable recreational fishery exists, especially in the Middle Atlantic and New England regions, but no statistics are available. There have been estimates of this fishery

TABLE 2
Quahog Clam Production in the U.S.

Regions	1981 t	1981 Value	1982 t	1982 Value	1983 t	1983 Value	1984 t	1984 Value	1985 t	1985 Value
New England Region										
Connecticut	—	—	136	998	114	828	205	1,485	333	1,288
Massachusetts	236	1,695	454	3,632	636	6,593	511	4,824	609	5,490
Rhode Island	2,045	13,000	1,682	10,730	1,906	10,123	1,763	12,952	1,911	15,011
Middle Atlantic Region										
New York	2,085	18,219	1,556	15,007	1,520	10,864	1,210	9,284	1,000	9,467
New Jersey	419	2,016	413	2,342	606	3,288	728	4,866	471	3,219
Delaware	12	43	18	68	12	58	14	71	9	49
Chesapeake Region										
Maryland	29	148	—	—	—	—	—	—	—	—
Virginia	2,339	8,525	285	1,427	530	2,492	336	1,837	258	1,446
South Atlantic Region										
North Carolina	663	5,387	774	6,606	610	5,402	631	5,506	633	5,654
South Carolina	117	580	233	1,165	155	1,206	183	1,366	82	618
Georgia	3	21	4	34	1	7	2	13	2	17
Florida	40	256	65	453	52	392	524	5,646	1,681	6,915
Gulf of Mexico Region										
Florida	11	47	−1	1	7	38	29	149	31	130
Total	8,599	49,938	5,620	42,463	6,149	41,291	6,136	47,999	7,020	49,304

Note: Metric tons (t) and value ($1,000) per year for each state, period from 1981 to 1985.

in some localities. In some areas in Massachusetts the recreational catch was estimated to be from 4 to 25% of commercial catch.[23] The recreational catch in Great South Bay, New York was estimated at 26 t in 1977.[24] An estimated 1 million quahog clams were taken by recreational fishermen in the Rehoboth Bay area of Delaware in 1957.[25]

Harvesting is by hand (raking, treading, and tonging) in the intertidal and shallow subtidal areas.

D. HARVESTING METHODS

Harvesting methods range from digging by hand (usually with a small pick) to using sophisticated power dredges. Most clammers work singly although larger boats with power dredges require a crew of two or more. Thousands of watermen harvest clams, either full- or part-time.

Hand methods account for the majority of the harvest: digging, treading, raking, and tonging. Hand harvest is limited to relatively shallow water, up to 6 or 7 m (by tongs). The so-called patent tongs, which are heavy tong heads with a trip mechanism operated by a power winch, can operate in deep water. Dredges similar to oyster dredges, but usually with a hose from a pump on ship deck to loosen the clams from the sediment are used in some areas. A modified version is the escalator dredge, in which the clams, after being uncovered with the pump hose, are carried upward on an endless belt leading up to the side of the ship. Escalator dredges can be used in water depth up to about 7 m.[3] A method used in

North Carolina is called ''kicking''; the propeller of the vessel washes the clams out of the bottom, and they are collected by a trawl net pulled behind.[26]

E. METHODS OF PROCESSING

Most quahogs are sold in the shell to a distributor, directly to stores and restaurants, or to individual customers. The smaller clams (45 to 75 mm long), called littlenecks and cherrystones, have the highest value per unit; individual small clams bring a higher price than large clams and are sold in the shell, either to be eaten raw on the half shell or lightly steamed. The long shelf life of the northern quahog is an asset for the shell trade. Large clams are tough and are shucked and usually diced or ground before preparation as chowders, fritters, or mixtures baked in the shell.

F. MANAGEMENT

All clam-producing areas are under strict surveillance and monitoring from health authorities to ensure that a safe product reaches the consumer. At any sign of pollution the area is closed to clamming. In some areas where domestic pollution occurs, clammers may harvest clams under permit to relay to ''clean'' waters for depuration or through depuration plants, using UV light. Some areas with industrial pollution are closed permanently because quahogs do not depurate heavy metals readily.[3] State agencies have relayed clams from domestically polluted areas where they are abundant to other areas for depuration. This accomplishes two goals: it utilizes the resource and it lessens the possibility of poaching the contaminated mollusks.[3]

Most states do not allow use of dredges on public clam beds, but only harvesting by hand. Some states allow dredging under a quota system by special permit. Quahogs are harvested year-round, but most states have minimum size laws (45 to 50 mm in length). Some localities have attempted to replenish depleted areas by planting hatchery-reared seed clams. These efforts have proved futile in most instances because of the high mortality rates of the seed, mainly from predation.

G. CONSTRAINTS

Like that of most shellfish harvests, the quahog harvest operates under many constraints. Some of these result from the vagaries of nature and some are man-made. Quahogs are noted for their erratic recruitment. In some areas and in some years very abundant set occurs. In other years hardly any clams from a spawning survive to commercial harvest size. The erratic recruitment is illustrated by the catch statistics from the several areas (see Tables 1 and 2). When an abundant recruitment and survival occurred in the 1980s in the Indian Lagoon on the lower east coast of Florida, experienced clammers from out of state moved in to harvest the bonanza. This benefited Florida in that the resource was exploited, but these clammers were unavailable to harvest clams in the areas from which they came. Another constraint, which all fishermen have to contend with, is inclement weather conditions. There have been winters during which Great South Bay, New York, froze and clammers were unable to work, except for a few hardy souls who chopped holes in the ice. Periodic outbreaks of paralytic shellfish poisoning occur in the New England area, and although quahogs are seldom contaminated, the demand for all shellfish decreases.

Pollution is one of the most serious constraints. The northern quahogs occur mostly in shallow estuarine waters, and these are often polluted.[27] In the 19th and early 20th century, New Jersey was one of the leading clam-producing states. Today, because of pollution, the harvest has been greatly reduced. The same is true for other areas. A decade or so ago, a serious hepatitis outbreak occurred in the eastern U.S. The outbreak was traced (in some cases) to eating contaminated clams. (New York City, a large market for shell clams, stopped all sale.) Many of the clam-eating public stopped buying clams, with serious financial loss to many clammers.

IV. RESEARCH AND DEVELOPMENT

A. BIOLOGY

Belding[29] made the first extensive study of quahogs (in Massachusetts). Since then, many studies have been made in areas from Canada to Texas. Many early attempts were made to rear clams from fertilized eggs, but were unsuccessful until methods were devised to provide adequate food for larvae. The first successful method, called the Wells-Glancy method after the investigators, was perfected in the late 1920s.[29] This method basically clarifies raw seawater by centrifuging or filtering it to remove silt and competing zooplankton, leaving the small phytoplankton as food for the larvae culture. Usually fertilizer is added to the clarified water, and it is incubated under light or in a greenhouse to enhance phytoplankton growth although some culturists use the water without enrichment. The Milford method, developed at the Milford, CT Laboratory of the National Marine Fisheries Services, uses purified water to grow monocultures of algae that have been determined to be good food for larvae. Marine laboratories as well as commercial hatcheries use one or both of these methods, each facility with its own modifications.

B. HATCHERIES

Hatcheries are essential to supply a reliable source of seed clams for culture. Operations of clam hatcheries have now become fairly routine. After metamorphosis the set clams are reared in the hatchery until they become large enough to plant in natural waters. It is possible to rear clams through to market size under controlled hatchery and growout conditions, but it is impracticable because of the amount of food, space, and labor needed. One method of rearing the seeds to planting size is to place them in troughs with continually flowing seawater, filtered to remove excess silt and predators. Another method that has proved successful and less expensive, called the upwelling method, is to force raw seawater up through several layers of clams.

C. NURSERY CULTURE AND GROWTH TO MARKET SIZE

Seed clams are planted in the natural habitat, but are still vulnerable to predation, especially by crustaceans, and must be given adequate protection. Various methods have been used, some more successful and less expensive than others. One method is to enclose the planted bed completely with small-mesh wire fences extending above the high water line, but it is very expensive. Some successful methods include covering the planted beds with plastic netting buoyed so as to be slightly above the bottom or using the netting over plantings in crushed shells. Others have used screened sand-filled boxes.[30] All of these methods are labor and equipment intensive. Quahogs can be grown up to about 15 mm at densities of 3000 to 4000/m^2, but then should be thinned to densities of about 400/m^2 to prevent stunting of growth.[8] Every culturist uses modifications, usually dictated by the locality.

D. SELECTION AND HYBRIDIZATION

Some hatcheries have attempted to domesticate quahogs to obtain better growth through selection of the fastest growing clams as brood stock. A hatchery in Massachusetts has developed a strain including three localities in the pedigree that is claimed to be adapted for good growth under a wider variety of conditions.[31] One hatchery has incorporated the *M. notata* markings, through selective breeding, for use as a trademark.[30]

The northern and southern quahogs hybridize readily, and F_1, F_2, and F_3 and backcrosses with either parent have been obtained.[32] The F_1 hybrid between the northern and southern species has been shown to have as good as or better growth rate in the semitropical waters of northwest Florida than the faster growing southern species. The growth rate of the hybrid was better than that of the northern quahog in this area even though the northern species had better annual growth than it did in the more northern latitudes. Preliminary observations found no better growth rates in the F_2, F_3, or backcrosses although the parents used were selected for high growth rates. Origin of the northern parents used for the hybrids ranged from Maine to Florida. Those from the Middle Atlantic region yielded the hybrid offspring with the best growth rates and shelf life.[32]

V. PRESENT AND FUTURE DEVELOPMENTS

By 1985 at least 20 companies were culturing clams for commercial sale, and additional companies have been started since then. Some of these have their own hatcheries, whereas others buy their seed from hatchery companies, which also may have grow-out operations. One company sold over 1 million clams for the shell trade in the past several years at prices as high as 20¢ each at the shipping point.[30]

The prognosis for future increase in commercial culture seems very encouraging. Techniques for producing seed from hatcheries are well developed. Successful nursery rearing and subsequent grow-out for commercial sale have been accomplished even though they are tedious and expensive. At the present time, the demand for the shell trade seems insatiable for this luxury food and the monetary value received justifies the expenses. Additional developments should lead to better techniques. Real domestication of this mollusk would appear to be possible in the near future. The excellent shelf life is a real asset, and the fact that, unlike many seafoods, small clams are in best demand offers an added incentive.

There is always the possibility that quahogs could be cultured so successfully that the demand would not keep up with the supply and the price would decrease. There are, however, many potential new customers who can be exploited with the proper promotion.

REFERENCES

1. **Abbott, R. T.,** *American Seashells,* 2nd ed., Van Nostrand Reinhold, New York, 1974, 1.
2. **McHugh, J. L., Summer, M. W., Flagg, P. J., Lipton, D. W., and Behrens, W. J.,** Annotated bibliography of the hard clam *(Mercenaria mercenaria), U.S. Department of Commerce, NOAA, NMFS,* Tech. Rep. NMFS SSRF-756, 1982.
3. **Menzel, W.,** The biology, fishery and culture of quahog clams, *Mercenaria,* in *Clam Mariculture in the United States,* Elsevier Science, Amsterdam, 1988, chap. 9, in press.
4. **Stanley, J. B. and Deweitt, R.,** Species profile: life histories and environmental requirements, North American hard clams, *U.S. Fish Wildl. Serv. Mar. Biol. Ser.,* FWS/OBS-82/11, 1983.
5. **Merrill, A. S. and Ropes, J. W.,** Distribution of southern quahogs off the middle Atlantic coast, *Comm. Fish. Rev.,* 29, 62, 1967.
6. **Chanley, P. E.,** Survival of some juvenile bivalves in water of low salinity, *Proc. Natl. Shellfish. Assoc.,* 48, 52, 1957.
7. **Davis, H. C.,** Survival and growth of clam and oyster larvae at different salinities, *Biol. Bull.,* 114, 296, 1958.
8. **Castagna, M. and Kraeuter, J. N.,** Manual for growing hard shell clams, *Mercenaria, Va. Institute Mar. Sci.,* Special Rep. No. 249, 1, 1981.
9. **Dow, R. L. and Wallace, D. E.,** A method of reducing winter mortalities of Venus *mercenaria* in Maine waters, *Natl. Shellfish. Assoc. Conv. Address,* 19, 1951.
10. **Loosanoff, V. L.,** Effects of temperature on shell movements in clams, *Venus mercenaria, Biol. Bull.,* 171, 76, 1939.
11. **Menzel, R. W.,** Seasonal growth of northern and southern quahog clams, *Mercenaria mercenaria* and *M. campechiensis,* and their hybrids in Florida, *Proc. Natl. Shellfish. Assoc.,* 53, 111, 1962.

12. **Sunderlin, J. B., Brenner, M., Castagna, M., Hirota, J., Menzel, R. W., and Roels, O. A.,** Comparative growth of hard clams *Mercenaria mercenaria* (Linne) and *Mercenaria campechiensis* (Gmelin) and their F_1 cross in temperate, subtropical, and tropical natural waters and in a tropical upwelling mariculture system, *Proc. World Maricult. Soc.*, 7th, 171, 1975.
13. **Haven, D. S. and Andrews, J. D.,** Survival and growth of *Venus mercenaria, Venus campechiensis* and their hybrids in suspended trays and on natural bottoms, *Proc. Natl. Shellfish. Assoc.*, 47, 43, 1956.
14. **Chestnut, A. F., Fahy, W. E., and Porter, H. J.,** Survival and growth of *Venus mercenaria, Venus campechiensis* and their hybrids, *Proc. Natl. Shellfish. Assoc.*, 47, 50, 1956.
15. **Loosanoff, V. L.,** Development of the primary gonad and sexual phases in *Venus Mercenaria, Biol. Bull.*, 72, 389, 1937.
16. **Loosanoff, V. L. and Davis, H. C.,** Rearing of bivalve mollusks, in *Advances in Marine Biology,* Russel, R. S., Ed., Academic Press, New York, 1963, 1.
17. **Carriker, M. R.,** Interrelation of functional morphology, behavior and autoecology in early stages of the bivalve *Mercenaria mercenaria, Elisha Mitchell Sci. Soc.*, 77, 168, 1961.
18. **Lutz, R. A. and Haskin, H. H.,** Some observations on the longevity of hard clams *(Mercenaria mercenaria), Annu. Meet. Natl. Shellfish. Assoc.*, 76 (Abstr.) 9, 1984.
19. **Ingersoll, E.,** The oyster, scallop, clam, mussel and abalone industries, in *The Fisheries and Fishing Industry of the United States,* Goode, G. B., Ed., U.S. Fish. Comm., Sect. 5, Vol. 11, 1887, 581.
20. **Schroeder, W. C.,** Fisheries of Key West and the clam fishery of southern Florida, *Rep. U.S. Fish. Comm. for 1923,* Append. 12, Doc. 962, 1924.
21. **Tiller, R. E., Glude, J. B., and Stringer, L. D.,** Hard-clam fishery of the Atlantic coast, *Comm. Fish. Rev.*, 14, 1, 1951.
22. *Fisheries Statistics of the United States,* U.S. Government Printing Office, Washington, D.C., numerous years.
23. **Conrad, J. M.,** Management of northeast clam resources: commercial and recreational considerations, in *Proc. Northeast Clam Industries: Management for the Future,* University of Massachusetts and MIT Sea Grant Program, SP-112, 121, 1979.
24. **Fox, R. E.,** An estimate of the recreational clam harvest from Great South Bay, New York, *N.Y. State Dep. Environ. Conserv.*, Proj. 3-263-R, 1978.
25. **Shuster, C. N., Jr.,** A biological evaluation of the Delaware River estuary, *Univ. Del. Mar. Lab., Inf. Ser. Publ.*, 3, 1959.
26. **Guthrie, J. F. and Lewis, C. W.,** The clam kicking industry of North Carolina, *Mar. Fish. Rev.*, 44, 16, 1982.
27. **Ritchie, T. P.,** A comprehensive review of the commercial clam industries in the United States, *U.S. Department of Commerce, NOAA, NMFS, Delaware Sea Grant Program,* Newark, DEL-SG-26-86, 1977.
28. **Belding, D. L.,** *A report on the Quahog and Oyster Fisheries of Massachusetts, Including the Life History, Growth and Cultivation of the Quahog (Venus mercenaria) and Observations on the Set of Oyster Spat in Wellfleet Bay, Boston,* Wright & Potter Printing Co., Boston, 1912.
29. **Wells, W. F.,** Reports of the experimental shellfish station, in *Early Oyster Culture Investigations by the New York State Conservation,* reprinted by the State of New York Conservation Department, Division of Marine and Coastal Research, 1917, 112.
30. **Chanley, P. C.,** personal communication, 1985-1987.
31. **Petrovits, E. J.,** personal communication, 1985.
32. **Menzel, R. W.,** Selection and hybridization in quahog clams, *Proc. World Maricult. Soc.*, 8, 507, 1977.

Chapter 6

MUSSEL AQUACULTURE IN MARINE AND ESTUARINE ENVIRONMENTS THROUGHOUT THE WORLD

Richard A. Lutz, Kashane Chalermwat, Antonio J. Figueras, Richard G. Gustafson, and Carter Newell

TABLE OF CONTENTS

I. INTRODUCTION

Mussels, represented by a variety of species within the bivalve mollusk family Mytilidae, are cultured in both marine and estuarine environments in numerous countries throughout the world. All current techniques rely on either the gathering or the natural settlement of seed mussels on various substrates. While technologies exist for the production of mussel larval and juvenile stages under controlled culture conditions, commercial hatcheries are not envisioned as contributing significantly to the total world production of mussels in the foreseeable future. While an exhaustive survey of mussel cultivation practices utilized in each of the countries involved is beyond the scope of the present chapter, we will attempt to provide a general overview of mussel culture and harvest throughout the world and to summarize in some detail the production status and techniques employed for the cultivation of these organisms in a number of selected areas that illustrate the range of aquacultural methods utilized throughout the oceans and estuaries of the world. Prior to such a discussion, we will present a "generalized mussel life cycle" in order that the reader might gain an appreciation of the reasons behind the development of certain cultivation practices.

II. LARVAL LIFE CYCLE AND SETTLING BEHAVIOR

The life cycle and settling behavior of numerous species of mussels, particularly *Mytilus edulis* Linnaeus 1758, have been described by numerous workers.[1-11] In general, within commercially important shallow-water species of the family Mytilidae, fertilization is external; the diameter of the fertilized egg is less than 100 μm (generally between 60 and 90 μm). The first cleavage division usually occurs within 1 hour of fertilization; subsequent embryological development proceeds by way of unequal, holoblastic spiral cleavage, with a ciliated trochophore stage being reached within 1 or 2 d after fertilization (Figure 1). Shortly after the trochophore stage is reached, the first signs of larval shell formation are apparent, and the organism soon develops a fully formed shell of the "veliger" stage with a length of approximately 100 μm (Figure 1).

The "planktonic" veliger state typically lasts approximately 1 to 4 weeks, during which time the larvae actively swim in the water column. When larvae reach approximately 150 μm in total length, a rounded umbo appears, marking the transition from the "straight-hinge" to the "umbo" stage of development (for shell morphology terminology, see Chanley and Andrews).[9] A well-defined foot is generally present in larvae with lengths between 200 and 300 μm, and in mature larvae this organ is capable of active extension (Figure 2). Larvae at this stage of development are referred to as "pediveligers" and are capable of metamorphosing; however, in the absence of a suitable substrate, the organisms have the ability to "delay" metamorphosis for substantial periods of time (up to 40 d at 10°C for *Mytilus edulis*).[6,7] Upon contact with a suitable substrate, such as mussel culture ropes, the pediveligers undergo a process known as "settlement", characterized by a crawling behavior that ends with secretion of byssal threads that anchor the larva to the substrate and signals the beginning of a sessile existence. It is this act of settlement upon which those engaged in mussel culture must capitalize.

III. COMMERCIAL PRODUCTION STATISTICS

At the present time, at least 17 species of mussels are harvested and/or cultured on a commercial scale in various regions. Table 1 provides a summary of the countries in which the various organisms are landed, as well as a detailed breakdown by species of production statistics for the year 1986. Worldwide mussel landings for the years 1970 through 1986 are graphically depicted in Figure 3A and are summarized by region for the years 1945 to

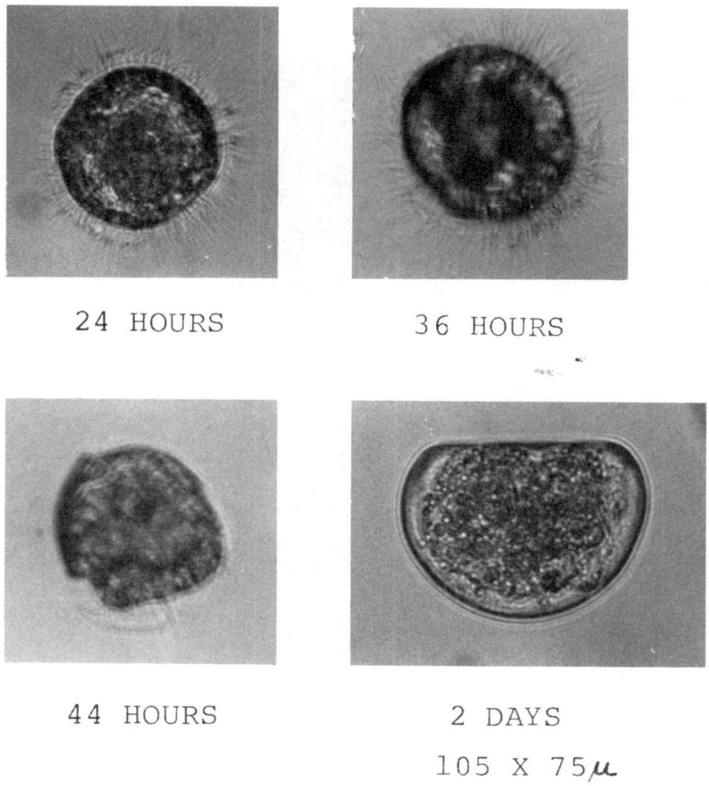

FIGURE 1. Photomicrographs depicting early larval development of *Mytilus edulis* from the ciliated morula stage (upper left) to the first shelled veliger stage (lower right). All micrographs were taken at the same magnification.

1986 in Figure 3B, by country for the year 1947 through 1986 in Figures 4 and 5, and by species for the years 1965 through 1986 in Figure 6. All data presented in Figures 3 through 6 are expressed in metric tons of total live mussel weight and represent figures obtained from either the Office of Data and Information Management of the National Marine Fisheries Service in Washington, D.C. or published statistics of the United Nations Food and Agriculture Organization.[12,13] It is important to note that the summarized figures represent minimum estimates of production totals for each region and country, as they reflect only officially reported landings. The landings for mussels are among the highest for any group of mollusks commercially cultured at the present time.

While few published data are available on the percentage of cultured vs. wild (natural harvest) landings in individual countries (see Figure 7 for a summary of the few countries or regions for which such data are available), the total mussel production values summarized by year and country in Figures 4 and 5 provide valuable insights into the relative importance of aquaculture in recent years in certain regions throughout the world. For at least the past 40 years, Europe and Asia have consistently been the leading regions of the world for both mussel culture and harvest (Figure 3B). The dramatic increase in production of *Mytilus edulis* and *M. galloprovincialis* (Figure 6A) since the mid-1960s is a marked reflection of increased mussel culture efforts in Spain (raft culture), France (bouchot culture), the Netherlands (bottom culture), Denmark (bottom culture), West Germany (bottom culture), and Italy (fixed platform hanging culture) (see Table 2). Cultivation of mussels on the bottom involves the transfer of juvenile, or seed, mussels from natural populations, where growth is poor due to lack of food or overcrowding, to bottom plots where conditions are better.

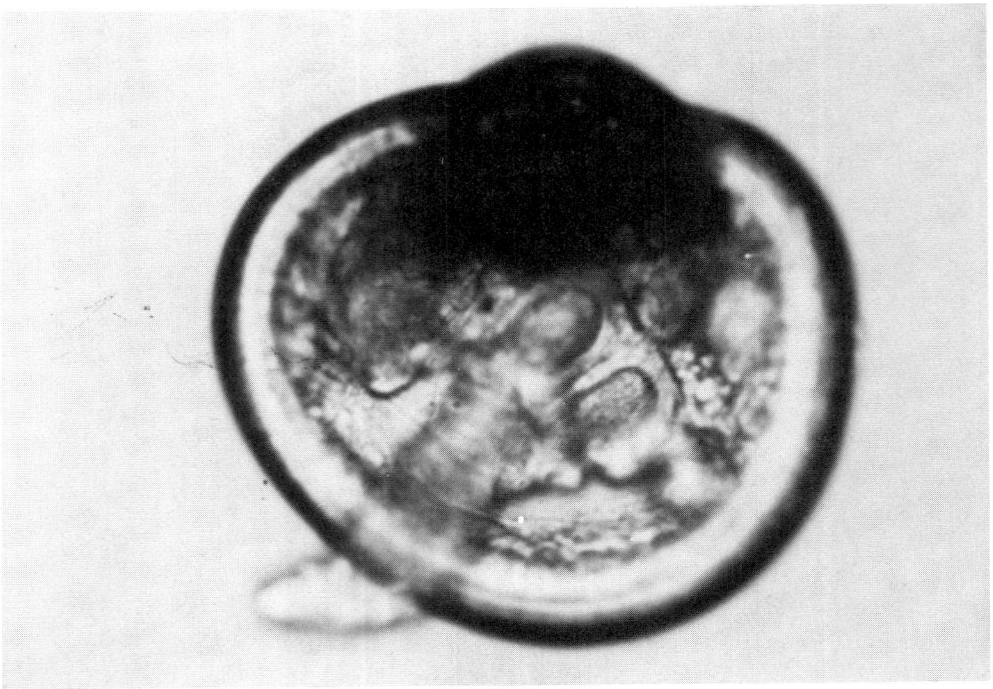

FIGURE 2. Pediveliger larva of *Mytilus edulis* with a shell length of 270 μm. Note the well-defined foot.

In the "bouchot" or pole culture method, seed mussels are allowed to set on wooden poles driven into the soft substrate and subsequently are either moved to grow-out poles or left to mature on the original poles. Raft culture involves the vertical suspension of mussel-covered ropes from a floating platform designed so that the ropes do not touch the bottom. In long-line culture, vertical mussel-covered ropes are suspended underwater from horizontal lines attached to a series of flotation devices. Fixed platform hanging culture is similar to raft culture, but the platform from which the ropes are suspended is immobile and attached directly to the shallow sea bottom. Among the Asian countries, Thailand, China, and Korea have been responsible for the vast majority of landings of *Perna viridis* (Thailand and China) and *M. crassitesta* (Korea) (Figures 4D and 6B, Table 1). The types of mussel cultivation systems utilized in these and other countries throughout the world are summarized in outline form in Table 2. While exhaustive surveys and descriptions of mussel cultivation practices utilized in each of these countries are beyond the scope of the present chapter, we will attempt to summarize the production status and techniques employed for the culture of mussels in a number of selected areas that illustrate the range of aquaculture methods employed throughout the world. The areas chosen for such illustrative purposes are (1) the U.S., (2) Spain, and (3) Thailand.

A. UNITED STATES

Figures 8 and 9 summarize the available published statistics for U.S. mussel landings (total and by state) over the past century. All data presented in these figures were obtained from the Office of Data and Information Management of the National Marine Fisheries Service in Washington, D.C. It is important to note that these statistics, which are annually collected (and ultimately published) by the National Marine Fisheries Service, are expressed in metric tons of wet mussel meats instead of metric tons of total live mussel weights, which is the unit used by the United Nations Food and Agriculture Organization (see Figures 7 A

TABLE 1
Compilation of the 17 Species of Marine Mussels which Contributed to the Worldwide Production of 829,497 t of Mussels (Whole Live Weight) Landed in 1986,[12,13] Together with Their Common Names, Countries Where Landings of Each Species Occurred in 1986, and Total 1986 Production for Each Species[12,13]

Species	Common names	Countries of occurrence	Production (t) (1986)
Mytilus edulis Linnaeus, 1758	Blue mussel	Canada, China, Denmark, F.R.G., France, Ireland, Netherlands, Norway, Portugal, Spain, Sweden, U.K., U.S.	437,684
M. galloprovincialis Lamarck, 1819	Mediterranean mussel	Bulgaria, Egypt, Italy, Morocco, Spain, Turkey, U.S.S.R., Yugoslavia	92,400
Perna viridis (Linnaeus, 1758) (= *M. smaragdinus* = *M. viridis*)	Green mussel	China, India, Malaysia, Philippines, Thailand	67,000
M. crassitesta Lischke, 1868	Korean mussel	Korea	43,000
Aulacomya ater (Molina, 1782)	Magellan mussel	Argentina, Chile, Peru	18,400
P. canaliculus (Gmelin, 1791) (= *M. canaliculus*)	New Zealand mussel	New Zealand	11,100
M. chilensis Hupe, 1854 (= *M. edulis chilensis*)	Chilean mussel	Chile	10,900
Musculus senhauseni (Benson, 1842)	Striped horse mussel	Thailand	7,500
M. planulatus Lamarck, 1819 (= *M. edulis planulatus*)	Australian mussel	Australia	1,400
M. platensis Orbigny, 1846 (= *M. edulis platensis*)	River Plata mussel	Argentina, Uruguay	1,000
P. perna (Linnaeus, 1758)	Rock mussel	Brazil, Uruguay, Venezuela	300
M. coruscus Gould, 1861	Thick-shell mussel	China	N.D.
M. californianus Conrad, 1837	California mussel	Mexico, U.S.	N.D.
P. indica Kuriakose and Nair, 1976	Brown mussel	India	N.D.
Modiolus metcalfei (Hanley, 1843)	Horse mussel	Philippines	N.D.
Modiolus philippinarum (Hanley, 1843)	Horse mussel	Philippines	N.D.
Choromytilus chorus (Molina, 1782)	Choro	Chile	N.D.
Unidentified marine mussels			138,813

Note: N.D. = no data available.

and B, 8, and 9). Furthermore, the unit of measurement generally used by the industry itself is a "bushel", which is a volumetric entity that creates a series of problems when one attempts to convert measurements to metric weight units. For formal conversion purposes in official published statistics (such as those presented in this chapter), the U.S. Government traditionally has assumed that 1 bushel weighs 60 lb (or approximately 27 kg), and that there are approximately 15 lb (or 6.8 kg) of wet mussel meat per bushel. In reality, the weight of a bushel of live cultured mussels can vary from 68 lb (30.8 kg) to 42 lb (19 kg),

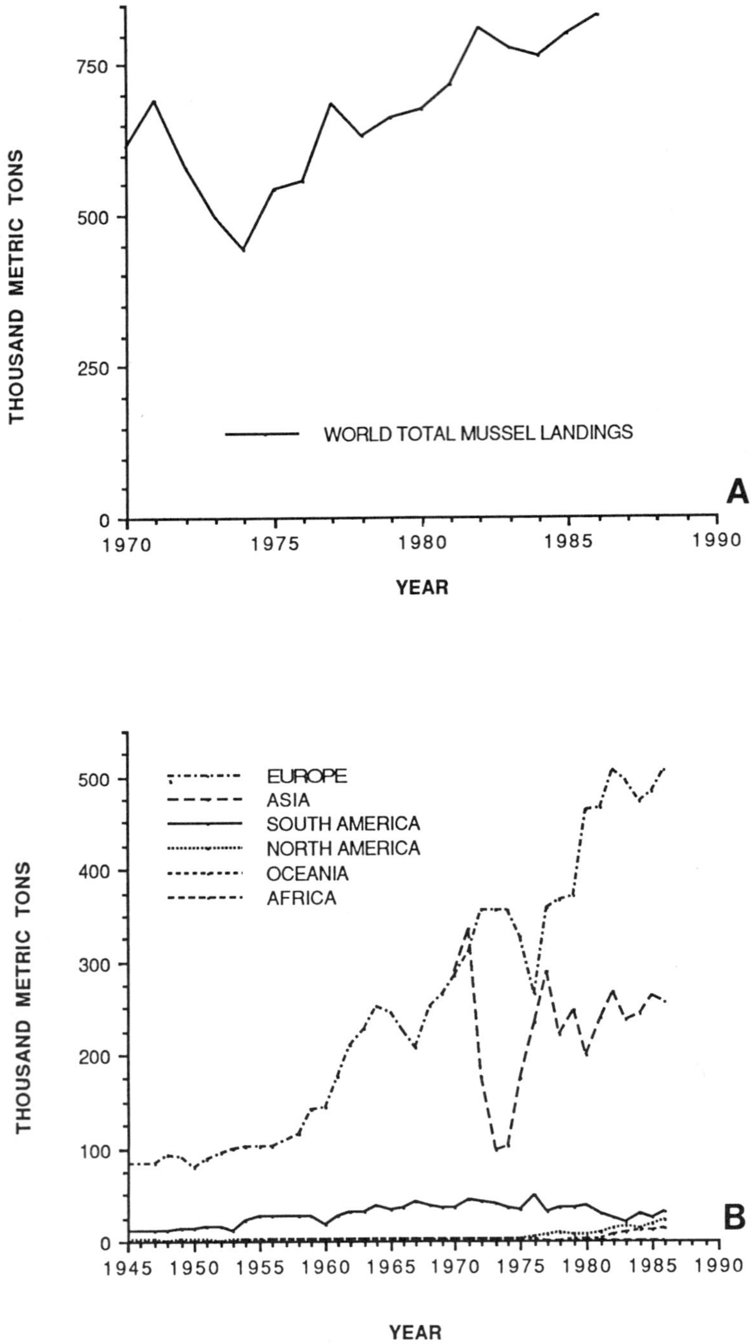

FIGURE 3. (A) Total world mussel production from 1970 to 1986.[12,13] (B) World mussel production by continent and year, 1945 to 1986. Values are expressed as whole live weight. Data are unavailable for Asia prior to 1970 and for Africa prior to 1979.[12,13] The precipitous decrease in both world and Asian mussel landings in 1972 and 1973 can be attributed largely to the collapse of the green mussel fishery in Thailand.[14]

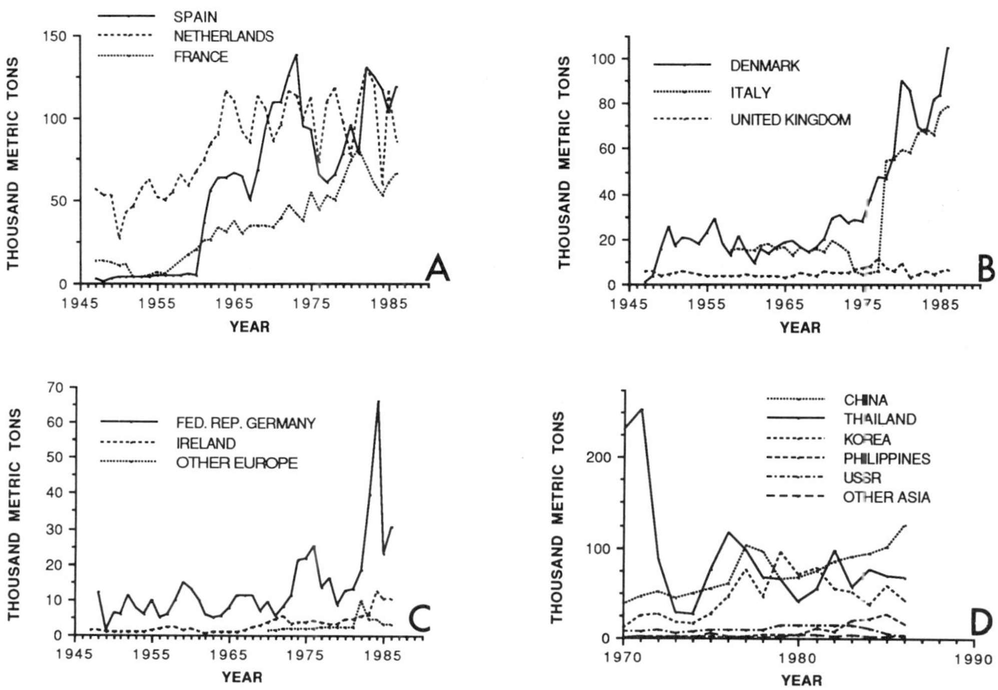

FIGURE 4. Total mussel landings, expressed as whole live weight, in Europe from 1947 to 1986 and in Asia from 1970 to 1986. Note the differing scales for the ordinate (y axis) in each graph. (A) Spain, the Netherlands, and France. The contribution from Spanish aquaculture was not reported by FAO prior to 1961, explaining the rapid increase shown for Spanish mussel landings in the early 1960s.[12,13] (B) Denmark, the U.K., and Italy. Data for Italy are unavailable prior to 1958 and have recently been revised for the years 1978 to 1986.[12,13] (C) Federal Republic of Germany, Ireland, and other mussel-producing countries of Europe (Albania, Bulgaria, Norway, Portugal, Sweden, the U.K., and Yugoslavia). (D) China, Thailand, Korea, Philippines, U.S.S.R., and other mussel-producing countries in Asia (Malaysia and Turkey). The precipitous decrease in Thailand landings for 1972 and 1973 can be attributed largely to the collapse of the green mussel fishery.[14]

depending on water loss and the weight of both meat and shell. Steamed meat weight in pounds per bushel varies from 6 to 16 with an average of 12 lb (5.4 kg) for cultured mussels and 8 lb (3.6 kg) for wild stocks. There are three obvious peaks in U.S. mussel production that we feel warrant explanation and discussion: (1) relatively high production during the late 1800s and early 1900s, including a dramatic production peak in 1908; (2) increased production during and immediately subsequent to World War II; and (3) striking increases in production over recent years.

1. Production During Early Years

Statistical sources of production data prior to the 1930s are poor, and, thus, the size of the commercial industry prior to that period may not be accurately reflected in the data presented in Figure 8. Despite this difficulty, certain trends in mussel production are apparent. Until the Great Depression of the 1930s, the only recorded years of significant mussel production for the East Coast were from the Middle Atlantic States for the years 1887, 1888, 1897, 1901, 1904, 1908, and 1921. For the West Coast, the only recorded years of significant activity were from 1888 to 1872, 1985, and 1899; however, as emphasized by Field,[2] production during a number of years prior to the early 1920s, for which production statistics are not available, may have been even higher than the large volumes reported for 1908. Field[2] points out that a single firm in New York reported handling 50,000 bushels of mussels 340 t of wet mussel meats) valued at $17,500 during the year 1912. Unemployment in the

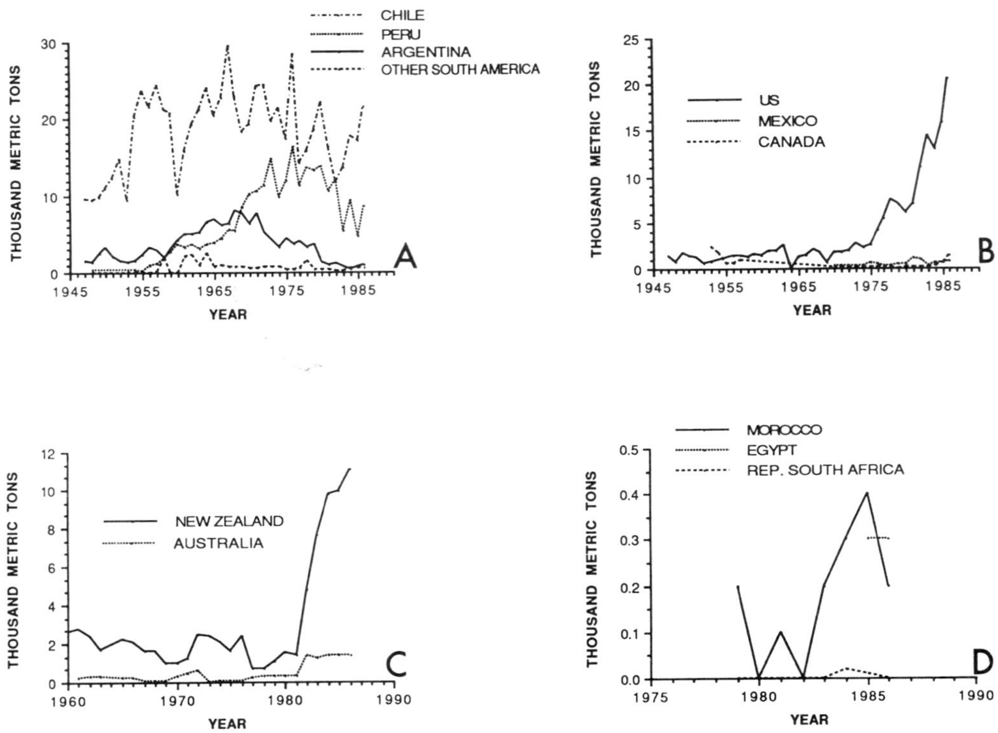

FIGURE 5. Total mussel landings, expressed as whole live weight, in North and South America from 1947 to 1986, in Oceania from 1960 to 1986, and in Africa from 1979 to 1986.[12,13] Note the differing scales for the ordinate (y axis) in each graph. (A) Chile, Peru, Argentina, and other mussel-producing countries in South America (Ecuador, Uruguay, and Venezuela). (B) Mussel production in North America (U.S., Mexico, and Canada). (C) Mussel production in Oceania (New Zealand and Australia). (D) African mussel production (Morocco, Egypt,[31] and the Republic of South Africa).

Middle Atlantic States was particularly high in 1908 following the "panic of 1907", and, during the third quarter of 1921, unemployment was the highest since the depression that followed World War I. Widespread wage cuts during these periods contributed to a low standard of living. As an inexpensive source of protein, mussels were substituted for more costly foods as the standard of living declined.[40]

2. World War II and More Recent Production

During World War II significant numbers of mussels were harvested and canned, as the species was considered a "marginal" food in the U.S. upon which no ration points were placed. Following the war, public demand for fresh or canned mussels declined sharply, and mussel production decreased to below prewar levels.[41] Since the early 1970s, there has been a renewed serious interest (with an accompanying dramatic increase in production; see Figure 8) in the utilization of this natural resource.[40,42-44]

Increased mussel landings over the past decade have led to considerable concern regarding the maximum sustainable yield (MSY) of the fishery. Estimation of MSY has proved extremely difficult due to the lack of quantitative estimates of available stocks, as well as the paucity of data accurately depicting growth dynamics of natural populations of this species;[40,44-46] however, the effect of increased production on fishery yields can be indirectly assessed through examination and interpretation of fishery statistics from the 1940s (see Figure 8). While increased production during the war years does not require an explanation, the sudden decrease in 1947 does. Dow and Wallace[41] attributed this sharp decline (at least

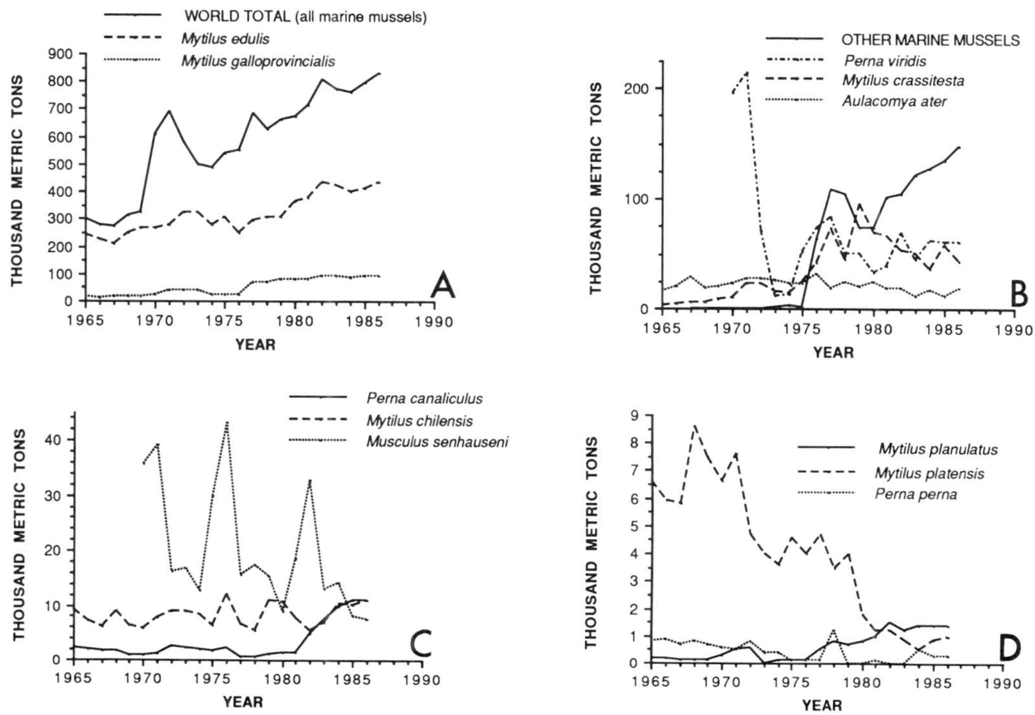

FIGURE 6. World mussel production by species and year from 1965 to 1986.[12,13] Data expressed as whole live weight. Note the differing scales for the ordinate (y axis) in each graph. (A) *Mytilus edulis, M. galloprovincialis,* and worldwide total production of all marine mussel species. (B) *Perna viridis, Mytilus crassitesta, Aulacomya ater*, and a combination of other marine mussel production not referred by FAO to a particular species. (C) *Musculus senhauseni, Mytilus chilensis,* and *Perna canaliculus.* (D) *Mytilus platensis, M. planulatus,* and *Perna perna.*

in Maine, where production during this period accounted for over two thirds of the total U.S. production) to two principal factors: (1) the availability of other protein foods considered to be more desirable and (2) the lack of available good-quality mussels. Concerning the second factor, they stated that "canning operations during the preceding war years had cropped almost completely the readily available supply."

In 1986, over 3000 t of wet mussel meats were harvested from natural populations within the U.S. (see Figure 8). Such a production level grossly exceeds that realized during World War II. If this high level of production is sustained or, as expected, increases during the next few years, the need for conservation of known resources and for additional sources of supply should be obvious. It has been, in fact, the problem of continual supply of high-quality mussels (in the face of growing demand) that has stimulated the recent sharp rise in industrial mussel culture activities in U.S. waters (see Figure 7A).

3. Recent Increases in Production and Culture Activity

In 1978, the U.S. National Academy of Sciences published a document entitled *Aquaculture in the United States: Constraints and Opportunities.*[47] In this publication the status of mussel culture in the U.S. was summarized in four brief sentences:

> The small demand for mussels *(Mytilus edulis)* in the United States has provided little incentive for mussel farming. Slowly increasing demand may accelerate development of a culture system. Mussels are a major aquaculture industry in some countries, such as Spain and the Netherlands. However, the labor-intensive methods used in these countries are not easily adaptable to the United States, and modification will have to be developed.

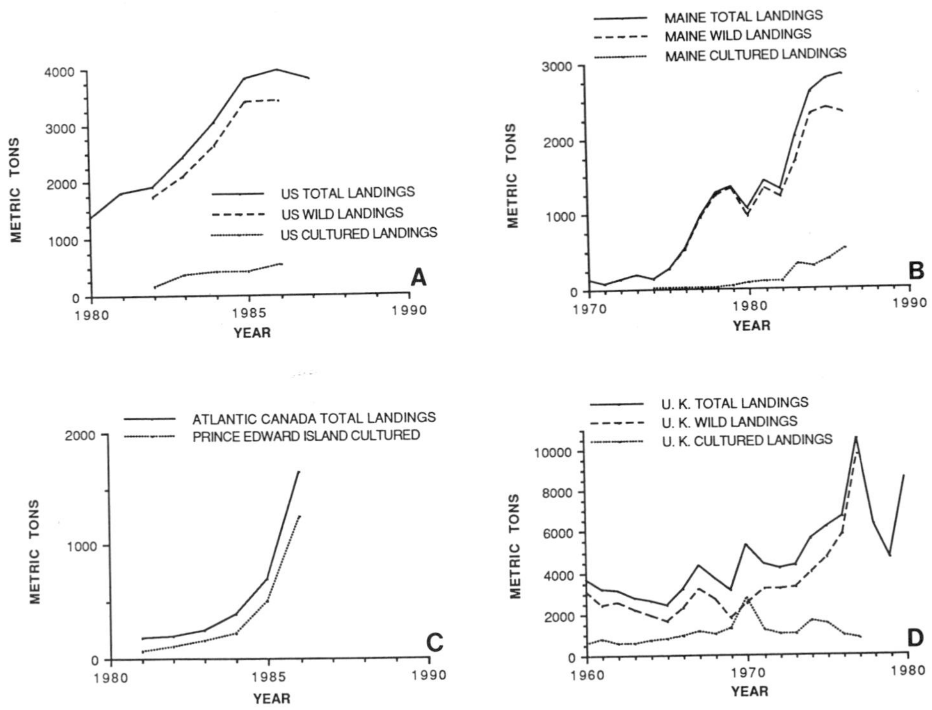

FIGURE 7. Mussel production resulting from natural and cultured harvest in the U.S., the state of Maine, Atlantic Canada, and the U.K. Note the differing scales for the ordinate (y axis) in each graph. (A) Total production of wild and cultured mussels, expressed as wet meat weight, in the U.S. from 1980 to 1987. Aquaculture data are available only for 1982 to 1986.[15] (B) Total wild and cultured production of mussels, expressed as wet meat weight, in the state of Maine from 1970 to 1986. Aquaculture data are available only for 1974 to 1986.[16] (C) Total mussel landings in Atlantic Canada and cultured landings in the province of Prince Edward Island, Canada from 1981 to 1986. Data are expressed as whole live weight.[17] (D) Total wild and cultured production of mussels, expressed as whole live weight, in the U.K. from 1950 to 1980. Cultured mussel data are available only from 1960 to 1977.[18]

Let it suffice to say that the above statements no longer adequately portray either the current status of mussel cultivation in the U.S. or the extensive progress made in this country over the past few years in successfully adapting Dutch bottom culture methods to various U.S. environments. As of the writing of this paper an updated aquaculture plan is being prepared by the U.S. Department of Agriculture (USDA) that should adequately address the status of, constraints to, and opportunities for the rapidly expanding mussel culture industry within the U.S.

Over the 5-year period from 1982 to 1986, both the volume and dockside value of blue mussel *(M. edulis)* landings in the U.S. have more than doubled. This dramatic rise can be traced in large part to a more than doubling of mussel landings in the state of Maine over this same period of time. While factors responsible for the increased Maine production are difficult to ascertain, there is an extremely intriguing correlation between the rising total mussel production and a striking increase in the percentage of cultured mussels being landed within the state (Figure 7B; also reflected in data available for total U.S. landings of wild vs. cultured mussels, Figure 7A).

In the early 1980s, bottom culture activities were initiated in various environments

TABLE 2
Countries of the World which Contributed to the 1986 Worldwide Production of 829,497 t (Whole Live Weight) of Marine Mussels, Together with Individual Production Totals[12,13] and Commercial and Experimental Scale Culture Methods Used in Each Country

Country	Production (t) (1986)	Culture techniques
China	125,440	Long-line, raft[19]
Spain	118,978	Raft[20], fixed platform hanging[21]
Denmark	105,012	Bottom[21]
Netherlands	85,896	Bottom[22]
Italy	78,561	Fixed platform hanging[23], open-sea long-line[24]
Thailand	67,510	Poles, fish traps, bottom, long-line[a], raft[a]
France	66,382	Bouchot (pole)[25], open sea long-line[a26], fixed platform hanging[21]
Korea	43,009	Long-line
Fed. Rep. Germany	30,684	Bottom[21]
Chile	21,604	Raft, long-line[27]
U.S.	20,709	Bottom, raft, long-line
Philippines	15,719	Hanging stake[28]
New Zealand	11,070	Raft, long-line[29]
Ireland	10,615	Bottom[21], raft, long-line
Peru	8,714	N.D.
U.K.	6,290	Bottom, raft[a 18]
Turkey	2,669	N.D.
Albania	2,100	N.D.
Canada	1,441	Bottom, raft, long-line
Australia	1,395	N.D.
U.S.S.R.	1,301	Fixed platform hanging[a30]
Yugoslavia	1,085	Fixed platform hanging[31]
Argentina	966	N.D.
Mexico	841	N.D.
Malaysia	709	Raft[a32]
Venezuela	350	Raft[a33]
Ecuador	215	N.D.
Morocco	184	N.D.
Uruguay	36	N.D.
Portugal	10	N.D.
Sweden	2	Long-line[a34]
Rep. South Africa	0	N.D.
Bulgaria	0	Open sea long-line[a35]
Norway	0	N.D.
Egypt	N.D.	Bottom[a36]
Greece	N.D.	Raft[a], long-line[a37]
India	N.D.	Raft[a38]
French Polynesia	N.D.	Off-bottom rack culture[a39]

Note: N.D. = no data available.

[a] Experimental culture.

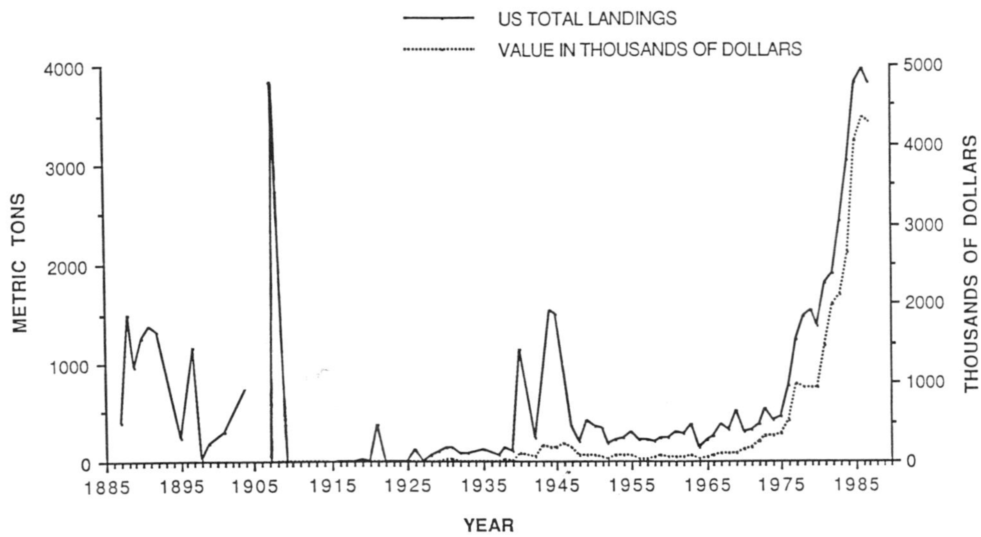

FIGURE 8. Quantity (by wet meat weight) and value (in nominal U.S. dollars) of annual U.S. mussel landings from 1887 to 1987.[40] For conversion purposes, the U.S. government considers meat weight to be equal to one fourth the organism's live weight in the shell.

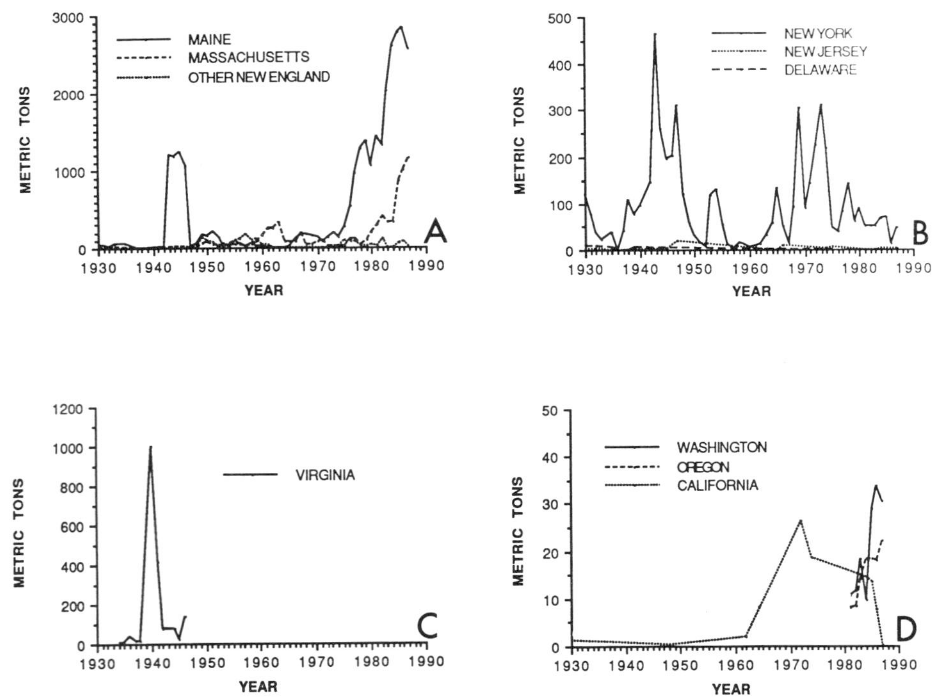

FIGURE 9. Quantity (by wet meat weight) of U.S. mussel landings by year and state[13,40] from 1930 to 1987. Data prior to 1930 are incomplete. Note the differing scales for the ordinate (y axis) in each graph. (A) Production of mussels in Maine, Massachusetts, and other New England mussel-producing states (Connecticut and Rhode Island). Data for Rhode Island are confidential from 1978 onward and therefore not reported. (B) Mussel production in the Middle Atlantic states of New York, New Jersey, and Delaware. (C) Mussel production in Virginia. Mussels were landed in the Chesapeake region only from 1934 to 1946.[40] (D) Mussel production on the West Coast of the U.S. in Washington, Oregon, and California.

throughout Maine. Coupled with these activities was an aggressive marketing strategy that appears to have had a profound impact upon consumer acceptability of mussel products. Before discussing the details of the bottom mussel culture practices which now form the basis of the majority of U.S. production, we will review briefly the history of mussel cultivation within U.S. waters.

The first mussel company, Abandoned Farm, Inc., was established in 1973. By 1982, there were four private industrial firms actively engaged in the aquaculture of blue mussels: (1) Abandoned Farm, Inc., 2 ha (5 Ac) in the Damariscotta River, Lincoln County, ME; (2) Penn Cove Mussels, 6 ha (15 Ac) in Puget Sound, Coupeville, WA; (3) Blue Gold Sea Farms, 24 ha (60 Ac) in Narragansett Bay, Middletown, RI and 24 ha (60 Ac) in Penobscot, ME; and (4) The Great Eastern Mussel Farms, 2 ha (5 Ac) in the Sheepscot River, Edgecomb, ME and 26 ha (65 Ac) in Stonington, ME. The 1982 combined production from the four commercial operations was approximately 165 t of wet mussel meats. By 1986, the total number of U.S. firms involved in mussel culture had increased to approximately 10, with a combined total production of 544 t. In Maine alone, there are presently 32 separate leased mussel sites consisting of a total of over 700 acres. Over 95% of this leased acreage is devoted to bottom culture, and this form of mussel culture accounts for over 95% of the cultured mussel landings within the state; the remainder of the acreage (less than 5%) is devoted to a variety of suspended mussel culture techniques (long-lines, rafts, etc.). From 1982 to 1986, there was an approximate fivefold increase in Maine's production of cultivated mussels (from 101 t of wet mussel meats in 1982 to 514 t in 1986; see Figure 7B). It might also be mentioned that off the coast of California a new form of "bouchot" (pole) culture is being practiced using offshore oil platforms. Here, mussels attach naturally to various structures associated with the platforms, and mussel growth rates are among the highest in the world. In 1986, approximately 1000 acres within the U.S. were under cultivation using various methods (rafts, long-lines, bottom culture, etc.), and the total value of cultured mussel landings was approximately $2 million.

4. Current Mussel Culture Practices

Prior to 1980, all U.S. firms involved in mussel aquaculture employed suspended culture systems. The designs and constructions of the employed systems, many of which are still in use, varied greatly from one firm to another and are reviewed in detail by Lutz.[44] In 1980, The Great Eastern Mussel Farms, Inc., initiated bottom culture operations adapting Dutch cultivation techniques to Maine waters. In these operations, "seed" mussels with minimum shell lengths of 20 mm are gathered from extensive intertidal or subtidal (as deep as 15 to 20 m) seed beds (Figure 10). Seed beds that are presently exploited commercially have average mussel seed densities of between 500 and 10,000 bushels/acre.

The initial density of seed planted on leased culture plots (the depth of which generally ranges from 1 to 5 m below mean low water) is extremely critical for optimal mussel growth and largely dependent on the flux of seston in any given area. As a general rule of thumb, transplanted seed mussels are planted (Figure 11) on bottom culture "farms" (Figure 12) at densities approximately equal to 1/10 of the density of the seed bed from which the mussel seed was gathered. Planting densities are thus generally in a range between 500 and 1000 bushels/acre (Figures 12B and 13A), although highly productive areas with high rates of seston flux are sometimes planted with densities as high as 15,000 bushels/acre (Figures 12C and 13B). The very process of seed transplantation at such reduced densities (relative to those encountered in natural populations) generally increases growth rates approximately fourfold over those encountered in natural populations. (Bottom-cultured mussels reach market size in approximately 2 years vs. 7 to 8 years in most natural populations.) In addition, meat yields of cultured mussels generally are significantly higher than those of mussels harvested from natural populations. When planting occurs at a density of 750 bushels/

A

B

FIGURE 10. Mussel (*Mytilus edulis*) seed beds. (A) Extensive intertidal mussel seed bed near Stonington, ME. (B) Extensive shallow subtidal mussel seed bed near Stonington, ME.

FIGURE 11. Work boat in the process of planting seed mussels (*Mytilus edulis*) on a bottom culture plot in the vicinity of Stonington, ME.

acre, the seed mussels are dispersed over the bottom in a series of mussel "patches" that cover approximately 20% of the bottom (see Figures 12B and 13A). Both patch size and percentage of bottom coverage increase with an increase in planting density (Figures 12C and 13B). The size of patches in areas with low current speed is extremely critical, as too large a patch may result in markedly reduced growth rates and an inconsistency in meat yield throughout the patch (Figure 14). In general, in areas with current speeds less than 15 cm/s, bottom patches should be less than 2 m in diameter; however, at current speeds greater than 15 to 60 cm/s, patch diameters may be as large as 5 m. Areas with current speeds in excess of 60 cm/s are generally considered unsuitable for farming activities due to problems associated with the instability of the associated mussel beds and sediments.

Mussel seed planting strategies involve the use of seed "counts" per bushel. There are approximately 1000 "market" (approximate shell length = 50 mm) mussels per bushel (i.e., 1000 "count"). When transplanting seed to growing areas, the following guidelines, which assume optimal planting densities (discussed below), are used by the industry: 4000 count seed mussels transplanted to bottom culture plots reach market size in approximately 1 year; 8000 to 12,000 count seed mussels are generally left on the culture plots for 18 months; and 2000 count seed mussels are left on the culture plots for considerably shorter periods of time, predominantly for fattening purposes. It is critical that the industry obtains high "seed to harvest" yields and high meat yields because of the high cost of seeding in relation to the exploitation of wild mussels. The industry currently grades mussels into four categories according to shell thickness, a measurement conveniently related to the width between steel grader bars through which a live mussel will pass: (1) large = >1 1/8 in. (>28.7 mm); (2) regular = 5/8 to 1 1/8 in. (15.9 to 28.7 mm); (3) half-grown = 1/2 to 5/8 in. (12.7 to 15.9 mm); and (4) seed = <1/2 in. (<12.7 mm). Market size mussels with a shell length of approximately 50 mm have a shell thickness of approximately 15.9 mm (5/8 in.).

A

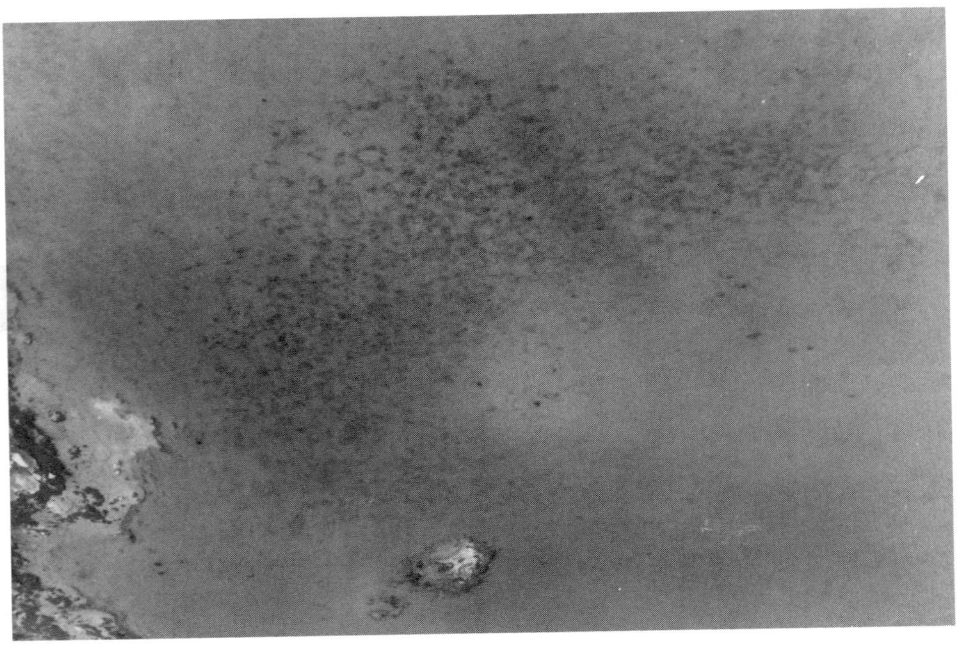

B

FIGURE 12. Aerial photographs of cultivated mussel (*Mytilus edulis*) farms near Stonington, ME. (A) Aerial view of two bottom culture mussel farms in shallow coastal Maine waters. The mussel farm to the left and directly above the small central island has been planted, in patches, at a density of 500 bushels of seed mussels to the acre, while the dense mussel farm near the mouth of the embayment in the upper center of the photograph has been planted, in patches, at a density of 1500 bushels of seed mussels per acre. This latter region is characterized by a high seston flux. (B) Closer view of 500 bushels per acre mussel farm shown in A. (C) Closer view of 1500 bushels per acre mussel farm shown in A.

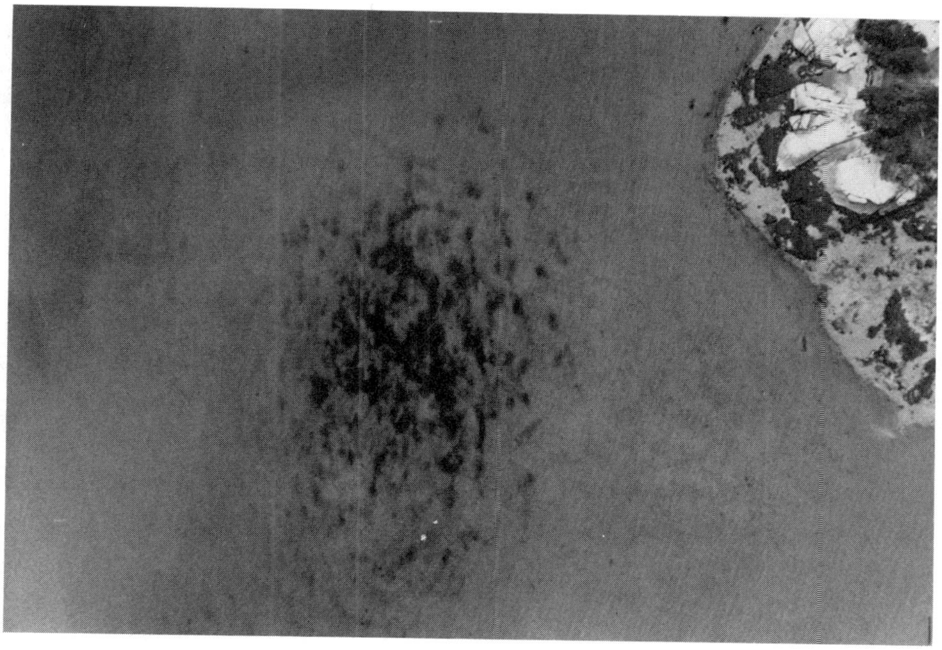

FIGURE 12C.

An interesting correlation has been found between the presence of extensive eelgrass beds (Figure 15) and the existence of productive mussel seed beds close by. The eelgrass itself and attached algal epiphytes offer an excellent substrate for primary mussel settlement, and up to 4000 mussel spat have been found on a single eelgrass blade in certain Maine environments.[48]

At the time of harvest, mussels are dredged from the bottom culture farm plots using metal dredges of various designs (Figures 16 and 17). After harvesting, the mussels generally are transported to shore-based facilities equipped with large purging tanks (Figure 18) and mechanical washer/grader devices for declumping, sorting, and cleaning the organisms for shipment (see Hurlburt and Hurlburt[49] for a photographic illustration of such a device). Mussels are primarily sold live in the shell, and significant quantities are transported in large refrigerated trucks and by air freight to various locations throughout the U.S. and Canada (Figure 19).

5. Constraints to Mussel Aquaculture

The seasonal and highly variable supply of mussels from natural populations has been a constraint to the development of stable mussel markets. In spite of this, the control over timing, quality, and volume afforded by mussel culture operations has allowed for significant market development.[50] A reputation for quality in mussels cultured within U.S. waters has resulted in significant economic benefit to the entire mussel industry (both the wild fishery and aquaculture operations) as shown in higher prices and increased market penetration. Further growth of the industry is limited by the availability of high-quality mussels from natural populations and is, thus, largely dependent upon the expansion of culture activities.

Predation by starfish, crabs, lobsters, and ducks continues to create major problems for firms engaged in mussel culture throughout the U.S. In New England, whelks appear to be a major predator on small mussel seed at some sandy and coarse substrate sites. Experimental studies conducted by Shumway et al.[51] on the effects of calcium oxide (''quicklime'') on ''nontarget'' organisms in mussel beds have suggested that large-scale liming of mussel

A

B

FIGURE 13. Underwater photographs of bottom culture mussel farms near Stonington, ME. (A) View of mussel farm seeded at a density of 500 bushels of seed mussels per acre. (B) View of mussel farm seeded at a density of 1500 bushels of mussels per acre.

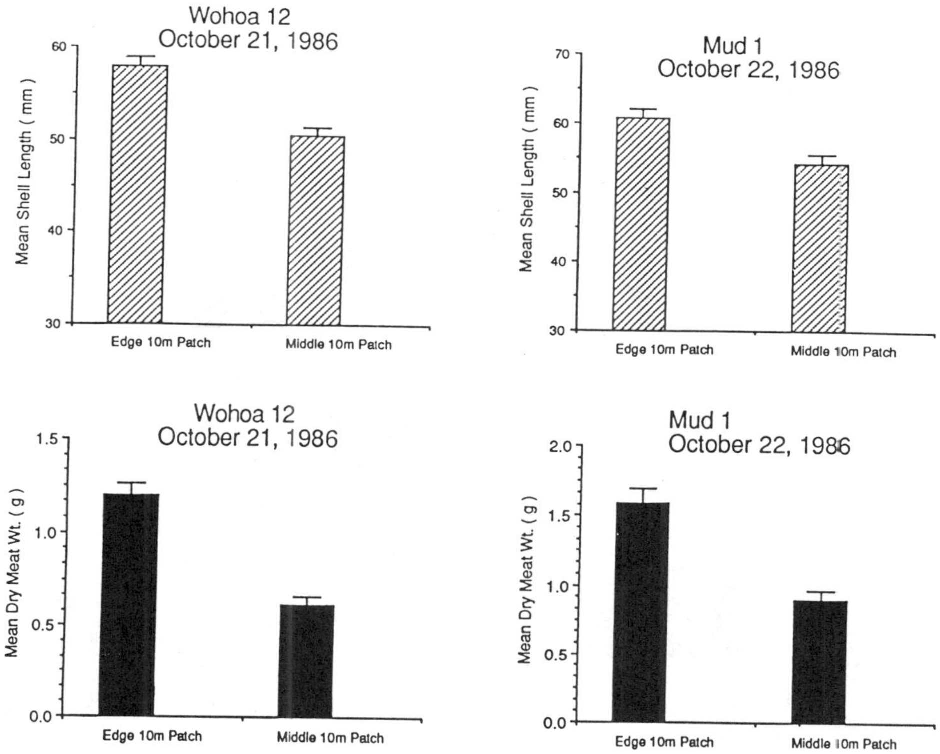

FIGURE 14. Shell length (mm) and dry meat weight (g) of mussels sampled in October 1986 from the edge and middle of 10-m patches 1 year after seeding of Wohoa Bay, ME, section 12, and Mud Cove, section 1. Values represent means plus one standard error.

FIGURE 15. Typical eelgrass (*Zostera marina* Linnaeus) bed in shallow coastal Maine waters; eelgrass and attached algal epiphytes provide an excellent substrate for primary settlement of the larvae of mussels (*Mytilus edulis*).

FIGURE 16. One of various types of metal dredges used to collect mussels from culture plots at the time of harvest. (From Hurlburt, C. G. and Hurlburt, S. W., *Mussel Culture and Harvest: A North American Perspective,* Lutz, R. A., Ed., Elsevier, Amsterdam, 1980, 69. With permission.)

bottom culture plots at concentrations of approximately 1 t/acre may represent an effective means of controlling starfish predation associated with bottom mussel culture farms. Such liming practices are currently employed on a large scale in Canadian waters (Figure 20) and by oyster culturists in Long Island Sound.

The consumption of large quantities of mussels by various species of sea ducks, particularly eiders and scooters, continues to be a large problem for those engaged in both suspended and bottom mussel culture activities throughout the U.S. In 1987 alone, one mussel firm lost approximately 25,000 bushels of mussels (170 t of wet mussel meats) located on 75 acres of leased culture bottom to eider duck predation. The use of mechanical sound devices (Figure 21) has been reported to have only limited success in controlling duck predation, and to date there exists no satisfactory means of controlling such predation.

Periodic outbreaks of "red tides" that result in the accumulation of algal biotoxins in mussels create continuing problems for mussel culture industries on both coasts of the U.S.[44,52,53] The consumption of contaminated mussels can result in paralytic shellfish poisoning in humans, and extensive monitoring and regulatory programs designed to prevent the sale of mussels containing biotoxins are currently under way on both the East and West Coasts. The species of dinoflagellates responsible for these periodic "red tides" are *Gonyaulax** *excavata* and *G. catenella* on the East and West Coasts, respectively, and considerable research is currently underway to ascertain factors responsible for such toxic algal blooms. For a detailed discussion of the accumulation of algal biotoxins in both natural and cultured stocks of mussels, see Yentsch and Incze.[53]

One of the serious limiting factors to the commercial use of large quantities of mussels within the U.S. is the presence of large (greater than 1 mm) and numerous pearls embedded

* The genus *Gonyaulax* has recently been renamed *Alexandrium*.

FIGURE 17. One of various types of metal dredges used to collect mussels from farm plots by the Maine mussel industry.

directly within the mantle epithelium of mussels in natural populations along the East Coast of the U.S. In the early 1940s, when large numbers of mussels were harvested, the pearls present in commercially important mussel populations along the northeastern coast from Eastport, ME[4] to Cape Cod, MA were quantitatively compared by Scattergood and Taylor.[54] As a result of their extensive survey, they recommended in 1942 the closing of six areas in Maine (areas that accounted for approximately one fifth of the total estimated Maine mussel production at that time) due to the presence of pearls. Canning operations were advised in January 1943 to avoid mussels taken from these areas.

More recently, in an extensive series of studies, Lutz[44,56,57] and Lutz and Hidu[55] have quantified the relationship between pearl incidence and age for a number of cultured and natural populations of mussels along the northeastern coast of the U.S. In these studies, marked differences in pearl incidence were observed between raft and shore populations of mussels of similar sizes, as well as between mussels of similar lengths obtained from rafts left in the water for various periods of time. The quantitative relationship found by these workers between number of pearls and age for two populations (raft and shore) located in Maine waters (Lutz and Hidu)[55] revealed that significant age-independent differences in pearl

FIGURE 18. Large holding tank at shore-based facility in Maine being used to purge mussels of mud and grit prior to shipment to market.

incidence exist between raft- and shore-based mussels from adjacent populations. They attributed the marked differences between the two populations with regard to the presence of pearls in organisms of similar lengths to differences in growth rate and age structure within the two population types. The implications of the results of these studies for both raft and bottom mussel culture are profound. As a result of the 1942 survey conducted by Scattergood and Taylor,[54] it was recommended that areas be closed where samples of mussels had either nine pearls with a diameter of 1 mm or greater, or more than three pearls with a diameter of 1.5 mm or larger in a 84-g sample of freshly drained meats. In the studies conducted by Lutz and Hidu,[55] no pearls with diameters greater than 1 mm were found in any mussels less than 5 years old, irrespective of whether the organisms were sampled from raft or shore locations. A marketable mussel (shell length >50 mm) can be obtained in periods ranging from 12 to 15 months in U.S. waters using suspended culture techniques and from 15 to 24 months using bottom culture. Therefore, problems associated with pearl incidence essentially are eliminated when such culture methods are employed.[29,42,44,57-63]

When bottom culture techniques are used, the situation is slightly more complex, since much of the mussel "seed" is of unknown age. This transplanted seed can consist of mussels of considerable size (shell lengths between 20 and 50 mm), and it is imperative that firms

A

B

FIGURE 19. Transport of mussels to market. (A) Great Eastern Mussel Farms, Inc. truck used to ship mussels to market. (B) View inside a large refrigerated truck being loaded with mussels prior to shipment to market.

FIGURE 20. Lime (calcium oxide) being applied to a bottom culture oyster farm in Atlantic Canada in an attempt to control starfish predation. Similar procedures are used on bottom culture mussel farms.

FIGURE 21. A mechanical sound device that is used by some mussel farm operations in an attempt to repel predation by dividing ducks on mussels.

engaged in bottom culture activities establish strict quality-control programs to ensure that the incidence of pearls in populations of seed mussels gathered for transplantation purposes is well below a level that will create consumer acceptability problems once the transplanted mussels reach market size. Along these same lines, it is important that adequate data are obtained and records kept to ensure that mussels harvested from bottom culture farm plots are less than 5 years old. Those firms actively engaged in culture activities at the present time recognize the importance of such efforts and devote considerable time and effort to quality-control programs aimed at preventing the transplantation or harvest of mussels with an undesirable incidence of pearls.

Mussel quality control for detectable pearls has been hindered by the lack of a rapid method for screening samples. At Great Eastern Mussel Farms, Inc., a new and rapid technique involves (1) steaming, removing, and weighing the meats from a 2-l sample of mussels; (2) homogenizing the meats in a high-speed blender for 30 s in an equal volume of water; (3) floating off the suspended tissue under gently running tap water in a beaker; and (4) rinsing the remainder into a sieve and counting the detectable pearls (>1 mm in diameter). An occurrence of 0.02 pearls per gram of mussel meat is suggested as a standard above which the mussels are considered undesirable.

Lack of an adequate understanding of the carrying capacity of open culture systems in relation to current flow, mussel density, and food availability represents a major constraint to the effective use of existing mussel culture leases. A basic understanding of factors controlling the yield of mussels from a given lease site will become increasingly important as fewer sites become available due to coastal development and competing interests and as the processing of mussel meats becomes a significant industry.

The identification and management of mussel seed beds will also become increasingly important as the industry expands. Mussel reproduction in relation to inshore productivity, mussel settlement on eelgrass beds, juvenile mussel migration, and predation on newly settled spat are all priority areas for increased research efforts aimed at developing more effective management strategies for seed beds.

It is estimated that over 35,000 acres of potential bottom culture sites exist in Maine alone and that current mussel production in this state is thus at only approximately 2% of its potential. At 1988 mussel prices and levels of mussel production on existing leases, the estimated potential value of landings ("dockside" value) associated with a Maine mussel culture industry expanded to its maximum would be approximately $75 million/year. It should be emphasized that this estimated value does not include any value-added processing potential. As emphasized by Wilson and Fleming,[50] realization of such an expanded mussel industry depends largely on the ability to "develop markets and resolve the problems of conflicting resource rights".

B. SPAIN

1. Background

On the northwest Atlantic coast of Spain, in a region known as Galicia, lie a series of bays or "rias" where raft culture of mussels has been practiced since the late 1940s (Figure 22). Culture on hanging ropes began in Spain in 1901 at Tarragona on the Mediterranean coast, spread to Barcelona by 1909, and began in Galicia in 1946. Raft culture is most intensive in the Ria de Arosa (about 2500 rafts), followed by the Ria de Vigo, the Ria de Pontevedra, the Ria de Muros (less than 100 rafts) (Figures 22 and 23) the Ria de La Coruna, and the Ria de Betanzas. Galician culturists have propelled Spain into the largest producer of cultured mussels in the world, with upward of 3500 mussel-growing operations accounting for between 250,000 and 300,000 t of mussels per year (whole live weight),[65] although the most recent figures issued by the Food and Agriculture Organization quote a figure of 119,000 for all of Spain in 1986[12,13] (Figure 4A). Prior to the 1940s, mussels in the Galician rias were used only as fishing bait and agricultural fertilizer.

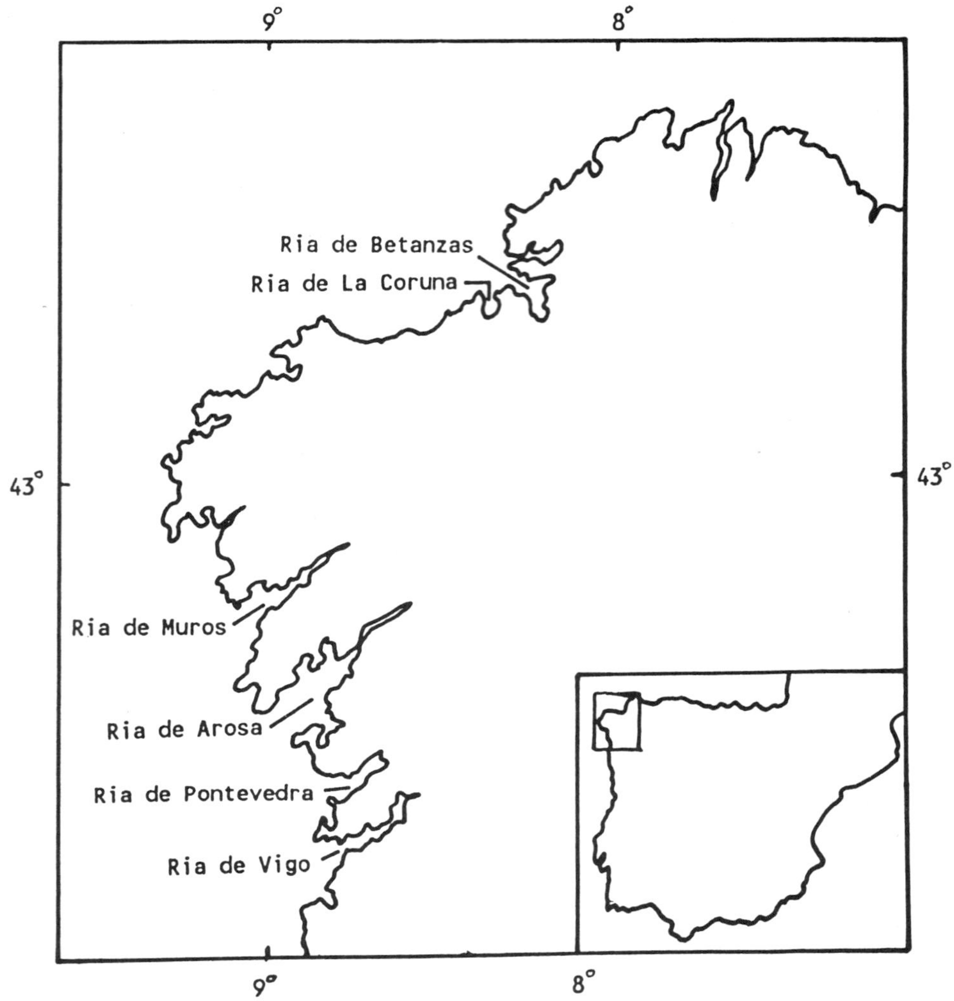

FIGURE 22. The Galician rias on the northwest coast of Spain.

2. Biology of the Species

Previous authors disagree over which mussel is farmed in Galicia, *Mytilus galloprovincialis*[66-68] or *M. edulis*.[20,65,69] The two forms coexist on Spain's Atlantic coast, and it is difficult to distinguish them, as hybrid forms may be present.[70,71] In Galicia, mussels reach market size (80 to 90 mm) in 1.5 years and obtain their maximum weight (with meat weight contributing upward of 50% wet weight) between September and December. There are two major periods of spawning: one in early spring with settlement occurring in April to June, and another in autumn with settlement in October to January. In the Ria de Vigo, mussels attain sexual maturity in 6 to 7 months at shell lengths between 40 and 50 mm.

Since a single mussel can filter from 2 to 5 l of water an hour, it has been estimated that mussels present on one rope can filter 90,000 l/h and that the mussels present on an entire raft can filter 70,000,000 l in a single day.[66] During an entire year, the mussels on one raft can ingest 180 ton of organic matter, 100 ton of which is returned to the sea floor where its effect on the bottom fauna can be catastrophic, as in the case of Europe's richest natural flat oyster bed, El Bao de Cambados in the Ria de Arosa. Tenore et al.[67] document the dramatic changes in the benthos of the Ria de Arosa over the past 40 years, from a diverse equilibrium assemblage to a low-diversity polychaete-nematode assemblage, as a result of the intense mussel cultivation.

FIGURE 23. Scores of mussel rafts of various construction moored in the Ria de Vigo, Spain. (From Myers, E. A., *Mussel Culture and Harvest: A North American Perspective*, Lutz, R. A., Ed., Elsevier, Amsterdam, 1980, 266. With permission.)

Fraga and Vives[72] estimated that 35 to 45% of the particulate organic matter is cleared during the passage of a given volume of water through a mussel raft. The proliferation of mussel rafts, coupled with the cumulative filtration of the expanded mussel biomass, has resulted in overall slower growth to market size in the Ria de Arosa than in the past. In the central part of this ria, seed mussels now require between 12 and 14 months to reach market size; whereas, in the early years of culture mussels reached this size in 8 or 9 months. Growth is still high close to the entrance of the Ria de Arosa; however, the rafts are less sheltered and are subject to storm damage.

3. Geography

The sheltered rias offer an ideal environment for suspended rope mussel culture on floating rafts. About 10% of the surface area of the Ria de Arosa is covered by polygonal arrays of 2200 rafts (200 of which are oyster rafts). These polygons are in the more protected inlets. The rias are deep (40 m in the Ria de Vigo, 60 m in the Ria de Arosa) and long (up to 25 km) and have a prolonged period of high phytoplankton production (with an annual average rate of photosynthesis of 10.5 μg C per liter per hour), strong tidal currents, relatively high salinity (35‰), and mild temperatures (9 to 21°C). The intermittent upwelling and intrusion into the rias of nutrient-rich North Atlantic water support the high primary production.[69]

4. Cultivation of *Mytilus edulis*

About 60 to 70% of the seed mussels transplanted to culture-growing ropes are collected from special ropes hung vertically or horizontally from the mussel rafts for this purpose. More seed mussels are available at the oceanic end of the rias, particularly near the Cies Islands in the Ria de Vigo, the Island of Ons in the Ria de Pontevedra, and the Island of Salvora in the Ria de Arosa. The best collection of seed mussels on ropes invariably occurs

FIGURE 24. Close-up of typical culture raft in the Ria de Vigo, Spain, showing the lattice framework with attached ropes and the method of wrapping the mussel seed onto the ropes. (From Hurlburt, C. G. and Hurlburt, S. W., *Mussel Culture and Harvest: A North American Perspective,* Lutz, R. A., Ed., Elsevier, Amsterdam, 1980, 69. With permission.)

in spring, with a single collector rope producing enough seed for thinning out onto 12 to 15 growing ropes.[21] A third minor source of seed comes from natural set on the growing ropes, although this requires much manual labor. Whenever possible, seed mussels are placed on the grow-out ropes the same day they are collected and are held in place with a special net or tissue (''tejido de malla'') that is wrapped around the rope and mussels (Figure 24). This net is manufactured in Vadalona-Barcelona and Galicia and is sold in 100- or 200-m reels, which is sufficient to load six to eight ropes with mussels. A skilled laborer can wrap a meter of rope with seed mussels in a minute; mechanization, however, has reduced this time to between 5 and 15 s. Although this mechanization process is expensive, 500 ropes can be wrapped in just 14 h with this machinery. The ropes are hung from the rafts by a loop made of thinner polyester rope (called a ''rabiza''). With a similar loop on either end of the rope, the whole rope can be inverted periodically, the aim being to produce mussels of equal size. Every 30 to 45 cm, a wood or plastic peg is inserted through the strands of each rope, which prevents the mussel clusters from sliding down the rope and being lost.

In the past, the mussel raft or parks were made from old sturdy wooden boats, ideal for exposed situations. Since these are no longer available, rafts are now constructed either as catamarans with long cubical wooden floats or well-tarred iron floats or as a single large centrally located float. The wooden floats may be coated with a layer of concrete as protection from shipworms. The floats support a superstructure of main beams, which help support the cross-beams to which the mussel ropes will be attached. An average raft, measuring 18 $\times$ 18 m, can support 600 to 700 ropes, while the largest may have 1000 ropes. Two or three ropes are hung per square meter of raft for a variety of purposes: (1) for catching seed; (2) for producing half-grown mussels; and (3) for producing market size mussels. The rafts, therefore, remain continuously productive. A medium-size wooden raft costs about 3 to 4 million pesetas ($24,000 to 32,000) and has a life span of 10 to 15 years, while a modern fiberglass raft costs 10 million pesetas ($80,000) but may last considerably longer.

After 3 to 4 months, or when the weight of mussels on the ropes reaches about 10 kg/m of rope, the mussels are removed from the ropes, cleaned, and redistributed onto two or three new ropes. This decreases the risk that clusters of mussels may drop off and increases the growth rate. Since the mussels are sensitive to light and tend to migrate up and down the rope until they reach a position with an optimal lighting regime, some fall off the ropes shortly after transplantation. The transplantation process is normally carried out on cloudy days and not in the summer because of this problem. For the next 8 to 12 months the mussels require little handling.

The absence of strictly seasonal spawning activity in Galicia's mussels means that a certain percentage of mussels harvested will have recently spawned, a condition that is not desirable. This is monitored by the grower daily during harvesting periods. At the time of harvest, the ropes are lifted in a basket and shaken loose by means of a crane located on either the raft or on a work boat. Market size mussels are either transported directly to canneries, or, if they are to be sold fresh in Spain or other countries, to depuration plants where they are cleansed for 48 h in water that has been purified by chlorination. Smaller mussels are rewrapped onto new ropes and placed back in the water. Under normal conditions, a meter of rope will produce approximately 10 kg of mussels.

Raft culture areas are controlled by the government and leased to the operators (usually a family business) for 10 years for a nominal fee. Where navigation is critical, as in the Ria de Vigo, the rafts are required to be moored with two anchors, and a distance of 100 m from other rafts must be maintained in all areas. When a raft concession is requested in an area outside designated "polygons", the Ministry of Agriculture, Fisheries, and Food must be assured that the placement of the raft in such an area does not compromise other interests (such as recreation, national defense, and navigation) and that the area in which the raft is placed is declared to be free of pollution by the national health department (Sanidad Nacional). In addition, officials must be satisfied that the planned increase in mussel production keeps pace with the market demand and that the area of the ria under consideration does not become overexploited.

5. Problems and Risks Associated with Mussel Farming

Excessive rainfall in the rias may lower the salinity to such a degree that mussels on the upper 0.5 to 1 m of the ropes will die. In addition, winter storms can damage and even sink the rafts, a situation that occurred in the winter of 1970—1971 when 25 mussel rafts sank in a gale in the Ria de Arosa. Little loss occurs due to predation by starfish and crabs, as the mussels are kept off the sea bottom, although predatory fish may occasionally eat the seed mussels attached to the culture ropes. Some farmers protect their rafts by constructing fencing around the rafts. The copepod parasite *Mytilicola intestinalis* Steuer has not greatly affected the mussels of Galicia, since more than two individuals are seldom found in a single mussel. Fouling organisms (such as algae, ascidians, bryozoans, barnacles, and polychaetes) are more of a problem, as they not only compete for food and space on the ropes, but also must be eliminated by hand during the harvesting process if the mussels are to be sold to the depuration plants.

Not all risks to the mussel farmers result from the vagaries of nature. In November 1970, the coastal vessel Erkowit, loaded with a cargo of pesticides including dieldrin and mercury compounds, went aground in Ria de La Coruna, releasing part of its cargo and causing all the cultured mussels in the ria to be declared unfit for human consumption. Mussel farming was halted for over a year as a result of this disaster.

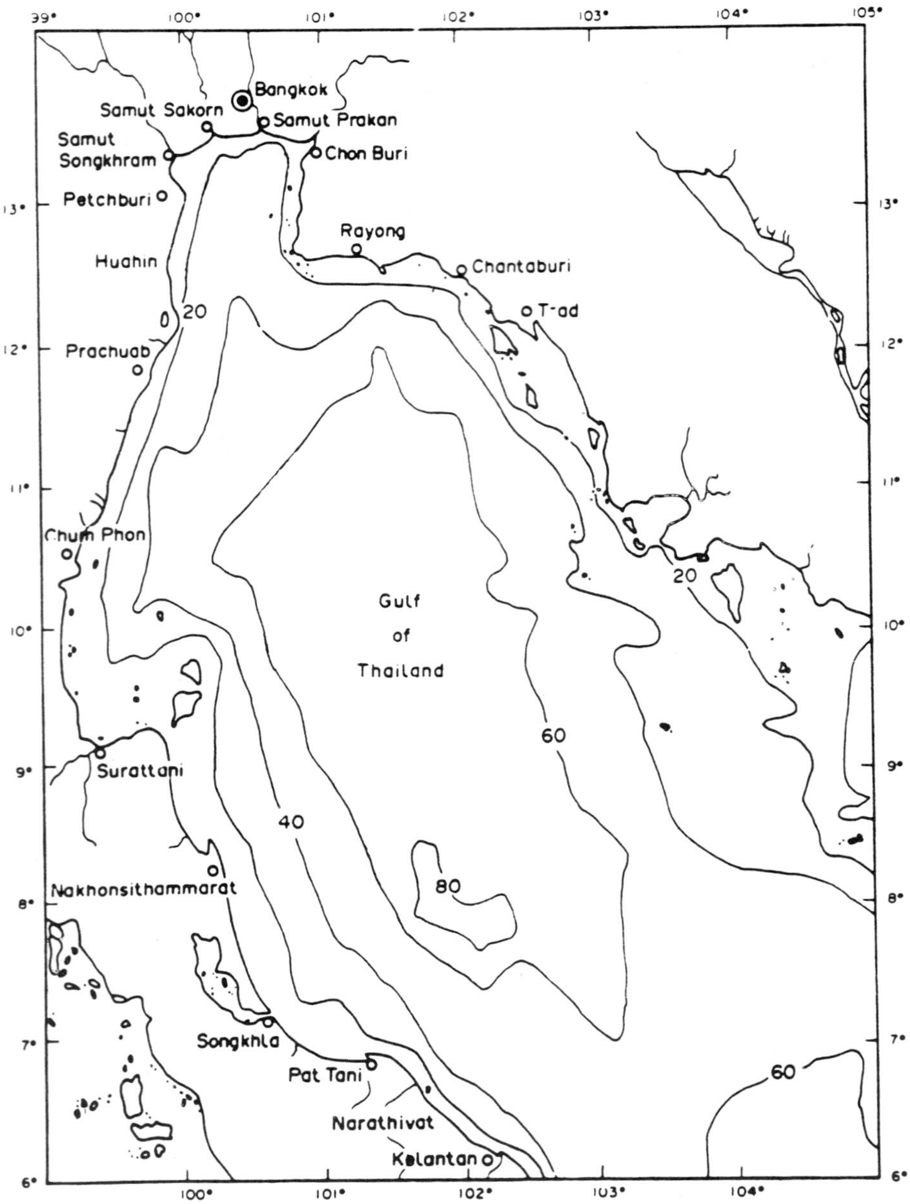

FIGURE 25. The Gulf of Thailand. After Hungspreugs and Yuangthon.[75]

C. THAILAND

1. Background

Mussel farming in Thailand can be described as a "semiculture," in which production of mussel seed and food for juvenile and adult mussels is left entirely to nature. With a 1986 production of 67,510 t of mussels (Table 2), Thailand is the second largest producer after China of mussels in Asia. Of the many shellfish industries that are of significance to the economy of Thailand, the mussel industry is the largest in terms of harvested biomass.[73,74] Although mussels are widely distributed in sheltered subtidal areas on both sides of the Gulf of Thailand, culturing activities are largely confined to the Upper Gulf provinces of Chon Buri, Chachoengsao, Samut Prakan Samut Songkram, and Petchburi (Figure 25).

2. Biology of the Species

Two species of mussels are cultivated: the green mussel, *Perna viridis* (Linnaeus, 1758), locally called "hoi malang-poo;" and the striped horse mussel, *Musculus senhauseni* (Benson, 1842), locally called "hoi ka-pong." Between the years 1975 and 1979, the area utilized for green mussel cultivation was 38% of the total area used for shellfish cultivation, and production was 45.5% of the total shellfish production. Corresponding values for the striped horse mussel were 5.6 and 17.5%, respectively.[74] Large natural populations of green mussels are found attached to jetties, submerged branches, and bamboo fish traps, whereas the striped horse mussel forms loose patches on the surface of soft muddy substrates.

It is apparent from the literature that the spawning period of green mussels is quite variable. Tiensongrassami[76] found that green mussels in the Gulf of Thailand spawned from September to November, while Chongpeepien[77] noted that this species spawned asynchronously from July to February in the province of Chachoengsao. In contrast, Tan[78] and Walter[79] demonstrated that spawning occurs year-round in *P. viridis* from Singapore and the Philippines. It is possible that the spawning cycle of green mussels has no distinct seasonality in the Upper Gulf of Thailand, particularly among those populations located near the mouths of large rivers where a large influx of freshwater occurs periodically. Green mussels in Thailand are sexually mature at a shell length as small as 2.5 cm and reach a maximum length of 16.5 cm.

Very little information is available on the biology of *M. senhauseni*. The preferred habitat of this species is on muddy surfaces at depths ranging from 0.5 to 2.0 m. The maximum shell length of this species is 5.5 cm.[80]

3. Geography and Hydrography

The upper Gulf of Thailand is a semienclosed square-shaped bay with a total area of approximately 8100 km^2. Four major rivers flow into the upper Gulf: the Mae Kong; the Ta Chin; the Bang Pakong; and the Chao Phraya, the longest, which has an average annual discharge of 21,000 m^3 (Figure 26). The climate within the Gulf of Thailand is tropical, with seasonally uniform yearly rainfall amounts of 250 cm. A strong salinity gradient exists between the near-shore environment and the mouth of the upper Gulf, ranging from 5 to 32‰ as a result of variations in the amount of river runoff. Air and water temperatures vary between 26 and 30°C, and there is evidence that upwelling occurs along the eastern Gulf during winter and along the western side during summer.[82]

4. Commercial Cultivation of *Perna viridis*

Prior to the initiation of mussel culture on bamboo poles in 1957, collection of mussels that secondarily settled on fish traps was the main source of green mussels.[83] The introduction of more modern fishing techniques after 1957 caused a gradual decrease in the number of fish traps, and, as a consequence, the harvest of green mussels also decreased. The collapse of the green mussel fishery in 1972 and 1973 led to a rapid expansion of cultivation activity in the mid-1970s.[14] At present, there are two methods of commercial production of green mussels: culture on bamboo poles (*Bamboosa* sp. and *Thyrsostachys* sp.) or palm stakes (*Licula* sp. and *Phoenix* sp.) that are driven into the muddy bottom in areas where natural mussel settlement occurs; and the incidental culturing of mussels on fish traps.

In 1986, areas used for culture were leased from the Royal Thai Government in the province of Chachoengsao at 500 baht per rai ($120 U.S./ha) and in the provinces of Chonburi and Petchaburi at 800 baht per rai ($192 U.S./ha). (1 baht = $0.04 in 1986; 1 rai = 0.16 ha). Total average costs (labor, boats, equipment, lease, etc.) per mussel farm in the three leading mussel-producing provinces of Chachoengsao, Chonburi, and Petchaburi are 75,599 baht ($2902); 131,916 baht ($5064); and 138,805 baht ($5328), respectively.[73]

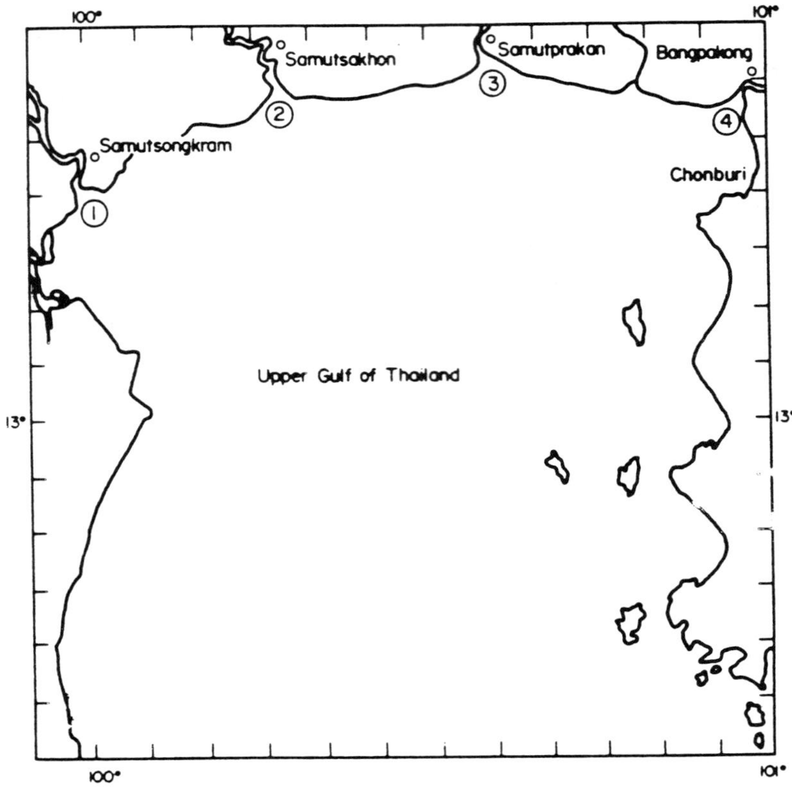

FIGURE 26. The upper Gulf of Thailand where most mussel farming activity is conducted. Numbered circles represent the mouth of (1) the Mae Kong River, (2) the Ta Chin River, (3) the Chao Phraya River, and (4) the Bang Pakong River. (After Menasveta, P. and Cheevaparanapiwat, V., *Mar. Pollut. Bull.*, 12, 19, 1981.)

a. Chachoengsao Province

Most mussel farms in this province are located in shallow water at distances between 2 and 3 km from shore. Bamboo poles 3 to 3.5 m in length are set from July to November during low tide. A work force of six or seven people is required to set the poles, which are transported to the site by boat. Farms in this province are typically small; the smallest farm has 2000 poles and the largest 8000 poles.[73] Harvesting begins in June, approximately 7 to 8 months after mussel settlement. During this period owners harvest only a specified quantity of mussels daily depending on the demands of the market as determined by the wholesalers. The farmer hires divers who pull out the poles (Figure 27) and also strip the mussels from the poles. Mussels are transferred to bamboo baskets on shore to be picked up by local collectors who transport them to the wholesalers (Figure 28). The organisms are delivered to market ungraded and uncleaned (Figure 29).

b. Chonburi Province

In Chonburi, the setting of poles is done at night from September to October during low tide. Approximately 3 months after mussels have set on the poles, the top 0.5 to 1 m of the pole is cut off, and the small mussels are removed and used as feed for domestic animals such as ducks. Market-size mussels for human consumption are harvested from February to July, with a certain percentage being held back by the growers for sale after October when these ''off-season'' mussels fetch a higher price. However, with such a harvesting strategy, the owners run the risk of missing the period of the maximum settlement

FIGURE 27. A green mussel farm in Chachoengsao Province. The poles are set approximately 1 m apart, while navigation lanes are approximately 10 m wide.

FIGURE 28. Harvested green mussels being taken to shore in a boat typical of those used in Chachoengsao Province.

FIGURE 29. Bamboo baskets used to transport mussels from Chachoengsao to the central market in Paknam, in Samut Prakarn Province, a journey of about 1 h.

of mussels that generally occurs between September and December. Typically, the farms in Chonburi are larger than those in Chachoengsao, with the number of poles ranging from 10,000 to 200,000 per farm.[73]

c. *Petchaburi Province*

In Petchaburi, the 4- to 6-m poles are set between August and December during low tide. Market-size mussels are harvested after 8 months with most harvesting activity occurring in July. The smallest farm in this province has 8000 poles, and the largest has 240,000 poles.[73]

d. *Samut Songkram Province*

This province is the only one with a significant production of green mussels harvested from fish traps. Mussels obtained from fish traps are usually larger in size than those from pole culture and therefore bring a higher price at market.[84] Fish-trap operators set poles with a length equal to the water depth at high tide. The main objective of the pole setting is to concentrate the fish within the confines of the trap where they are easily caught. A single fish trap may require from 10,000 to 60,000 poles and can produce up to 48 t of mussels in the shell.[83] The poles are set in the western Gulf in April, and larval mussels attach to subtidal portions of the poles (Figure 30). These mussels are harvested 7 to 8 months later when the structures are disassembled for reconstruction. When fish-trap owners sell their mussels to local collectors, the amount of money received is usually a prearranged lump sum of between 40,000 to 50,000 baht ($1535 to $1919 U.S.).[85] Most mussels harvested from fish traps are sold in a central shellfish market in Paknam in the province of Samut Prakarn.

5. Experimental Cultivation of *Perna virdis*

Chaitanawisuti and Menasveta[86] have recently shown that green mussel seed can be

FIGURE 30. A portion of the wings of a fish trap in Chachoengsao Province at low tide. Notice green mussels attached below the low-water mark.

successfully collected on bamboo stakes and subsequently transferred to synthetic ropes or cylindrical plastic mesh nets. These ropes and/or nets are, in turn, transported to spat-deficient areas with better growing conditions. These experimental studies have involved the use of long-lines and rafts and have demonstrated many benefits associated with the long-line method of culture from a standpoint of durability and maintenance. Although the results of these experimental studies show promise for the future, obstacles remain in transferring the new technology to mussel growers. Government-sponsored extension agents, who are today few in number, may prove to be a critical factor for the successful introduction of such new technologies to the involved mussel culture operations.

6. Cultivation of *Musculus senhauseni*

The striped horse mussel is not popular as a human food source because of its small size. However, in 1982, 95% of the total production of 32.9 t of this species was sold to local collectors who, in turn, sold 82% of their stock to duck farms.[85] At Bangsai, in the province of Chonburi, striped horse mussels are gathered from natural populations when they are about 0.5 cm long and sown on designated plots a few hundred meters from shore. These mussels, destined to be used as animal feed, are harvested 2 or 3 months later at a length of about 2 cm and are loaded directly onto small trucks for shipment (Figure 31). Mussels large enough for human consumption (4 cm in length) are harvested further out to sea, packed in baskets, and transported to the wholesale market at Paknam in the province of Samut Prakarn. Usually, a whole family is involved in the harvest of horse mussels, each person gathering the mussels in a tubular net (2 m long and 1 m in diameter) made of fine plastic mesh, the size of which varies with the size of the mussels. Harvesting is done at high tide in about 0.5 m of water. With one end of their tubular net tied off (Figure 32) and the other end held open by a D-shaped iron frame, the harvesters scoop the mussels into the open end of the net by hand and, when the net is full, step into the net and clean

FIGURE 31. Striped horse mussels being loaded on a small truck for transportation to local duck farms.

FIGURE 32. Striped horse mussel harvesters tying shut the unsupported end of plastic tubular nets before harvesting begins.

FIGURE 33. Hauling a plastic tubular net full of striped horse mussels on board a small work boat after the contained mussels have been cleaned by trampling under water.

the mud off the mussels, simultaneously separating the clumps with a trampling action underwater. The mussels are extremely hard to clean by hand. When the mussels are considered ''white'', the net containing about 100 kg of washed mussels is hauled aboard (Figure 33) and placed in the boat, which has a 2-mm fine mesh net laid inside it to help contain the mussels (Figure 34). Most fishermen engage in striped mussel harvesting as a secondary source of employment, as many are also fish trappers or green mussel farmers.

7. Problems and Risks Associated with Mussel Farming

Although economic data for 1978 indicate that green mussel farming had the highest net return over expenditures of any aquatic animal culture in Thailand,[84] a large number of operators have had to negotiate loans from banks or relatives on a yearly basis to keep their farms in operation.[85] Up to 42% of the set poles may be lost per farm during the monsoon season, which lasts from August to October. Mussel farmers on the eastern seaboard are faced with the added problem of obtaining culture poles from the western provinces of Kanchanaburi and Petchaburi. In addition, the wholesale shellfish marketing system is nearly a monopoly among a small number of individuals, and this results in both producers and consumers being subject to unfair pricing of green mussel products.[73,74,85] Joint efforts begun in 1981 among the International Center for Living Aquatic Resources Management (ICLARM), the Fisheries Department of Agriculture Economics at Kasetsart University, and the German Agency for Technical Cooperation (GTZ) have resulted in studies of the following: (1) shellfish economics and marketing, (2) economics and biology of the culture and harvest of Thai commercial bivalves, (3) product standardization, (4) production costs, and (5) assessment of public health aspects of bivalve consumption.[87]

The upper Gulf waters are increasingly being used for disposal of industrial wastes, for recreation, and for fishing by both commercial and sport fishermen. The increasing industrialization near traditional mussel farming regions may be responsible for a steady decrease

FIGURE 34. A net made of fine mesh lining the bottom of a boat in preparation for harvesting striped horse mussels at Muang District, Chonburi Province.

in mussel production from 1978 to the present,[73,88] although this is not reflected in published statistics from the Food and Agriculture Organization (Figure 4D).

Of the 520,000 ha of intertidal flats available for mollusk culture in the Gulf of Thailand and on the Andaman sea, only 7200 rai (1152 ha) are presently being used for this.[89] Therefore, with careful planning and funding on a national scale, expansion of mussel culture in areas away from the upper Gulf may yield rewarding results.

REFERENCES

1. **Stafford, J.,** On the fauna of the Atlantic coast of Canada, third report—Gaspe, 1905—1906, *Contrib. Can. Biol. Fish.*, 1906—1912, 45, 1912.
2. **Field, I. A.,** Biology and economic value of the sea mussel, *Mytilus edulis, U.S. Bur. Fish. Bull.*, 38, 127, 1922.
3. **Werner, B.,** Über die Entwicklung und Artuntersheidung von Muschellarven des Nordseeplanktons, unter besonderer Beruschsichtigung der Schalenentwicklung, *Zool. Jahrb. Abt. Allg. Zool. Physiol. Tiere*, 66, 1, 1939.
4. **Sullivan, C. M.,** Bivalve larvae of Malpeque Bay, P.E.I., *Bull. Fish. Res. Board Can.*, 77, 1, 1948.
5. **Bayne, B. L.,** Primary and secondary settlement in *Mytilus edulis* L. (Mollusca), *J. Anim. Ecol.*, 33, 513, 1964.
6. **Bayne, B. L.,** Growth and the delay of metamorphosis of the larvae of *Mytilus edulis* (L.), *Ophelia*, 2, 1, 1965.
7. **Bayne, B. L.,** The biology of mussel larvae, in *Marine Mussels: Their Ecology and Physiology*, Bayne, B. L., Ed., Cambridge University Press, Cambridge, 1976, 81.
8. **Ockelmann, K. W.,** Developmental types in marine bivalves and their distribution along the Atlantic coast of Europe, in *Proc. 1st Eur. Malacological Congress, London, 1962*, Cox, L. R. and Peake, J. F. Eds., Conchological Society of Great Britain and Ireland and the Malacological Society of London, London, 1965, 25.

9. **Chanley, P. E. and Andrews, J. D.,** Aids for identification of bivalve larvae of Virginia, *Malacologia,* 11, 45, 1971.
10. **de Schweinitz, E. H. and Lutz, R. A.,** Larval development of the northern horse mussel *Modiolus modiolus* (L.), including a comparison with the larvae of *Mytilus edulis* L. as an aid in planktonic identification, *Biol. Bull. (Woods Hole, Mass.),* 150, 348, 1976.
11. **Lutz, R. A. and Hidu, H.,** Hinge morphogenesis in the shells of larval and early post-larval mussels *Mytilus edulis* L. and *Modiolus modiolus* (L.), *J. Mar. Biol. Assoc. U.K.,* 59, 111, 1979.
12. Yearbook of Fishery Statistics, Food and Agriculture Organization, Rome, 1947—1986.
13. Office of Data and Information Management, National Oceanographic and Atmospheric Administration, National Marine Fisheries Service, U.S. Department of Commerce, personal communication, 1988.
14. ICLARM, Mollusc fisheries of Thailand, *ICLARM Newsl.,* January, 9, 1981.
15. **Drucker, B.,** Office of Science and Information, Prediction, Assessment and Monitoring Division, National Oceanographic and Atmospheric Administration, National Marine Fisheries Service, U.S. Department of Commerce, personal communication, 1988.
16. **Lewis, R.,** Department of Marine Resources, Augusta, ME, personal communication, 1988.
17. Fisheries Research Branch, Department of Fisheries and Oceans, Ottawa, Ontario, Canada, personal communication, 1988.
18. **Dare, P. J.,** Mussel cultivation in England and Wales, Lab. Leafl. MAFF Direct. Res., Lowestoft, 1980, 50.
19. **Zhang, F.,** Mussel culture in China, *Aquaculture,* 39, 1, 1984.
20. **Korringa, P.,** *Farming Marine Organisms Low in the Good Chain. A Multidisciplinary Approach to Edible Seaweed, Mussel and Clam Production,* Elsevier, Amsterdam, 1976, chap. 4.
21. **Mason, J.,** Cultivation, in *Marine Mussels: Their Ecology and Physiology,* Bayne, B. L., Ed., Cambridge University Press, Cambridge, 1976, chap. 10.
22. **Korringa, P.,** *Farming Marine Organisms Low in the Food Chain. A Multidisciplinary Approach to Edible Seaweed, Mussel and Clam Production,* Elsevier, Amsterdam, 1976, chap. 2.
23. **Korringa, P.,** *Farming Marine Organisms Low in the Good Chain. A Multidisciplinary Approach to Edible Seaweed, Mussel and Clam Production,* Elsevier, Amsterdam, 1976, chap. 5.
24. **Fabi, G., Fiorentini, L., and Giannini, S.,** Growth of *Mytilus galloprovincialis* Lamk on a suspended and immersed culture in the bay of Portonovo (central Adriatic Sea), in Report of the Technical Consultation on Open Sea Shellfish Culture in Association with Artificial Reefs, Food and Agriculture Organization, Rome, 1986, 144.
25. **Korringa, P.,** *Farming Marine Organisms Low in the Food Chain. A Multidisciplinary Approach to Edible Seaweed, Mussel and Clam Production,* Elsevier, Amsterdam, 1976, chap. 3.
26. **Raimbault, R.,** Historique de l'implanation de la conchyliculture en mer ouverte (Méditerranée francaise), in Report of the Technical Consultation on Open Sea Shellfish Culture in Association with Artificial Reefs, Food and Agriculture Organization, Rome, 1986, 65.
27. **Winter, J. E., Toro, J. E., Navarro, J. M., Valenzuela, G. S., and Chaparro, O. R.,** Recent developments, status, and prospects of molluscan aquaculture on the Pacific Coast of South America, *Aquaculture,* 39, 95, 1984.
28. **Korringa, P.,** *Farming Marine Organisms Low in the Food Chain. A Multidisciplinary Approach to Edible Seaweed, Mussel and Clam Production,* Elsevier, Amsterdam, 1976, chap. 8.
29. **Skidmore, D. S. and Chew, K. K.,** *Mussel Aquaculture in Puget Sound,* Washington Sea Grant Program, University of Washington, Seattle, 1985.
30. **Korringa, P.,** *Farming Marine Organisms Low in the Food Chain. A Multidisciplinary Approach to Edible Seaweed, Mussel and Clam Production,* Elsevier, Amsterdam, 1976, chap. 7.
31. **Korringa, P.,** *Farming Marine Organisms Low in the Food Chain. A Multidisciplinary Approach to Edible Seaweed, Mussel and Clam Production,* Elsevier, Amsterdam, 1976, chap. 6.
32. **Sivalingam, P. M.,** Aquaculture of the green mussel, *Mytilus viridis* Linnaeus in Malaysia, *Aquaculture,* 11, 297, 1977.
33. **Lizarraga, M.,** Tecnicas aplicadas en el cultivo de moluscos en America Latina, in Simp FAO/CARPAS Sobre Acuicultura en America Latina, Food and Agriculture Organization, Rome, 1974, 1.
34. **Loo, L. O. and Rosenber, R.,** *Mytilus edulis* culture: growth and production in Western Sweden, *Aquaculture,* 35, 137, 1983.
35. **Popov, V. and Zlatanova, S.,** Development of mussel culture along the Bulgarian Black Sea Coast, in Report of the Technical Consultation on Open Sea Shellfish Culture in Association with Artificial Reefs, Food and Agriculture Organization, Rome, 1986, 74.
36. **Ettewa, I.,** Report on intensive mussel culture in Egypt, in Report of the Technical Consultation on Open Sea Shellfish Culture in Association with Artificial Reefs, Food and Agriculture Organization, Rome, 1986, 155.
37. Report of the Technical Consultation on Open Sea Shellfish Culture in Association with Artificial Reefs, Food and Agriculture Organization, Rome, 1986, 30.

38. **Nakamura, R.,** Aquaculture development in India: a model, *Bioscience,* 35, 96, 1985.
39. **Coeroli, M., De Gaillande, D., Landret, J. P., and Aquacop (Coatena, D.),** Recent innovations in cultivation of molluscs in French Polynesia, *Aquaculture,* 39, 35, 1984.
40. **Lutz, R. A., Incze, L. S., Chew, K. K., Clifton, J. A., Haley, R., Miller, B. A., Brownell, W., Chaves-Michael, L., Blumenstock, M. W., Hayes, K. P., Weldon, A., Dearborn, R. K., and Lloyd, D.,** A Comprehensive Review of the Commercial Industries in the United States, Stock No. 003-020-00133-5, U.S. Government Printing Office, Washington, D.C., 1977, 1.
41. **Dow, R. L. and Wallace, D. E.,** Blue mussels *(Mytilus edulis)* in Maine, Department of Sea and Shore Fish. Bull., Maine Department of Marine Resources, Augusta, ME, 1954, 1.
42. **Lutz, R. A.,** Raft cultivation of mussels in Maine waters: its practicability, feasibility and possible advantages, Mar. Sea Grant Bull. No. 4, University of Maine, Walpole, 1974, 1.
43. **Lutz, R. A., Ed.,** *Mussel Culture and Harvest: A North American Perspective,* Elsevier, Amsterdam, 1980, 1.
44. **Lutz, R. A.,** Mussel aquaculture in the United States, in *Crustacean and Mollusk Aquaculture in the United States,* Huner, J. V. and Brown, E. E., Eds., AVI Publishing, Westport, CT, 1985, 311.
45. **Lutz, R. A.,** Annual growth patterns in the inner shell layer of *Mytilus edulis, J. Mar. Biol. Assoc. U.K.,* 56, 723, 1976.
46. **MARITEC,** An Inventory of the Blue Mussel *(Mytilus edulis)* Stocks from the Damariscotta River Estuary to Jonesport, Maine, New England Fisheries Development Program, National Marine Fisheries Service, Gloucester, MA, 1978, 111.
47. Aquaculture in the United States: Constraints and Opportunities, National Research Council, U.S. Government Printing Office, Washington, D.C., 1978, 1.
48. **Newell, C.,** unpublished data, 1988.
49. **Hurlburt, C. G. and Hurlburt, S. W.,** European mussel culture technology and its adaptability to North American waters, in *Mussel Culture and Harvest: A North American Perspective,* Lutz, R. A., Ed., Elsevier, Amsterdam, 1980, 69.
50. **Wilson, J. and Fleming, D.,** *The Economics of the Maine Mussel Industry,* University of Maine, Orono, 1987.
51. **Shumway, S., Card, D., Getchell, R., and Newell, C.,** Effects of calcium oxide (Quicklime) on non-target organisms in mussel beds, *Bull. Environ. Contam. Toxicol.,* 40, 503, 1988.
52. **Lutz, R. A. and Incze, L. S.,** Impact of toxic dinoflagellate blooms on the North American shellfish industry, in *Toxic Dinoflagellate Blooms,* Taylor, D. L. and Seliger, H. H., Eds., Elsevier/North-Holland, New York, 1979, 476.
53. **Yentsch, C. M. and Incze, L. S.,** Accumulation of algal biotoxins in mussels, in *Mussel Culture and Harvest: A North American Perspective,* Lutz, R. A., Ed., Elsevier Scientific, Amsterdam, 1980, 223.
54. **Scattergood, L. W. and Taylor, C. C.,** The mussel resources of the North Atlantic region. II. Observations on the biology and the methods of collecting and processing the mussel, *Commer. Fish. Rev.,* 11, 8, 1949.
55. **Lutz, R. A. and Hidu, H.,** Some observations on the occurrence of pearls in the blue mussel, *Mytilus edulis, Proc. Natl. Shellfish. Assoc.,* 68, 17, 1978.
56. **Lutz, R. A.,** Pearl incidence in *Mytilus edulis* and its commercial raft cultivation implications, *Proc. World Mariculture Soc.,* 9, 509, 1978.
57. **Lutz, R. A.,** Pearl incidence: mussel culture and harvest implication, in *Mussel Culture and Harvest: A North American Perspective,* Lutz, R. A., Ed., Elsevier, Amsterdam, 1980, 193.
58. **Lutz, R. A.,** Bivalve molluscan mariculture: a *Mytilus perspective, Proc. World Mariculture Soc.,* 10, 595, 1979.
59. **Lutz, R. A.,** Introduction: mussel culture and harvest in North America, in *Mussel Culture and Harvest: A North American Perspective,* Lutz, R. A., Ed., Elsevier, Amsterdam, 1980, 1.
60. **Chaves, L. A.,** *Experimental Culture of Mussels at Seabeck Bay,* Sea Grant Publ. No. WSG-TA-75-14, University of Washington, Seattle, 1975.
61. **Chaves, L. A. and Chew, K. K.,** Mussel culture studies in Puget Sound, Washington, *Proc. World Mariculture Soc.,* 6, 185, 1975.
62. **Lutz, R. A. and Porter, B.,** Experimental culture of blue mussels *Mytilus edulis* in heated effluent waters of a nuclear power plant, *Proc. World Mariculture Soc.,* 8, 427, 1977.
63. **Incze, L. S., Porter, B., and Lutz, R. A.,** Experimental culture of *Mytilus edulis* in a northern estuarine gradient: growth, survival and recruitment, *Proc. World Mariculture Soc.,* 3, 523, 1978.
64. **Myers, E. A.,** The evolution of a commercial mussel culture operation, in *Mussel Culture and Harvest: A North American Perspective,* Lutz, R. A., Ed., Elsevier, Amsterdam, 1980, 266.
65. **Conrad, J.,** Mussel production plant in Spanish Galicia, *Aquaculture Mag.,* 12, 57, 1986.
66. **Andreu, B.,** Fishery and culture of mussels and oysters in Spain, in *Proc. Symp. on Mollusca III,* Marine Biological Association of India, 1970, 835.
67. **Lubet, P.,** Exposé synoptique des données biologique sur la moule *Mytilus galloprovincialis* (Lamarck 1819), Synopsis FAO sur les peches, No. 88, Food and Agriculture Organization, Rome, 1973.

68. **Figureas, A. and Figueras, A. J.,** Variabilidad ecomorfica del mejillon silvestre y cultivada en Espana (gen. *Mytilus*) y relacion con su posicion sistematica, *Invest. Pesq.*, 47, 57, 1983.
69. **Tenore, K. R., Boyer, L. F., Cal, R. M., Corral, J., Garcia-Fernandez, C., Gonzalez, N., Bonzalez-Gurriaran, E., Hanson, R. B., Iglesias, J., Drom, M., Lopez-Jamar, E., McClain, J., Pamatmat, M. M., Perez, A., Rhoads, D. C., De Santiago, G., Tietjen, J., Westrich, J., and Windom, H. L.,** Coastal upwelling in the Rias Bajas, NW Spain: contrasting the benthic regimes of the Rias de Arosa and de Muros, *J. Mar. Res.*, 40, 701, 1982.
70. **Yonge, C. M.,** The mussel form and habit, in *Marine Mussels: Their Ecology and Physiology,* Bayne, B. L., Ed., Cambridge University Press, Cambridge, 1976, chap. 1.
71. **Brock, V.,** Immuno-electrophoretic studies of genetic relations between populations of *Mytilus edulis* and *M. galloprovincialis* from the Mediterranean, Baltic, East and West Atlantic, and East Pacific, in *Proc. 19th Eur. Mar. Biol. Symp.*, Gibbs, P. E., Ed., Cambridge University Press, Cambridge, 1984, 515.
72. **Fraga, F. and Vives, F.,** Retencion de particulas organicas por el mejillon en los viveros flotantes, *Reun. Prod. Mar. Expl. Pesq. Inst. Invest. Pesq.*, 4, 71, 1960.
73. **Natathanapat, N.,** An Economic Analysis of Green Mussel Culture in Chachoengsao, Chonburi, and Petchaburi Provinces of Thailand, 1982, (in Thai), unpublished Master's thesis, Kasetsart University, 1984.
74. **Poosiripinya, R.,** The Green Mussel (*Perna viridis*) Marketing System of Provinces in the Inner Part of the Gulf of Thailand, 1982, unpublished Master's thesis, Kasetsart University, 1984.
75. **Hungspreugs, M. and Yuangthong, C.,** A history of metal pollution in the Upper Gulf of Thailand, *Mar. Pollut. Bull.*, 14, 465, 1983.
76. **Tiensongrassami, B.,** Principles of Sea Farming (in Thai), Kasetsart University, 1974.
77. **Chongpeepien, T.,** Reproductive Cycle of Adult Mussel at Samaekao, Chachoengsao (in Thai), Brackishwater Division, Department of Fisheries, Ministry of Agriculture and Cooperatives, 1980.
78. **Tan, W. H.,** Egg and larval development in the green mussel, *Mytilus edulis* Linnaeus, Veliger, 18, 151, 1975.
79. **Walter, C.,** Reproduction and growth in the tropical mussel *Perna viridis* (Bivalvia: Mytilicae), *Kalikasan, Philipp. J. Biol.*, 11, 83, 1982.
80. **Tiammeth, J.,** Edible Molluscs of Bangkok (in Thai), Faculty of Fisheries, Kasetsart University, 1960, 111.
81. **Menasveta, P. and Cheevaparanpiwat, V.,** Heavy metals, organochlorine pesticides and PCBs in green mussels, mullets and sediments of river mouths in Thailand. *Mar. Pollut. Bull.*, 12, 19, 1981.
82. **Emery, K. O. and Niino, H.,** Sediments of the Gulf of Thailand and adjacent continental shelf, *Geol. Soc. Am.*, 74, 541, 1963.
83. **Tiensongrassami, G.,** Principles of Green Mussel Culture (in Thai), Faculty of Fisheries, Kasetsart University, 1977, 25.
84. **Jinpinpet, R.,** Economic Analysis of Green Mussel *(Perna viridis)* Production in Association with Fish Traps in Selected Provinces of Thailand, 1982, unpublished Master's thesis, Kasetsart University, 1984.
85. **Tokrisna, R., Tugsinavisuitti, S., Kaoian, S., and Ghuntungkul, P.,** Marketing System of Shellfish (in Thailand), (in Thai), Asian Fisheries Social Science Research Network: Thailand, Department of Agriculture Economics, Faculty of Economics and Business Administration, Kasetasart University, 1985.
86. **Chaitanawisuti, N. and Menasveta, P.,** Experimental suspended culture of green mussel *Perna viridis* (Linn.) using spat transplanted from a distant settlement ground in Thailand, *Aquaculture*, 66, 97, 1980.
87. **MacLean, J. L. and Dizon, L. B., Eds.,** ICLARM Rep., International Center for Living Aquatic Resources Management, Manila, 1983.
88. **Menasveta, P.,** Effects of water pollution on aquaculture development in Thailand, *Water Qual. Bull.*, 12, 116, 1987.
89. **Saraya, A.,** Bivalve culture country report, in *Bivalve Culture in Asia and the Pacific: Proc. Workshop Held in Singapore,* Graham, F. B. and Graham, M., Eds., Publ. No. 2003, Internation Development Research Center, 1983.

Chapter 7

CULTURE OF *HINNITES* AND RELATED SCALLOPS ON THE AMERICAN PACIFIC COAST

David L. Leighton

TABLE OF CONTENTS

I. INTRODUCTION

The purple-hinge rock scallop, *Hinnites multirugosus* (Gale)* (= *H. giganteus*)[1], is a relatively large pectinid that cements to rock or other firm surface to mature after a free-living juvenile existence. The rock scallop is well known and highly esteemed by sport divers in California for its flavorful adductor muscle; however, the shellfish has not been the subject of commercial fisheries in California, since harvests by other than sport interests has been precluded by fish and game regulation.[2,3] Throughout its range from Sitka, AK to Magdalena Bay, Baja California, Mexico, it is generally found from the lower intertidal to depths as great as 35 m.[4] Offshore reefs are typically populated by the rock scallop, but concrete pier pilings and jetty rock at entrances to bays in southern California have become favored habitats.[2,3]

Cementing species of scallops are generally restricted to tropical waters (e.g., *Spondylus* and *Plicatula*. *Hinnites*, however, inhabits temperate-boreal ocean waters.[5] Early development of *Hinnites* proceeds much as in *Pecten maximus* (L.);[6] metamorphosis is completed about 2 months after fertilization. Larvae are in the plankton for about 4 weeks. After settlement, the pediveliger changes gradually in the course of metamorphosis to the early juvenile. Juveniles may swim in short bursts, but generally cling to shaded rock surfaces by byssal threads. Cementation occurs after about 6 months of growth, when the juvenile shell is 2 to 3 cm in diameter. Permanent attachment is achieved by deposition of shell matter by the right shell valve mantle in close conformity with the microrelief of the substrate surface.[2,3] Adults exceeding 20 cm are occasionally found, but the average adult is 10 to 15 cm in diameter (Figure 1). *Hinnites* produces a relatively large adductor muscle comprising from 30 to 55% of the soft body weight; adductor muscles typically weigh 50 to 80 g in adults 13 to 15 cm in diameter.

Interest in culture of the purple-hinge rock scallop arose in the early 1970s.[3,7] *Hinnites* was found to be especially hardy and to grow relatively rapidly (6 cm/year). Adults could be induced to spawn readily in the laboratory, and larvae proved amenable to culture using simple procedures and commonly available algal foods. Furthermore, the cementing habit was shown to be of extraordinary value to aquaculture because the species' spatial distribution in grow-out could be controlled and densities optimized for maximum growth. Research from 1975 to 1982 was supported by the University of California Sea Grant Program and the San Diego State University.[3] Most of the following information resulted from that program.

II. BIOLOGY

Gravid adult rock scallops from southern California populations proved responsive to spawn-induction stimuli chiefly during late spring and mid-fall.[3] Major periods of maximum gonadal bulk and advanced gametogenesis were found in late spring and late fall in the San Diego area.[8]

Hinnites multirugosus in the northern Puget Sound area is reported to spawn only in late spring;[9] however, nearby populations in the Hood Canal spawn twice each year.[10] In northern California, rock scallops appear to be reproductively active through most of the year.[11]

In this dioeceous pectinid, the ripe gonad is distinctive—red-orange in females and cream-white in males. Hermaphrodites are rarely found, but it has been suggested that the relative paucity of males among adult *Hinnites* from northern Puget Sound indicates protandry.[9] Among samples of adult rock scallops from the San Diego area, the ratio of females to males increased with size as well. For young adult scallops (5 to 10 cm) from Mission Bay and Point Loma, the ratios were 1.45:1.00 and 1.00:1.00, respectively (n = 130 for

* *H. giganteus* has been a synonym for *H. multirugosus*[6] but recently the genus and species have been changed to *Crassadoma gigantea (Gray)*.[35]

FIGURE 1. Adult rock scallop opened to reveal adductor muscle, gonad, and viscera.

Mission Bay, 60 for Point Loma). Larger adults (10 to 19 cm) from these locations were predominantly female (2.61:1.00 for Mission Bay, n = 65, and 2.95:1.00 for Point Loma, n = 75).[3] A single hermaphrodite was found in a total of 303 scallops. The high relative abundance of females may indicate protandry, but it may also reflect greater growth rate, survival, and longevity among females. No evidence of sex-related differential growth rate or mortality was found in our studies.

III. CULTURE

A. BROODSTOCK AND SPAWN INDUCTION

Sampling from field populations of *Hinnites* during times of major reproductive activity yielded gravid adults that spawned readily when given appropriate stimulation in the laboratory.[3] Scallops approaching maturity were brought into the laboratory during other seasons for conditioning and eventual spawning. Scallops held under spring conditions (14 to 18°C, 16 h light, 8 h dark) and fed *Tetraselmis suecica* could be induced to spawn 2 to 4 months thereafter.[12]

Although gravid *Hinnites* has been induced to spawn by a number of methods (thermal shock, short-period desiccation, and addition of testis fluids to the medium), the use of seawater exposed to short-wavelength ultraviolet (UV) light (peak, 254 nm) has proven most reliable and suitable to hatchery procedures.[3] The spawn-inducing effect of UV-irradiated seawater was discovered accidentally by Japanese biologists working with both abalone and scallops.[13,14] Other methods reported to cause spawning in *Hinnites* are the addition of hydrogen peroxide to the spawning medium[15] and the intragonadal injection of serotonin.[11]

Routinely, we pass filtered seawater (5 μm) at 15 to 20°C through a simple UV sterilization apparatus. Flow rate to containers holding ripe scallops is maintained initially at 0.5 l/min, but stopped entirely at first signs of spawning. Under these conditions, gamete release

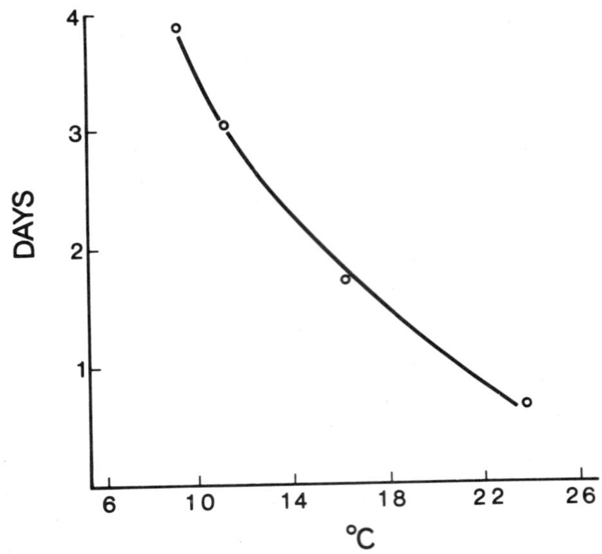

FIGURE 2. Temperature dependency of hatching in developing eggs of *Hinnites multirugosus*.[2]

by males begins within 50 to 60 min, preceded by about 10 min of "valve-clapping" behavior (rapid gaping and closure of valves). Females begin to respond by valve clapping in 60 to 75 min and spawn by 80 min. Sex products are ejected vigorously. A 12-cm female may release 10^7 ova in the course of one spawning.[2,3]

B. Fertilization and Incubation of Eggs

Freshly released ova are placed in clean plastic pails (10 l) at a concentration of about 100/ml and fertilized with sperm at about 10^3/ml. After ½ h, the fertilized eggs have settled to the bottom of the containers, and superficial water is decanted. The eggs are then resuspended in 5-μm-filtered and UV-sterilized seawater. The process is repeated twice more to ensure removal of most excess sperm and other substances. Antibiotics (penicillin G potassium and streptomycin sulfate, each at 25 to 50 ppm) are added prophylactically. Temperature is held in the range 15 to 18°C. No aeration is supplied.

C. CARE OF LARVAE

Hatching rate is strongly temperature dependent (Figure 2). At 18°C, larvae hatch from the eggs in about 32 h. Development is halted at 10°C and accelerated beyond normal control above 22°C in southern California *Hinnites* larvae; the optimal range for early development is 16 to 18°C.[3]

Larvae are cultured in 10-l polyethylene pails at concentrations of 20,000 to 40,000/pail. During the first week, water is changed daily; larvae are retained on an immersed 45-μm frame-supported mesh and washed into freshly filtered (5 μm) and UV-sterilized seawater. Antibiotics are administered with each water change during the first week at the dose given above. Antibiotics are administered only once weekly during the following 3 weeks of larval culture.[3]

Algal foods are provided after the third day, as veligers advance to the straight-hinge stage (110 μm). Generally, only two algae, *Isochrysis galbana* (or the Tahitian strain of *Isochrysis*) and *Pavlova* (= *Monochrysis*) *lutheri*, are supplied at 20,000 to 40,000 cells per milliliter during daily feedings. As larvae reach 150 μm (age 15 to 18 d), other algae, including *Tetraselmis suecica* and *Phaeodactylum tricornutum*, are added to the diet.[3]

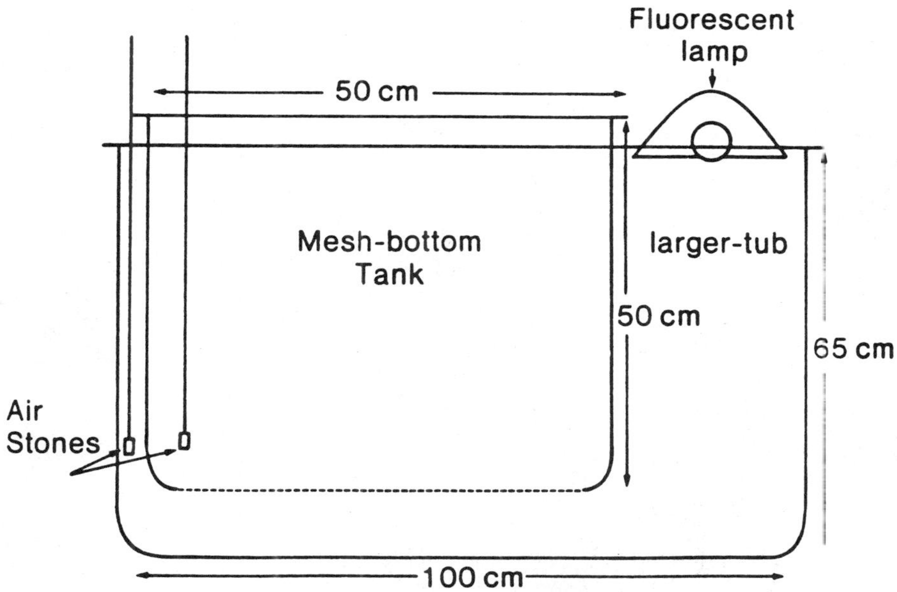

FIGURE 3. System employed for containment and rearing of late-stage *Hinnites* larvae through metamorphosis and early juvenile stages. (A mesh-bottom tank is immersed in a larger tank holding seawater and algae. Developing scallops occupy the darkened interior of the mesh-bottom tank. Illumination is provided to the outer medium. Aeration mixes both tanks.)[2]

D. METAMORPHOSIS

Metamorphosis in the rock scallop begins at an age of about 1 month, when the pediveliger stage is reached (200 μm).[2] Pediveligers can be carried through metamorphosis when fed *Isochrysis* and *Pavlova* alone, but growth and survival are improved when the alga *Rhodomonas* sp. is included.[3] Metamorphosis is gradual, requiring about 2 weeks. At an age of 2 months, the young rock scallop is about 1 mm long.

The culture containers yielding highest survival of pediveligers and early juveniles were black-surfaced fiberglass tanks with mesh bottoms (130 μm) immersed in larger tanks containing food algae and supplied illumination and continuous mixing by aeration (Figure 3). Larvae approaching metamorphosis were introduced into the system. As the pediveligers began to produce byssal threads, they held to the darkened vertical container walls with valves oriented toward the aeration current.[3] Young rock scallops can be carried through metamorphosis in a simple, round tank with an air-lift arrangement to maintain a gentle rotary flow.[10,16]

E. REARING OF JUVENILES

Although juveniles have been reared in the laboratory on a mixed diet including *Isochrysis*, *Pavlova*, *Rhodomonas*, and *Tetraselmis*, growth rate was superior in juveniles placed in fine mesh (2 mm) cages suspended in natural waters to feed on wild plankton.[3,17] Juvenile *Hinnites* have proved remarkably hardy; little mortality occurred among groups held under a range of conditions in bay, ocean, and laboratory environments. Scallops retained in cages in Mission Bay reached 1 to 2 cm at an age of 6 months (Figure 4).[3] In our studies, early juveniles produced in the laboratory were routinely transferred to cages secured at a depth of 2 to 3 m in phytoplankton-rich bay waters for growth to cementing maturity. However, results of preliminary attempts to maintain juveniles in laboratory tanks throughout the precementing period suggest large concentrations (to 10/l) could be effectively reared in expanded rotary-flow round-tank systems receiving natural phytoplankton supplemented with

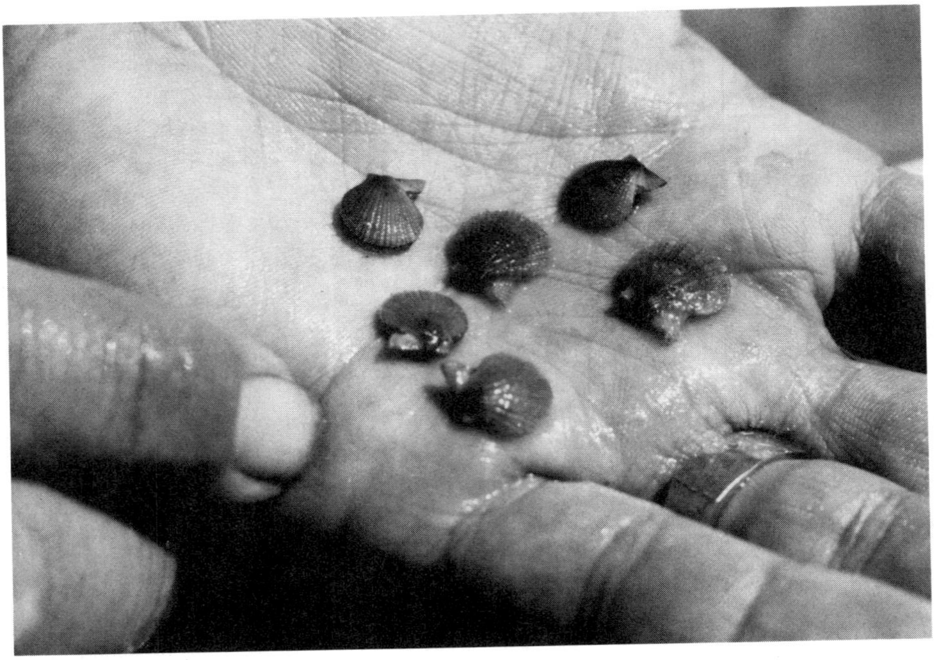

FIGURE 4. Juvenile rock scallops nearing cementing maturity (age 4 months).

cultured algae such as *Tetraselmis*. Large-scale culture of juvenile *Hinnites* has yet to be undertaken.

F. GROWTH TO MARKET SIZE

Experimentally, juvenile rock scallops have been reared to adulthood in both bay and ocean environments while held in cages or attached to panels or sheets of asbestos construction board, concrete, or plastic. Cage confinement was least satisfactory because scallops often aggregated, resulting in reduced growth. Furthermore, scallops cemented either to one another or to the cage material itself and were consequently broken or otherwise damaged during harvest. Cage netting was torn, as some individuals had incorporated the mesh into their shells. Conversely, rearing of scallops attached to vertically suspended panels was entirely satisfactory and had none of the problems associated with cage rearing. The early experimental use of asbestos board and thin concrete sheet led to the development of a plastic panel to which juvenile scallops could be induced to attach in a manner and density yielding maximum growth.

Panel rearing of rock scallops requires that juveniles nearing cementation be held in position on the planar surface for a period of 2 to 3 months until cementation has progressed sufficiently to hold them in place. In our initial studies, cup-shaped cagelets formed of 4-mm polyethylene mesh were glued in rows to sheets of concrete or asbestos board. Precementing scallops 5 to 15 mm in diameter were introduced singly into cagelets through small slits cut in the mesh. The panels were then suspended vertically in food-rich bay waters. The scallops soon cemented to the board, and cagelets could be removed. Growth thereafter through the 1 to 2 year period to market size was unobstructed.[3]

On the basis of this preliminary experience, a plastic panel was developed that was vacuum-formed (of ABS or styrene) to provide a series of depressions or wells spaced about 10 cm apart. A single juvenile scallop was placed in each well and held in place with flat plastic mesh taped to the panel. Once cementation had occurred, the overlying mesh was removed. Perforations in the bottom of each well provided attachment points for cementing

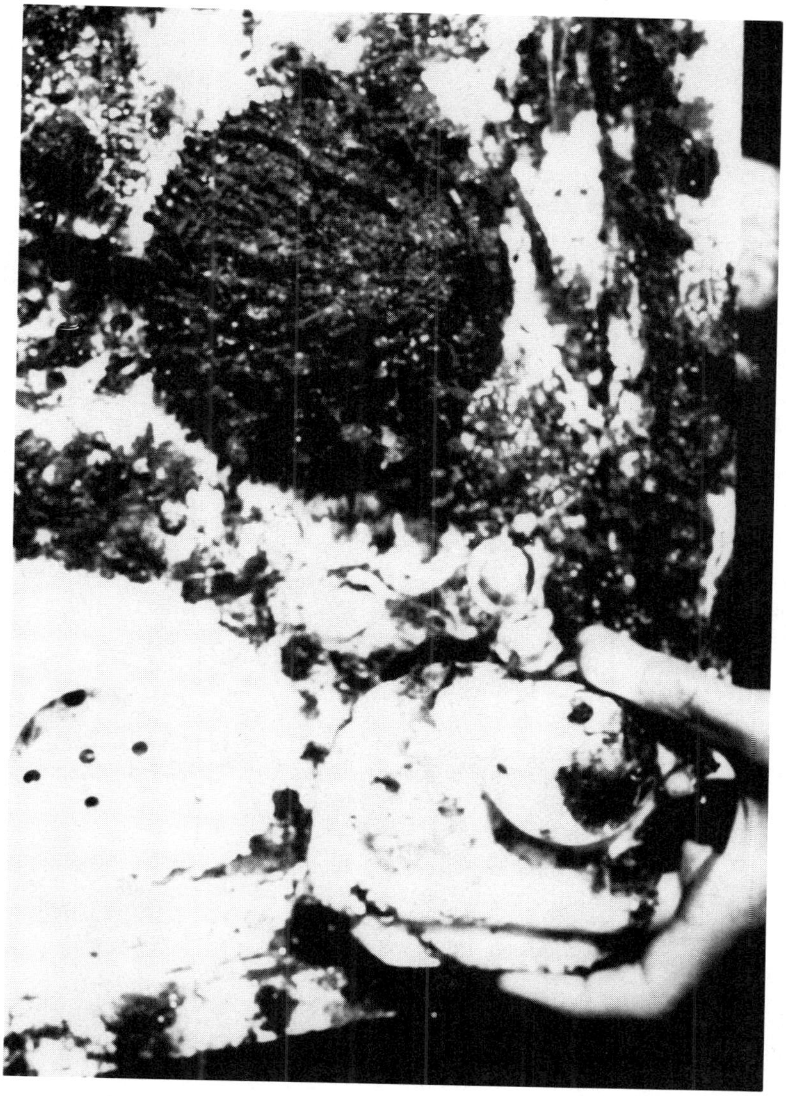

FIGURE 5. Market-size rock scallop as removed from specially designed rearing panel. (Note cementation attachment points matched with perforations on well bottom.)

scallops (Figure 5). At harvest, scallops were simply removed, and the panels cleaned and restocked for the next rearing period.[18,19] A spatial distribution of 40/m^2 allowed growth to maturity with minimal effects of crowding and competition.

G. OBSERVATIONS ON GROWTH RATE

Growth from juvenile to adult stages usually occurred at 5 to 6 cm/year.[2] Closely matched growth rates were found in groups reared in protected marine waters of Mission Bay and in the open ocean near San Diego (Figure 6).[20] Where scallops were held in cages at different depths, growth was greatest at intermediate levels—2 to 6 m in the bay and 6 to 18 m in coastal waters nearby. Groups of young scallops held at six depths to 120 m in the open sea 5 km from shore at La Jolla, CA, averaged 6.1, 8.1, and 6.9 mm/month at depths of 8, 15, and 30 m, respectively. Those at 60, 90, and 120 m averaged 4.8, 2.7, and 2.5 mm/month, respectively. Reduced growth at depths greater than 30 m was due to paucity of phytoplankton and low temperatures.[21]

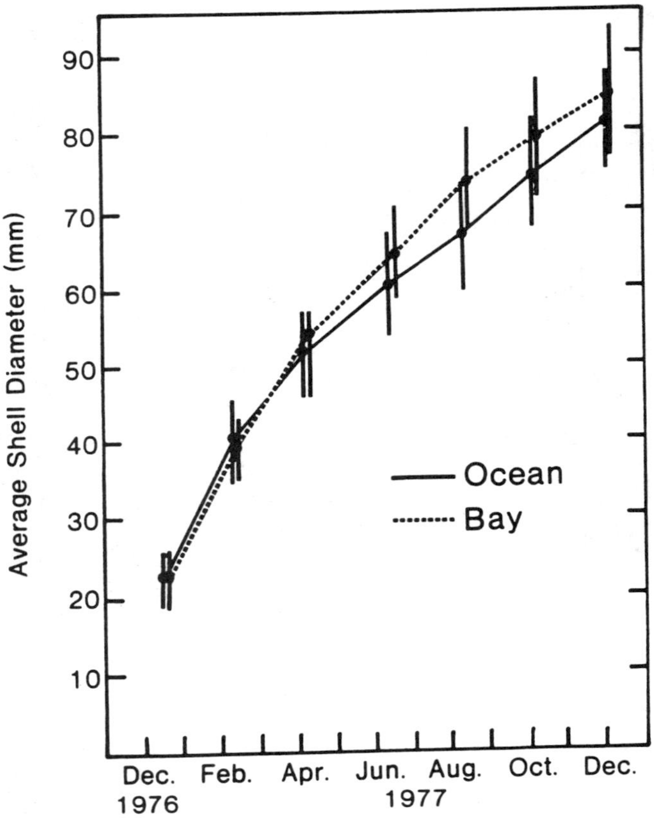

FIGURE 6. Growth curves for scallops reared for 1 year in bay and ocean environments. (Groups of scallops (n = 30 to 32) affixed to asbestos-board slats were held at intermediate depths (bay, 3 m; ocean, 9 m). Average shell diameter (height + width/2) was obtained each 2 months.[2,21]

Growth rate varied by season as well as with depth. In Mission Bay, scallops averaged 7.9 mm/month during summer (20 to 22°C), 4.0 and 5.1 mm/month during fall and winter (13 to 21°C), and 3.0 mm/month during spring (14 to 19°C) when held at a depth of 4 m.[3] As phytoplankton was abundant during the spring, it appears temperature was the major factor limiting growth at that season.

Scallops held in Mission Bay for 2.5 years reached 10 to 12 cm. At an age of 3.0 years, these scallops were 12 to 15 cm long. A fully marketable scallop can therefore be produced after about 2 years of growth. Adductor muscles in scallops 2.5 years old averaged about 40 g (Figure 7).[3]

IV. ENVIRONMENTAL REQUIREMENTS

The distribution of *Hinnites multirugosus* over a rather broad range of latitude and depth along the Pacific coast of North America suggests the species is relatively flexible in its requirements for temperature and salinity. However, its absence from innermost reaches of bays and its only temporary population of other areas noted for wide variation of these factors suggest limitations to distribution may be imposed by seasonal fluctuations in salinity and temperature. Studies of survival and growth of caged *Hinnites* held for 1 year at six stations throughout Mission Bay have supported this contention. Growth and survival were excellent in scallops placed in the fore-bay and poor in back-bay locations.[3]

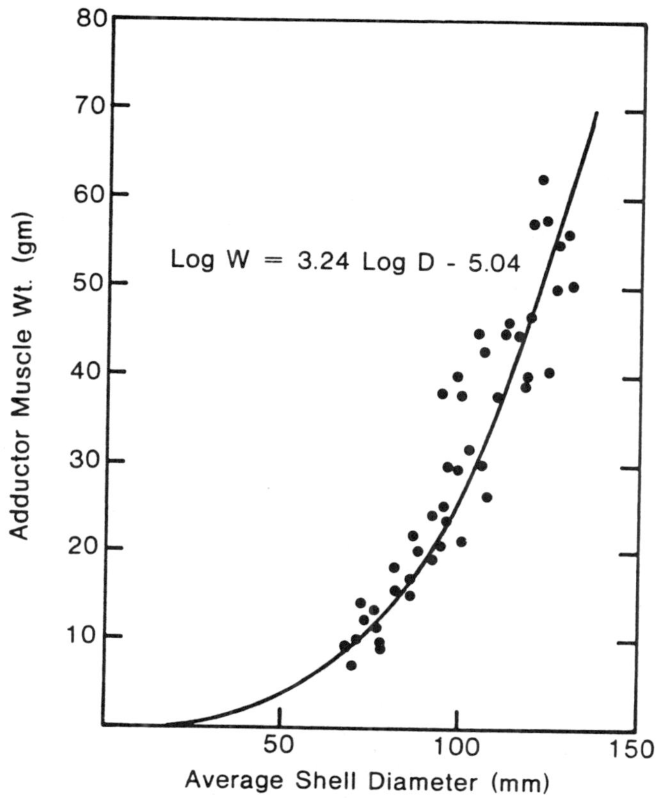

FIGURE 7. Adductor muscle weight (g) vs. size (mm) for rock scallops from Mission Bay, San Diego.[2]

The temperature range for normal physiological processes in larvae from southern California *Hinnites* was found to be 12 to 18°C. Thermal tolerance of juveniles and adults encompassed a somewhat broader range; near the upper limit (24°C) growth rate declined, as it did below 14°C. The practical thermal range for culture of the rock scallop in southern California appears to be 14 to 22°C. *Hinnites* from Puget Sound, WA, exhibit tolerance to lower temperatures.[10]

Hinnites appears to be relatively intolerant to moderate dilution of normal seawater. Southern California larvae and juveniles do not survive salinities below 24.5‰. Salinity reduction following heavy winter rains and excessive runoff has been most often implicated in occurrences of mass mortality to rock scallops in southern California bays.[3] Puget Sound *Hinnites*, in contrast, seems tolerant to greater dilution. Juvenile rock scallops reared in the Hood Canal, WA, survived salinity drops to 18‰.[10]

V. QUALITY AND COMPOSITION OF ADDUCTOR MUSCLE

The adductor muscle in adult *Hinnites*, the portion normally marketed, forms 9 to 10% of the total in-shell fresh weight of scallops 6 to 12 cm in diameter.[3] For 49 scallops (10 to 15 cm) from Mission Bay, adductor muscles comprised about 41% of the soft body weight. In females, the adductor muscle fraction was found to be about 2% less than in males. Seasonal variation in relative weight of adductor muscle reflects chiefly temporal changes in gonadal bulk.

Flavor profile analyses were conducted in which *Hinnites* adductor muscle was compared to commercially available "bay and sea scallops" by taste panel evaluation. The four major categories of aroma, taste, texture, and aftertaste were subdivided into their constituent sensory elements, and *Hinnites* excelled in almost all cases.[3,22]

Chemical studies were made of *Hinnites* adductor muscle in samples from the San Diego area. Proximate analyses indicated adductor muscle consists of 70% protein, 24% carbohydrate, 2% lipid, and 4% ash (by dry weight).[3,22]

VI. SPAT COLLECTION IN THE NATURAL ENVIRONMENT

Experimental observations on settlement of young stages of *Hinnites* and other scallops on the Pacific coast on artificial substrates suggest a feasible alternative to hatchery production for the procurement on young. Nylon netting and other materials effective as spat collectors for *Patinopecten yessoensis* (Jay)[23] and *Placopecten magellanicus* (Gmelin)[24,25] are also attractive to *Hinnites*, although its normal habitat and mode of living are very different from those of its free-living relatives. Onion bags packed with nylon monofilament (often old gill netting) have been found effective in trapping *Hinnites*, presumably as late-stage veligers and early pediveligers.[26,27] Yields vary by season, immersion time, depth, location, surface area, and composition of synthetic material. Harvestable juvenile scallops are limited by predation and competition within the collector bags. Numbers of *Hinnites* spat recovered from bags immersed for 3 months have exceeded 1000/bag.[28]

Spat collectors are rather nonspecific. Other species of scallops, mussels, clams, and crustaceans are commonly trapped, and separation of scallops by species when young is often difficult.[27] *Pecten latiauritus* Conrad appeared with *Hinnites* in spat collectors used in our San Diego studies, and two species of *Chlamys* (*C. rubida* and *C. hastata*) occurred with *Hinnites* in Puget Sound.[26,27] Closely related species can, in some cases, be distinguished and sorted by differences in attachment behavior.

VII. STRATEGIES IN GROW-OUT

Best growth of rock scallops resulted in groups held off the bottom (avoiding siltation, benthic predators, etc.) and below surface waters (avoiding excessive wave action, salinity reduction, and heavy fouling). Vertically suspended panels holding scallops at a density of 4/0.1 m^2 proved most successful for *Hinnites* rearing. Cage-enclosed arrays of panels suspended by long-lines and buoys should prove effective for mid-water-column culture in the open ocean. Protected waters permit use of a wide variety of grow-out structure designs. In our studies, we made wide use of redwood posts set in sedimentary bottom. Cages holding scallops were held off the bottom. A concrete block stabilized the structure (Figure 8). In another arrangement, more appropriate for use in coastal water, rock scallops are planted on surfaces of hollow-core (2-cm polyvinyl chloride pipe) concrete pilings or posts set in sedimentary bottom.[16] Prototype structures of this and other designs have yet to be tested.

VIII. PRESENT COMMERCIAL CULTURE OF ROCK SCALLOPS

A single aquaculturist in California is engaged in culture of *Hinnites* as a secondary crop. The Johnson Oyster Company nurtures juvenile set recovered from shells of harvested oysters; the young scallops are held to maturity in pens supported off the bottom in an enclosed bay (Drakes Estero, Point Reyes, CA). Adult scallops are marketed at $1.00 each and yield adductor muscles averaging about 50 g. Scallop meat is therefore valued at about $19.80/kg ($9.00/lb).[3]

IX. BIOLOGICAL CONSIDERATIONS

Hinnites appears highly resistant to attack by most predators. In the adult, vigorous

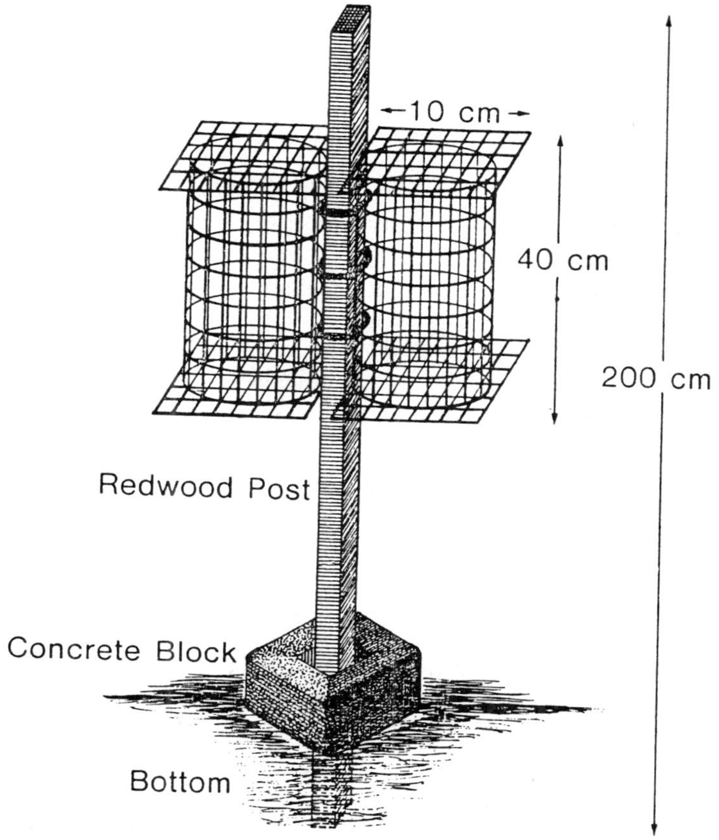

FIGURE 8. Cages attached to weighted and stabilized redwood stake used on sedimentary bottom in experimental rock scallop rearing.[2]

contractions bring sharp-edged valves together at the slightest disturbance. Among experimental animals reared from juvenile to adult stages in the sea and in well-circulated bay environments, survival was close to 100%.

Contaminants, including heavy metals and human pathogens, have not been of concern in the marketability of the rock scallop. In general, only the scallop adductor muscle is consumed, and concentrations of contaminants in that tissue would be considerably less than in viscera.

A single (and fatal) case of paralytic shellfish poisoning (PSP) has been reported, associated with consumption of *Hinnites* collected during a ''red tide'' in northern California. In that instance, viscera (not adductor muscle) estimated to contain 26,000 μg saxitoxin/100 g was consumed by a diver who simultaneously consumed alcohol.[29] The incidence of PSP was due to concentration of saxitoxin from the dinoflagellate *Gonyaulax catanella*. The toxicity hazard level for saxitoxin is set at 80 μg/100 g by the California State Department of Public Health. This episode has brought some concern to potential scallop farmers hoping to operate in waters seasonally populated by *G. catanella* (north of Point Conception, CA). *Hinnites* reared in cold-temperate waters may have to be checked periodically for saxitoxin, especially during the mussel quarantine months (May to October) in northern California and the Pacific Northwest.

X. CULTURE OF OTHER SCALLOP SPECIES ON THE PACIFIC COAST

In North America, most efforts to culture scallops have been experimental. The relatively small spiny and pink scallops (*Chlamys hastata* and *C. rubida*) and the larger weathervane scallop, *Patinopecten caurinus*, have been subjects of aquaculture research in Puget Sound and the Strait of Georgia. *Chlamys* juveniles are collected on artificial substrates and reared to maturity in suspended lantern nets.[27,30]

Few attempts to produce scallops by aquaculture have been made in Latin America. Several species occur in waters of Mexico, but *Argopecten circularis* appears to stand alone as the one receiving attention in that country. In suspended pens, this species was reared from juvenile stages to adult size in commercial numbers in the late 1970s near the harbor of La Paz, Baja California. A hurricane destroyed most of the containment structures and associated moorings in 1982. Further attempts seem not to have been made to reestablish the project.[31,32]

The South American bay scallop, *Chlamys purpurata* Lamarck, is the subject of a new aquaculture effort in Peru and Chile. Juveniles collected in shallow water by divers are reared in bottom culture within net-enclosed sections of the Bay of Paracas, Peru. The rapidly growing scallops reach harvest size (75 mm) after a 5-month grow-out period. An approach is proposed that would collect spat with increased efficiency by adding artificial substrates to ''recruitment plots'' in areas of known high settlement rates. Experimental culture of *Chlamys purpurata* in Chile has shown the species to be highly suitable for controlled mariculture operations. According to DiSalvo, Universidad del Norte, Coquimbo, this scallop is easily matured in captivity and induced to spawn. Its larvae grow and settle well with minimum mortality during metamorphosis. Juveniles (1 mm) placed in cages in natural environments and held to maturity (9 cm, 14 months) are hardy and experience little mortality.[33] This shellfish appears to have special potential for large-scale aquaculture. The hatchery production of juvenile stock may supplement natural set during periods when recruitment is below normal.[34]

REFERENCES

1. **Roth, B. and Coan, E. V.,** Nomenclatural notes on *Hinnites giganteus* (Gray), *Veliger,* 20, 297, 1978.
2. **Leighton, D. L. and Phleger, C. F.,** The purple hinge rock scallop: a new candidate for marine aquaculture, *Proc. Annu. Meet. World Mariculture Soc.,* 8, 457, 1977.
3. **Leighton, D. L. and Phleger, C. F.,** The Suitability of the Purple-Hinge Rock Scallop to Marine Aquaculture, Rep. No. T-SCSGP 001, California Sea Grant College Prog. Publ., Center for Marine Studies, San Diego State University, San Diego, 1981.
4. **Grau, G.,** *Pectinidae of the Eastern Pacific,* Rep., Allan Hancock Pacific Exped. 23, University of Southern California Press, Los Angeles, 1959.
5. **Yonge, C. M.,** Studies on Pacific Coast mollusks. II. Observations on *Hinnites multirugosus* (Gale), *Univ. Calif. Publ. Zool.,* 55, 409, 1951.
6. **Gruffydd, L. D. and Beaumont, A. R.,** A method for rearing *Pecten maximus* larvae in the laboratory, *Mar. Biol.,* 15, 350, 1972.
7. **Leighton, D. L. and Phleger, C. F.,** Preliminary studies on the aquaculture potential of the Pacific coast purple-hinge rock scallop, *Proc. Annu. Meet. World Mariculture Soc.,* 7, 213, 1976.
8. **Jacobsen, F. R.,** The Reproductive Cycle of the Purple-Hinge Rock Scallop, *Hinnites multirugosus* Gale 1928, Master's thesis, San Diego State University, San Diego, 1977.
9. **Lauren, D. J.,** Oogenesis and protandry in the purple-hinge rock scallop, *Hinnites giganteus,* in upper Puget Sound, Washington, USA, *Can. Zool.,* 60, 2333, 1982.
10. **Olsen, S.,** Completion Report on Invertebrate Aquaculture, Shellfish Enhancement Project, Point Whitney Shellfish Lab., Washington Department of Fisheries, 1984, 309.

11. **Malachowski, M.,** Gametogenesis, induced spawning and hermaphroditism in rock scallops, *Hinnites*, from Humboldt Bay, California, paper presented at the 16th Annu. Meet. World Mariculture Soc., Reno, NV, January 1986.
12. **Leighton, D. L. and Phleger, C. F.,** Project completion report, 1982—1983, University of California Sea Grant Program, Scripps Institute of Oceanography, La Jolla, 1983.
13. **Uki, N. and Kikuchi, S.,** On the effect of irradiated seawater with ultraviolet rays on inducing spawning in the scallop, *Patinopecten yessoensis* (Jay), *Bull. Tohoku Reg. Fish. Res. Lab.*, 34, 87, 1974.
14. **Kikuchi, S. and Uki, N.,** Technical study on artificial spawning of abalone, genus *Haliotis*. II. Effect of irradiated seawater with ultraviolet rays on inducing to spawn, *Bull. Tohoku Reg. Fish. Res. Lab.*, 33, 79, 1974.
15. **Morse, D. E., Hooker, N., and Morse, A.,** Chemical control of reproduction in bivalve and gastropod molluscs. III. An inexpensive technique for mariculture of many species, *Proc. Annu. Meet. World Mariculture Soc.*, 9, 543, 1978.
16. **Leighton, D. L.,** Cultivo experimental de la venera *Hinnites multirugosus,* paper presented at the 1st Symp. Latin American Aquaculture Assoc., Maracay, Venezuela, November 1977.
17. **Cary, S. C., Leighton, D. L., and Phleger, C. F.,** Food and feeding strategies in culture of larval and early juvenile purple-hinge rock scallops, *Hinnites multirugosus* (Gale), *J. World Mariculture Soc.*, 12, 156, 1981.
18. **Leighton, D. L. and Phleger, C. F.,** U.S. Patent 4,317,429, 1982.
19. **Leighton, D. L.,** Rock scallop grow-out, *Aquaculture Mag.*, July 1985.
20. **Monical, J. B.,** Comparative studies on growth of the rock scallop, *Hinnites multirugosus* (Gale), in marine waters of southern California, *Proc. Natl. Shellfish. Assoc.*, 70, 14, 1980.
21. **Leighton, D. L.,** A growth profile for the rock scallop *Hinnites multirugosus* held at several depths off La Jolla, California, *Mar. Biol.*, 51, 229, 1979.
22. **Phleger, C. F., Holtz, R. B., Grimes, P. W., Jacobsen, F. R., and Leighton, D. L.,** Chemical and sensory analysis of the purple-hinge rock scallop *(Hinnites multirugosus), J. Food Sci.*, 43, 1793, 1978.
23. **Taguchi, K.,** Japanese scallop culture techniques boost yield, *Aust. Fish.*, March, 20, 1976.
24. **Naidu, K. S. and Scalpen, R.,** Settlement and survival of the giant scallop, *Placopecten magellanicus*, larvae on enclosed polyethylene film collectors, in *Advances in Aquaculture,* Pillay, T. and Dill, W., Eds., Fishing News Books, England, 1979, 379.
25. **Naidu, K. S., Cahill, F. M., and Lewis, D. B.,** Relative efficacy of two artificial substrates in the collection of sea scallop (*Placopecten magellanicus*) spat, *J. World Mariculture Soc.*, 12, 165, 1981.
26. **Phleger, C. F., Cary, S. C., and Leighton, D. L.,** Spat collection of the purple-hinge rock scallop on artificial collectors, paper presented at the 1982 Meet. Western Soc. Naturalists, Long Beach, CA, 1982.
27. **Bronson, J., Bettinger, T., Goodwin, L., and Burge, D.,** Investigations of spat collection on artificial substrates for the weathervane scallop (*Patinopecten caurinus*) and the rock scallop *(Hinnites multirugosus)* in Puget Sound, Washington, Rep., Washington State Department of Fisheries, August 1984.
28. **Shaw, W. N.,** The purple-hinge rock scallop: a mariculture species of the future?, *Aquaculture Mag.*, January/February 1985.
29. Anon., Fatal paralytic shellfish poisoning from scallops—a rarity, California Morbidity Rep. 39, Department of Health Services, State of California, October 1980.
30. **Mottet, M. G.,** A review of the fishery biology of scallops, Tech. Rep. 39, Washington Department of Fisheries, Olympia, 1979.
31. **Cardenas, E. B.,** Status of mulluscan aquaculture on the Pacific Coast of Mexico, *Aquaculture,* 39, 83, 1984.
32. **Sievers, A. K.,** Commercial mariculture of the bay scallop *Argopecten circularis* in the Bahia de La Paz, Baja California, Mexico, unpublished manuscript, 1981.
33. **Costello, T.,** Peru has thriving scallop industry, *Aquaculture Mag.*, May 1985.
34. **Winter, J. E., Torro, J. E., Navarro, J. M., Valenzuela, G. S., and Chaparro, O. R.,** Recent developments, status, and prospects of molluscan aquaculture on the Pacific coast of South America, *Aquaculture,* 39, 95, 1984.
35. **Bernard, F. R.,** *Crassadoma* gen. nov. for "*Hinnites*" *giganteus* (Gray, 1825) from the Northeastern Pacific Ocean (Bivalvia: Pectinidae), *Venus,* 45(1), 70, 1986.

Chapter 8

CULTURE OF *CRASSOSTREA CORTEZIENSIS* IN MEXICO

Erik Baqueiro C.

TABLE OF CONTENTS

I. INTRODUCTION

Crassostrea corteziensis (Hertlein) has been described as an allopatric species of *C. virginica* (Gmelin) of the Atlantic coast.[1] The similarity of the species and their habitats has led aquaculturists on the west coast of Mexico to apply or adopt the techniques used by their Atlantic coast colleagues.

II. BIOLOGY

A. SPECIES

Crassostrea corteziensis was originally placed in the genus *Ostrea* by Hertlein,[2] but presence of a proximal chamber and nonincubatory larval development place it in *Crassostrea*.[3,4] Castillo[5] and Harry[6] consider it a junior synonym of *Ostrea columiensis* (Hanley), and Rodriquez et al.[7] of *Ostrea chilensis* (Philippi). Menzel[8] readily crossed *C. corteziensis* with *C. virginica*, obtaining F_1, F_2, and backcrosses with no abnormal chromosomal behavior, and tentatively considers it a subspecies of *C. virginica*.

B. MORPHOLOGY

The shell is of variable shape, elongated with rounded irregular borders to subtriangular, from 75 to 120 mm in length. It is compressed and relatively thin. The hinge is small with a wide ligament area. The beak is laterally curved, and the lower valve is deep, usually attached over its entire surface. The upper valve is smaller; its external proximal surface has a variable number of radiating weak scales, and its distal border has several concentric lamellae which form foliaceous flutings. The inner margin is smooth. The external color is dark purple, blackish blue, yellowish green, or white with purple bands. The inside is white with purple or olive green borders. The adductor muscle scar is purple.[3]

C. RANGE AND DISTRIBUTION

The species has been reported from Baja California, Mexico to Paita, Peru[5] or with a more restricted range from the Gulf of California to Panama.[9] In Mexico the species is known by the common name of "ostión de mangle" (mangrove oyster). Attempts to culture the species have been made in the Mexican States of Sonora, Sinaloa, Nayarit, and Colima, where an important natural fishery is slowly being replaced by aquaculture practices.

Crassostrea corteziensis inhabits estuaries and coastal lagoons at intertidal levels and is found on mangrove roots and other solid substrates available for attachment. It occurs at salinities from 3 to 39‰,[10] with heavy mortalities at lower salinities. Larval settlement is reportedly limited to 21 to 37‰[10] or to 25 to 36‰.[4] Optimum growth is attained from 23 to 28‰.[11]

Temperature does not appear to limit distribution of *C. corteziensis*. It endures a winter minimum of 12°C at the northernmost limit in San Quintin Bay, Baja California to a maximum of 35°C on the Mexican Pacific coast.

D. REPRODUCTION AND SEXUALITY

Stuardo and Martinez,[4] Orbe,[10] Ruiz,[12] and Cuevas and Martinez[13] have described the reproductive cycle, and Rodriguez et al.[17] the karyotype of *C. corteziensis*. The species has 20 diploid chromosomes, 7 metacentric and 3 submetacentric.

Gametogenic activity has been detected throughout the year, with peak activity occurring just before spawning in March. Spawning takes place from June to October, with peak activity occurring in September. A postspawning period is reported from October to November in 70% of the population. Eggs are reported to be up to 57.6 μm in diameter, with an average of 45.7 μm. Hermaphroditism occasionally has been detected.

TABLE 1
Oyster Production for the Periods 1940—1953 and 1952—1963 and the Years 1980—1984

	1940—1953	1952—1963	1980	1981	1982	1983	1984
Baja California[a]	—	10	—	—	60	144	247
Baja California Sur[b]	—	—	1,034	905	999	401	510
Sonora[a]	451.5	300	—	—	78	256	178
Sinaloa	173	600	—	360	224	204	126
Nayarit[c]	130	30	163	427	141	43	60
Jalisco	4	30	28	359	141	13	16
Colima	—	30	—	—	12	—	—
Michoacán	—	20	146	246	49	201	241
Guerrero[c]	—	10	847	1,101	2,082	1,017	671
Oaxaca	—	10	—	31	26	19	3
Chiapas	—	—		—	—	—	
Pacific coast total			2,218	3,351	2,634	3,038	2,052
National total			43,525	39,134	31,660	36,544	1,957,981
Value per t (pesos)			$35,655	$35,655	$77,410	$77,500	$90,000
($1 U.S. = pesos)			72	72	125	300	600

Note: Volume in metric tons (t).

[a] Production made up of wild and cultured *C. gigas*.
[b] Production made up of wild *C. corteziensis* and cultured *C. gigas*.
[c] Production made up of wild and cultured *C. corteziensis*.

Cortez and Martinez[14] reported larvae from the estuary of Nayarit from October to March, with the highest density in November, and again in July at a salinity from 24 to 33‰ and temperature from 24 to 34°C. Settlement on experimental collectors was highest in August.

Oysters grown on different types of collectors have reached the commercial size of 75 mm in 7 months,[4,10,14] with a maximum length of 110 mm. The fastest recorded growth was 20 mm/month for the first 2 months. An average of 7.5 to 7.7 mm/month was found for all localities.

Even though the larvae are regularly cultured in the hatchery at San Blas, Nayarit, a formal description of larval development has not been published.

III. FISHERY

The oyster fishery on the Pacific coast of Mexico comprises *C. corteziensis, C. iridescens* (Hanley, 1854), and, in the past five years, *C. gigas* (Thunberg), which has been introduced for culture from hatcheries in the U.S. Table 1 presents the annual mean production for the periods 1940 to 1953 and 1952 to 1963[3] compared to years 1980 to 1984.

C. corteziensis is sold whole or shucked. Of the total production, standardized to oyster with shell (1 kg shucked = 10 kg with shell), 35% is sold shucked. The price paid to fishermen for a kilogram of whole oysters rose from $27.55 pesos in 1979 to $150 pesos in 1984 ($1.10 U.S. in 1979 to $6.00 U.S. in 1984).

Oysters are eaten fresh on the shell or in cocktails and cooked in a variety of ways. Their market is restricted to Mexico. Most fishing and culture areas have been certified free of pollution. Oyster production and culture are reserved to cooperative fishermen only.

When it is considered that 92% of Mexican oyster production comes from the Gulf of Mexico coast and that over 55% of the Mexican Pacific coast consists of coastal lagoon area, it is clear there is great potential for development of an oyster fishery through aquaculture and proper management of natural resources on the Pacific coast of Mexico.

IV. AQUACULTURE

In spite of the abundance and diversity of mollusk species found in Mexican waters and the commercial use and high value of many of them, only *Crassostrea corteziensis* Ramirez ad Sevilla, 1965, is being cultured commercially.[10] The first efforts at modern aquaculture in Mexico were made with *C. corteziensis* at Guaymas, Sonora in 1965. Different types of spat collectors were tried. Shell bags and shell collars (wire loops through which oyster shells are inserted) proved to be the cheapest and most efficient. Successful results aided the promotion of the first cooperative for oyster culture in Mexico.

Clay tiles, mangrove branches, and oyster shells have been used as spat collectors. Shell bags and collars have been adopted as standard collection procedures for both research on abundance and commercial culture. Such bags or collars are hung from long lines or rafts. When spats are 3 to 5 mm long, the shells are laid on the bottom or hung from rafts and long lines on wire spaced with plastic hose.

A hatchery at San Blas, Nayarit produces *C. corteziensis*, but natural settlement is so abundant that laboratory-reared larvae are insignificant to aquaculture production. Culture of *C. corteziensis* is becoming widespread in coastal lagoons on the Pacific coast of Mexico.

REFERENCES

1. **Castillo, R. Z. and Garcia-Cubas, A.,** Taxonomia y anatomia comparada de las ostras en las costas de Mexico, *An. Inst. Cienc. del Mar y Limnol.*, Univ. Nac. Auton. Mexico, 13(2), 1986.
2. **Hertlein, G.,** Description of two species of marine pelecypods from west Mexico, *Bull. Soc. Calif. Acad. Sci.*, 50, 68, 1951.
3. **Ramirez, R. G. and Sevilla, M. L.,** *Las ostras de Mexico,* Sec. Ind. y Com. Dir. Gral. de Pesca e Ind. Con., 1965, 1.
4. **Stuardo, J. and Martinez, A.,** Relacion entre algunos factores ecologicos y la biologia de poblaciones de *Crassostrea corteziensis* Hertlein, 1951, de San Blas, Nayarit, Mexico, *An. Cent. Cienc. del Mar y Limnol.*, Univ. Nac. Auton. Mexico, 2, 84, 1975.
5. **Castillo, R. Z.,** Contribución al estudio taxonomico de algunas especies Mexicanas de la Familia Ostreidae, Tesis profesional, Fac. Cienc., Univ. Nac. Auton. Mexico, 108, 1977.
6. **Harry, H. W.,** Synopsis of the supraspecific classification of living oysters (Bivalvia Gryphaeidae and Ostreidae), *Veliger*, 21, 121, 1985.
7. **Rodriguez, R. F., Uribe, M. A., and Laguarda, F. A.,** The karyotype of *Crassostrea corteziensis* Hertlein (Mollusca; Ostreidae), *An. Cent. Cienc. del Mar y Limnol.*, Univ. Nac. Auton. Mexico, 6, 15, 1979.
8. **Menzel, W.,** Hybridization of oysters and clams, *Proc. World Symp. in Selection, Hybridization and Genetic Engineering in Aquaculture,* Bordeaux, 27—30 May 1986, Berlin II, 47, 1987.
9. **Keen, M. A.,** *Sea Shells of Tropical West America, Marine Mollusks from Baja California to Peru,* Stanford University Press, California, U.S., 1971.
10. **Orbe, M. A.,** Estudio ostricola realizado en el Pacifico por la Dirección de Aguacultura, Sec. Rec. Hidr., presented at *5th Nac. Cong. Oceanogr.*, Mexico, 1974.
11. **Sevella, M. L.,** Datos biologicos para el cultivo del ostión de Guaymas, Son. Sec. Ind. y Com. Dir. Gral. de Pesca e Ind. Con., Mexico, 13, 159.
12. **Ruiz, D. F.,** Estudio histologico comparativo de los ciclos gonadicos de *Ostrea corteziensis* H., *Crassostrea virginica* G., y *C. iridescens* H. Simp., Carpas en Aguacultura en America Latina, F.A.O. Carpas/6/74.SE/8, 1974.
13. **Cuevas, G. C. and Martinez, G. A.,** Estudio gonadico de *Crassostrea corteziensis* Hertlein, *C. palmula* Carpenter, y *C. iridescens* Hanley de San Blas, Nayarit, Mexico (Bivalvia: Ostreidae), *An. Cent. Cienc. del Mar y Limnol.*, Univ. Nac. Auton. Mexico, 6, 37, 1979.
14. **Cortez, G. A. and Martinez, G. A.,** Identificatión y cuantificación de larvas pediveliger de *Crassostrea corteziensis* Hertlein y balanidos de los esteros de San Blas, Nayarit, Pacifico de Mexico, *An. Cent. Cienc. del Mar y Limnol.*, Univ. Nac. Auton. Mexico, 6, 81, 1979.

Chapter 9

BIOLOGY AND CULTURE OF THE CARIBBEAN OR MANGROVE OYSTER, *CRASSOSTREA RHIZOPHORAE* GUILDING, IN THE CARIBBEAN AND SOUTH AMERICA

Anibal Vélez

TABLE OF CONTENTS

I. INTRODUCTION

Crassostrea rhizophorae (Guilding, 1828) is in the family Ostreidae. Dall and Simpson[1] classified the common oyster of Puerto Rico as *C. virginica* (Gmelin); however, Mattox[2] in a later paper considered it to be *C. rhizophorae* (Guilding).* It is very similar to *C. virginica* but differs in that the lower left valve is less plicated than that of *C. virginica* and in that the muscle scar is more rounded and often unpigmented. Ahmed[3] believes that *C. rhizophorae* may have descended from *C. virginica*.

The common name ''mangrove oyster'' originated from its association with the mangrove trees, *Rhizophora mangle* (L.), that are found bordering lagoons and bays of the coastal waters of the Caribbean Sea and adjacent islands and along the north Caribbean coast of South America. It is a small oyster, not exceeding 6 cm in length, whose shell is often ribbed radially, light, thin, and foliaceous. The bottom valve is deeply cupped, and the smaller, flat upper valve fits into the lower one (Figure 1).

II. BIOLOGY

A. ENVIRONMENT

The mangrove oyster is only commonly found in the narrow intertidal zone (tidal amplitude seldom exceeds 60 cm in the Caribbean), attached to the aerial roots of the red mangrove.[4-6] Other organisms also attach in this zone, but only oysters, some species of barnacles, and red algae survive.[7] Barnacles apparently are not a serious competitor because they have been found overgrown by oysters.[5] Just below the intertidal band of oysters are found the drunk oyster (*Isognomon alatus* [Gmelin]), mussels, bryozoans, the ascidians *Ascidia nigra* and *Ecteinascidia* sp., and tubicolous polychaetes. Four bivalves often attach subtidally at lower levels: *Spondylus americanus* Hermann, *Chama macrophylla* (Gmelin), *Pinctada radiata*, and *Ostrea (Lopho) frons* (Linneaus). The mangrove oyster lives in a wide variation of salinities; different reports give ranges of 5 to 20‰,[4,8] 18 to 32‰,[9] 22 to 40‰,[10] and 27 to 47‰.[2,11-13] The pH of the water in which oysters live has been found to vary from 7.7 to 8.3 and dissolved oxygen from 2.0 to 4.4 ml/l.[14,15]

B. SEXUAL DEVELOPMENT, SPAWNING, AND SPATFALL

The mangrove oyster is a protandric hermaphrodite,[16] predominantly male when small and female when large. As seen from Figure 2, males may develop gonadal tissue when they are as small as 5 mm, but probably do not spawn until they are about 10 mm long. Female gonads, on the other hand, may be seen in oysters as small as 10 mm but probably do not function until they are 25 mm, about 8 weeks old. A large proportion of the oysters under 20 to 25 mm are undifferentiated. A smaller portion are always undifferentiated, even in the larger sizes; these are probably spawned out oysters in which new gonadal tissue has not yet developed.

About 25% of the population develop as males first, and about 20% develop initially as females. Approximately 35% of the young oysters have initially bisexual gonads; they develop as males, then after the first spawning change to females,[10] beginning when the individuals are about 25 mm long. Sexual differentiation appears to be rapid in individuals but slow and asynchronous in the cohort.[17] In older oysters the bisexual phase is seen in about 3% of the population. Release of both sperm and eggs at the same time does not occur; the gametes do not reach maturity at the same time. The overall sex ratio is 1:1.

* Editor's Note: Menzel, W. (Hybridization of oysters and clams, in *Proc. World Symp. on Selection, Hybridization, and Genetic Engineering in Aquaculture*, Bordeaux, May 27—30, 1986, Berlin, II, 47, 1987) readily obtained F_1, F_2, and backcrosses with *C. virginica* with normal meioses and mitoses.

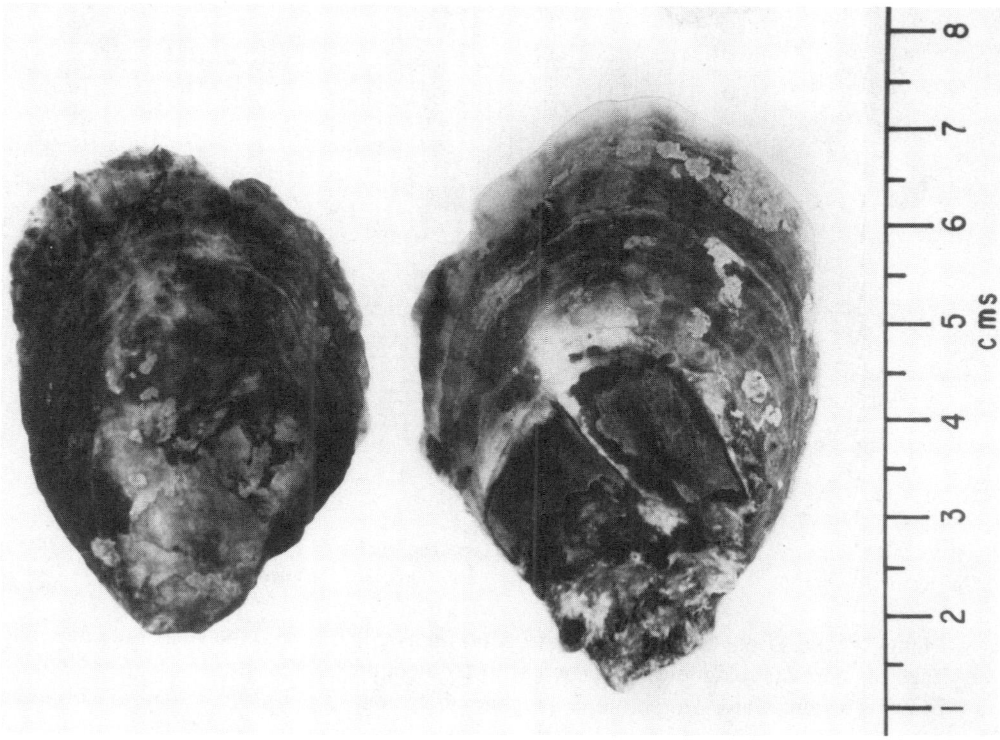

A

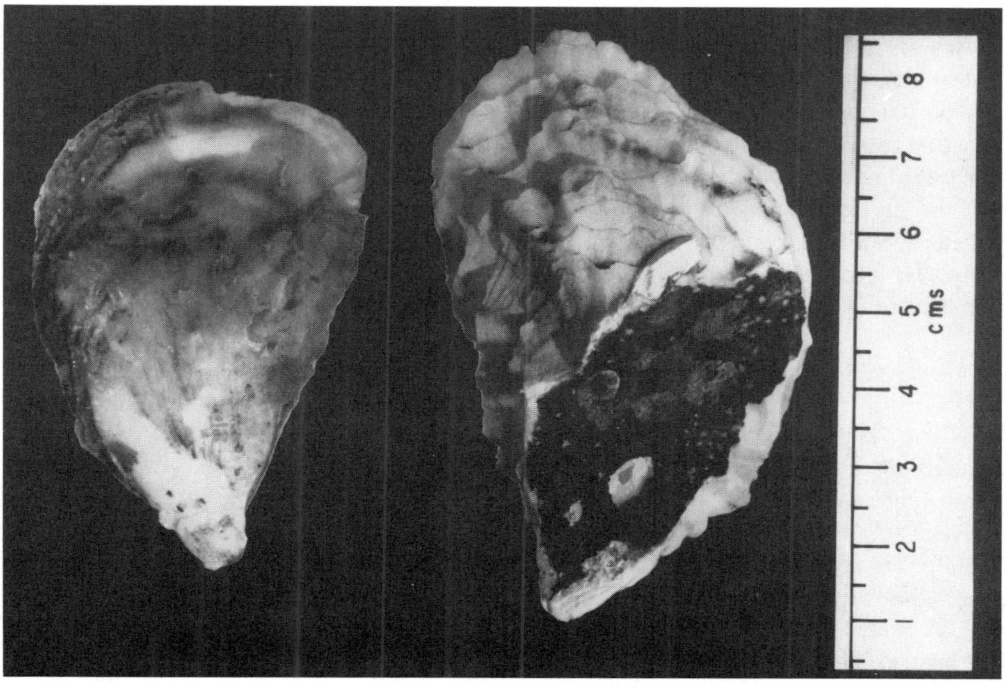

B

FIGURE 1. Photograph of right and left shells of *C. rhizophorae*.

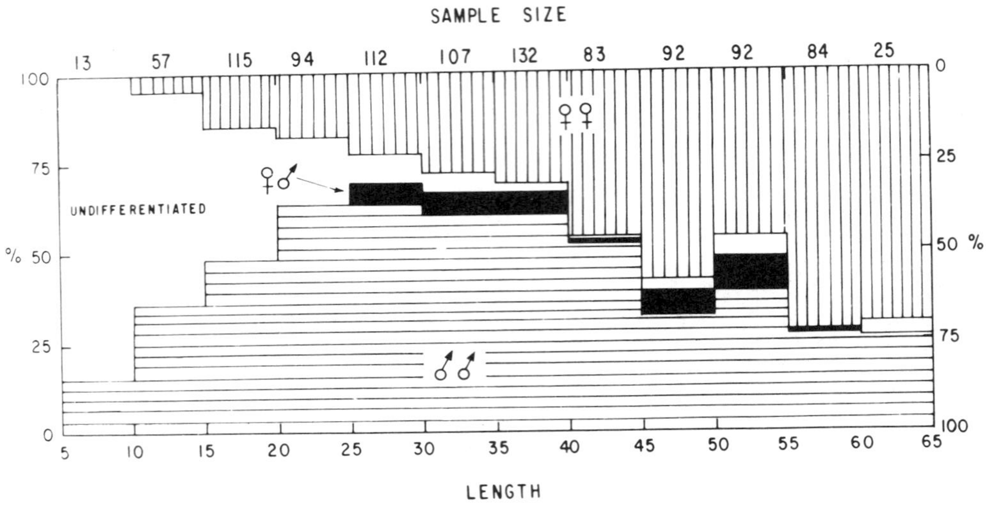

FIGURE 2. Sex ratios of *C. rhizophorae* in each size class. The number of oysters in each size class is shown at the top of the diagram. (From Vélez, A., *Aquaculture*, 21, 129, 1982. With permission.)

Mangrove oyster populations reproduce continuously throughout the year, but there may be peaks in seasonal spawning associated with temperature and salinity.

In areas where water temperature fluctuates between 23 and 29°C, there are bimodal peaks, lower from January to June and higher from July to December.[5,18] During the high reproductive period, the condition index, glycogen, and total lipids are low.[19,20] High reproduction has been associated with upwelling, which causes an increase in nutrients for phytoplankton growth and a decrease in temperature.[18,21,22] Spawning and spatfall are continuous in lagoons with temperatures higher than 25°C. The spawning process in mangrove oysters is like that in other members of the genus *Crassostrea*. The sperm and eggs are released into the water column, where fertilization takes place. After development the mature larvae (late umbo stage) attach to firm, clean surfaces, such as mangrove roots or stones. Spatfall can occur at depths of up to 8 m, but mortality is high subtidally, especially in the first 45 d or so.[7]

In the laboratory, stripping mature male and female gonads and mixing the gametes yields fertilized eggs.[22] Spawning can be readily induced by the addition of hydrogen peroxide (H_2O_2) to alkaline seawater.[23] This method is inexpensive, simple, and reliable for the control of synchronous spawning of the sexes.

III. FISHERY

The oysters are collected by hand picking in coastal mangrove swamps. The oysters are placed in seawater until they are sold in local tourist places, where they are consumed fresh. Sometimes the fishermen sell their catch to oyster dealers who distribute in large cities. The oysters are usually placed in sacks and kept in cold storage until sold. Most of the catch is sold to restaurants and oyster bars. Although oysters may be consumed at any time of the year, they are in the best condition from January to May.[20]

Although mangrove oysters do not represent a large fishery, they are an important source of income for the poorer communities of the Caribbean. Their easy accessibility from small boats enables fishermen to harvest them with little investment and only simple equipment, such as knives and machetes. However, the high price the oysters bring in some Caribbean countries frequented by tourists has led some fishermen to harvest them indiscriminately, cutting the mangrove roots and destroying the oysters' natural substrate.[5,24] This practice has resulted in declining populations in certain areas. In some Caribbean countries, management efforts have been directed toward closure of certain areas for the protection of juveniles. Unfortunately, these measures are often not implemented because they are unpopular among the fishermen.

IV. CULTURE

Because it has been difficult to prevent the progressive decline of natural mangrove oyster production, some affected countries such as Cuba, Jamaica, Venezuela, Nicaragua, the Bahamas, Colombia, Puerto Rico, and Brazil have considered intensive farming as an alternative that would protect the resource and at the same time increase production. Suitable mangrove oyster culture areas are found in inshore regions along the shoreline of the Caribbean Sea, the Guayanas, and Brazil, but only Cuba, Venezuela, and Jamaica have developed pilot and commercial mangrove oyster culture programs. Annual oyster production for the period 1969 to 1973 was about 30 ton in Venezuela and 26 ton in Cuba. Unfortunately, the commercial attempts at oyster culture failed in Venezuela and Nicaragua because of problems that will be discussed later. Jamaica is developing a pilot program whose production is unknown but reports high demand from hotels and restaurants.[25] Presently, Cuba is the only country with a working commercial program. Nineteen farms are in operation, with an estimated potential annual production between 900 and 1100 t of unshucked oysters.[25]

A. SPAT COLLECTION

Spat collection of mangrove oysters has been found to depend on the combined characteristics of the natural stock, the coastal hydrographic features of the area, and larval behavior.[6,13,15] The best places for spat collection are in semienclosed waters at times of high temperature and salinity. The highest larval fixation and spat survival occur under horizontal surface collectors in the intertidal zone.[7]

In pilot as well as commercial operations, successful spat collection has been carried out in Cuba with mangrove branches, in Venezuela with mussel shells and asbestos sheets, and in Jamaica with pieces of old car tires and frames.[7,13,15,25] The terminal mangrove branches without leaves are trimmed so that the majority of them will be between 30 and 40 cm long (Figure 3). Mussel shells are placed in chickenwire bags with the inner faces down. The pieces of old car tires are cut into squares and drilled in the center. They are tied together with monofilament line. The collectors are suspended from wooden intertidal racks for larval settlement.

B. GROWING AREAS

After spat collection, the collectors are left on the intertidal racks or suspended from rafts for growing (Figure 3). The oysters on racks reach the commercial size of 50 mm in 5 to 6 months, whereas those on the rafts reach 70 to 80 mm in length in the same period.[7,15,27] However, the oysters harden better on the intertidal racks and also avoid fouling organisms such as ascidians, bryozoans, sponges, and some species of barnacles whose habitats are below the intertidal zone. The production costs of this method are lower.

Suitable areas for growing mangrove oysters naturally vary from place to place but must meet some basic requirements, such as water depth, salinity, water currents, and the absence of predators.[7,15,27] Mangrove oysters grow faster in salinities between 28 and 47‰, but a small freshwater inflow can help eliminate some competitors and predators. Land drainage

FIGURE 3. Terminal branches of mangrove used as collectors of *C. rhizophorae* in Cuba. (From Nikolić, M., Bosch, A., and Alfonso, A., *Aquaculture*, 9, 1, 1976. With permission.)

brings in nutrients, which promote the growth of microscopic food organisms. Racks should be placed at a depth of 2 m at low tide. Oyster growth is faster where water currents constantly replenish the food supply.[15]

C. CULTURE PROBLEMS

Most of the problems of mangrove oyster culture are related to predation, parasites and fouling, and marketing. Some gastropods, predators such as *Chicoreus breviformis*, *C.*

(Marer) breviformis (Lamarck), *Thais haemastoma* (Linnaeus), and *Cymatium (Monoplex) parthenopeneum* (Von Salis), and some crustaceans, such as *Callinectes danae* (Smith) and *Menippe nodiformis* (Simpsons), do not represent initial problems, but could do so with the increase of culture. Probably the most serious potential predator is the gastropod *C. parthenopeneum*.

Several species of sponges and mudworms and a mussel, *Lithophaga* sp., bore the shell of the mangrove oyster. These are not true parasites, but they excavate tunnels in the calcareous shell, and the oysters must devote a portion of their energy to repairing the shell damage they cause. Generally, the heaviest infections are found in older individuals. The crab *Pinnotheres ostreum* San lives within the shells of *Crassostrea rhizophorae* but is not common in Venezuela and Cuba.

In Venezuela, the most serious problems are the competitors of the oyster: the ascidians, sponges, and barnacles. Barnacles may cause problems for very young oysters, and ascidians and sponges are serious competitors in cultures on floating rafts. The high cost of removing fouling and of marketing has been the main cause of failure of culture efforts in Venezuela, where the market for the *C. rhizophorae* is replaced by that for another species, *C. virginica*.

Fouling organisms and diseases are the main problems of Jamaican oyster culture. The major problems of Cuban mangrove oyster culture are the wide fluctuation in climate and low spat settlement. A hatchery under construction under the supervision of experts from the Food and Agriculture Organization may overcome the latter problem.[25]

D. THE FUTURE

Most of the commercial and pilot culture operations of mangrove oysters have had low production capacity, all of which has been absorbed by internal markets. Further expansion of the industry will have to be based on the culture practice of using intertidal racks to reduce the operational cost, and on the improvement of distribution channels. Unfortunately, *C. rhizophorae* is one of the smallest species in the genus and does not compete well with other larger species in demand in the North American and European markets. However, most Caribbean islands and countries have high rates of seafood consumption.[28] Researchers should be encouraged to produce seed under controlled conditions and to try to develop a new, more competitive variety through hybridization. The Instituto Oceanografico of Venezuela and the National Council for Scientific Research (CONICIT) are presently directing their efforts toward obtaining a hybrid of *C. rhizophorae* and *C. virginica*.

REFERENCES

1. **Dall, W. H. and Simpson, C. T.,** The molluscs of Puerto Rico, *Bull. U.S. Fish. Comm.*, 20, 1, 351, 1902.
2. **Mattox, W. T.,** Studies on the biology of the edible oyster *Ostrea rhizophorae* (Guilding) in Puerto Rico, *Ecol. Monogr.*, 19, 339, 1949.
3. **Ahmed, M.,** Speciation in living oysters, *Adv. Mar. Biol.*, 13, 357, 1975.
4. **Martínez, R.,** Aspectos Bioecologicós de *Crassostrea rhizophorae* en Laguna Grande del Obispo, *Publ. Espec. Univ. Oriente,* 1962, 1.
5. **Saenz, B. A.,** El Ostión Antillano, *Crassostrea rhizophorae* Guilding y su cultivo experimental en Cuba, *Inst. Nal. Pesca Cuba,* 6, 1, 1965.
6. **Saint-Felix, C.,** Les girements huitrieres de *Crassostrea rhizophorae* en Martinique, *Sci. Pêche, Bull. Inst. Pêches Marit.*, 214, 1972.
7. **Vélez, A.,** Algunas observaciones sobre la ostricultura en el oriente de Venezuela, *Laguna,* 35/36, 9, 1975.
8. **Bacon, P. R.,** Studies on the biology and cultivation of the mangrove oyster in Trinidad with notes on other shellfish resources, *Trop. Sci.*, 12, 265, 1971.
9. **Nascimento, I. and Lunetta, J. E.,** Ciclo sexual da ostra de mangle e sua importancia para el cultivo, *Bol. Fisiol. Anim. Univ. Sao Paulo,* 2, 63, 1978.

10. **Nascimento, I., Simoes, M. I. and Dos Santos, A.**, Sex-ratio occurencia de hermafroditismo en *Crassostrea rhizophorae* (Guilding, 1828), in *Proc. Int. Symp. Brasileiro Aquaculture*, Recife and Pe, Brazil, 1981, 385.
11. **Angel, C.**, Maduración gonádica y filjación de *Crassostrea rhizophorae* en una laguna hipersalina del nororiente de Venezuela, *Mem. Soc. Cienc. Nat. La Salle*, 23, 215, 1972.
12. **Margalef, R.**, Composición y distribución del fitoplancton, *Mem. Cienc. Nat. La Salle*, 26, 1, 1965.
13. **Vélez, A.**, Ensayos del cultivo del ostion *Crassostrea rhizophorae* en el oriente de Venezuela, *Laguna*, 19/20, 11, 1968.
14. **Jmeliova, N. N. and Sanz, J.**, Respiración y algunas particularidades de la alimentación del ostión *Crassostrea rhizophorae* Guilding, 1818, *Ser. Oceanol. Acad. Cienc. Cuba*, 3, 1, 1969.
15. **Nikolić, M., Bosch, A., and Alfonso, A.**, A system for farming the mangrove oyster (*Crassostrea rhizophorae*) Guilding, 1828, *Aquaculture*, 9, 1, 1976.
16. **Vélez, A.**, Hermafroditismo del ostión *Crassostrea rhizophorae*, *Aquaculture*, 21, 129, 1982.
17. **Vélez, A.**, Crecimiento, edad y madurez sexual del ostión *Crassostrea rhizophorae* de Bahía de Mochima, *Bol. Inst. Oceanogr. Univ. Oriente*, 15, 65, 1976.
18. **Vélez, A.**, Ciclo sexual de reprodución del ostión *Crassostrea rhizophorae* (Guilding) de Bahía de Mochima, *Bol. Inst. Oceanogr. Univ. Oriente*, 15, 87, 1977.
19. **Bonilla, J., Benítez, J., Okuda, T.**, Variación estacional de la composición química del ostión *Crassostrea rhizophorae* (G) en la Laguna Grande y la Bahia de Mochima, *Bol. Inst. Oceanogr. Univ. Oriente*, 8, 46, 1969.
20. **Vélez, A. and Bonilla, J.**, Variacion estacional del engorde del ostión *Crassostrea rhizophorae*, de Bahia de Mochima y Laguna Grande, *Bol. Inst. Oceanogr. Univ. Oriente*, 11, 39, 1972.
21. **Mandelli, E. and Acuna, A.**, The culture of the mussel, *Perna perna*, and the mangrove oyster, *Crassostrea rhizophorae*, in Venezuela, *Mar. Fish. Rev.*, 37, 15, 1975.
22. **Menzel, W.**, Possibilities of molluscan cultivation in the Caribbean, *FAO (F.A.O.U.N.) Fish Rep.*, 71, 2, 1983.
23. **Morse, D. E., Hooker, N., and Morse, A.**, Chemical control of reproduction in bivalve and gastropod molluscs. III. An inexpensive technique for mariculture of many species, *Proc. World Mariculture Soc.*, 9, 543, 1978.
24. **Chipman, W. and Thompson, P.**, Possibilities for oyster culture in Puerto Rico and the Virgin Islands, *U.S. Bur. Fish. Bull. Spec. Sci. Rep. Fish.*, 9, 1, 1950.
25. **Jory, D. E. and Iversen, E. S.**, Molluscan mariculture in the greater Caribbean: an overview, *Mar. Fish. Rev.*, 47, 1, 1985.
26. **Vélez, A.**, Fijación de la larva de la ostra de los bancos naturales de Bahía de Mochima y Laguna Grande, *Bol. Inst. Oceanogr. Univ. Oriente*, 11, 97, 1972.
27. **Angel, C.**, Crecimiento y mortalidad de la ostra de mangle cultivada *(Crassostrea rhizophorae), Bol. Inst. Oceanogr. Univ. Oriente*, 33, 154, 1973.
28. **Olsen, D. A., Nellis, D. W., and Wood, R. S.**, Ciguatera in the eastern Caribbean, *Mar. Fish. Rev.*, 46, 21, 1984.

Chapter 10

CRASSOSTREA RHIZOPHORAE (GUILDING) AND *C. BRASILIANA* (LAMARCK) IN SOUTH AND CENTRAL AMERICA

Iracema Andrade Nascimento

TABLE OF CONTENTS

I. BIOLOGY

A. CHARACTERISTICS

Along the almost 8000 km of Brazilian coastline, most of it located in tropical and subtropical areas, the presence of the mangrove environment is a normal feature. The most representative animal species of these mangrove areas are *Crassostrea rhizophorae* (Guilding) and *C. brasiliana* (Lamarck). The first is more conspicuous at lower latitudes, with broad distribution from the West Indies and the Caribbean to Santa Catarina in southern Brazil,[1] but southward from Espirito Santo, in Brazil, *C. brasiliana* gradually increases in density to complete predominance in the more temperate areas.

Crassostrea brasiliana, the Brazilian oyster, like *C. rhizophorae*, the Caribbean or mangrove oyster, is an oviparous cup-like species (Figure 1) having a promyal chamber and small gill ostia.[2] *Crassostrea rhizophorae* has a thin, foliaceous, and deeply cupped right valve; the upper (left) valve is small and flat and fits into the lower one (Figure 2). The beak is twisted dorsally, and the muscle scar, often unpigmented, is near the dorsal margin of the shell. Adults may reach 10 cm in height,[2] but normally, in their natural environment, are stunted, growing to a maximum size of 5 cm.[3]

Crassostrea brasiliana is very similar to *C. rhizophorae*. It is less cup-like than *C. rhizophorae*, and the muscle scar is generally slightly blue or brown.[4] The biggest size reached by *C. brasiliana* in Cananéia (25°01′S 47°55′W) is 14.0 cm, but they normally reach only 10.5 cm in natural populations.

B. ENVIRONMENT

Crassostrea rhizophorae, like *C. brasiliana*, is commonly found in aggregations attached to the aerial roots of the mangrove *Rhizophora mangle* Linnaeus or to those of *Avicennia* and *Laguncularia* trees. Both species are adapted to life in waters with considerable variations in salinity (from 0‰ during the annual rains, to 40‰ in the dry season), but the optimum salinity range is roughly from 7.2 to 28.8‰ for *C. rhizophorae*[5] and from 8.0 to 34‰ for *C. brasiliana*.[4] Growth and survival under the higher salinity limits are lower for *C. rhizophorae*,[6,7] but low salinities (less than 18‰ seem to be deleterious for gonadal and larval development and spatfall.[7] A salinity range of 25 to 37‰ is the best for the embryonic development of *C. rhizophorae*, whose embryos do not develop at salinities lower than 19‰.[8] *Crassostrea rhizophorae* in Cuba tolerates salinities from 22 to 40‰ for a number of days, but it is not found naturally at salinities below 12‰ or above 40‰.[9] In Brazil, *C. rhizophorae* withstands violent salinity fluctuations of short duration during the rainy seasons.

Crassostrea brasiliana grows and fattens better in lower salinity zones (15 to 25‰), but spat collection for this species is better in areas of higher salinities.[4] In Cananéia (São Paulo, Brazil), the spatfall is best at depths of 7 to 8 m, where the currents promote a higher salinity than is found in the surface water.[10] These findings lead to the conclusion that, for both *C. rhizophorae* and *C. brasiliana*, although oyster settlement is heaviest in hypersaline conditions (up to 37‰), growth to market size (6 to 7 cm) is fastest in brackish water, upriver locations, more favorable to the proliferation of mangrove vegetation.[11]

In tropical and subtropical areas the temperature remains near 28°C through the year. In temperate areas, the variation is greater but, in most of the mangrove areas along the Brazilian coast, is similar to that at Todos os Santos Bay, where the temperature varies from 20 to 32°C.[12] The higher temperatures occur around January and February; the lowest occurs from June to August. Generally, oysters withstand large temperature variations,[2] but the optimum temperature for *C. rhizophorae* is below 30°C. At 25°C a high proportion (58.5%) of normal D-larvae is found 24 h after fertilization; at 30°C the proportion does not surpass 7%.[8]

Distribution of the mangrove oyster is determined mainly by the availability of sub-

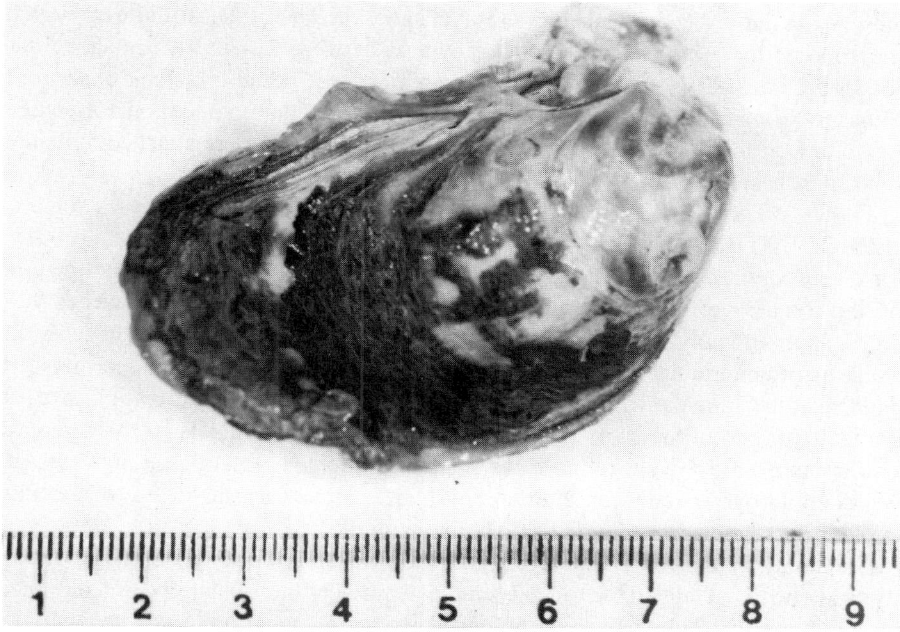

FIGURE 1. *Crassostrea brasiliana* collected from natural populations at Cananéia, São Paulo, Brazil.

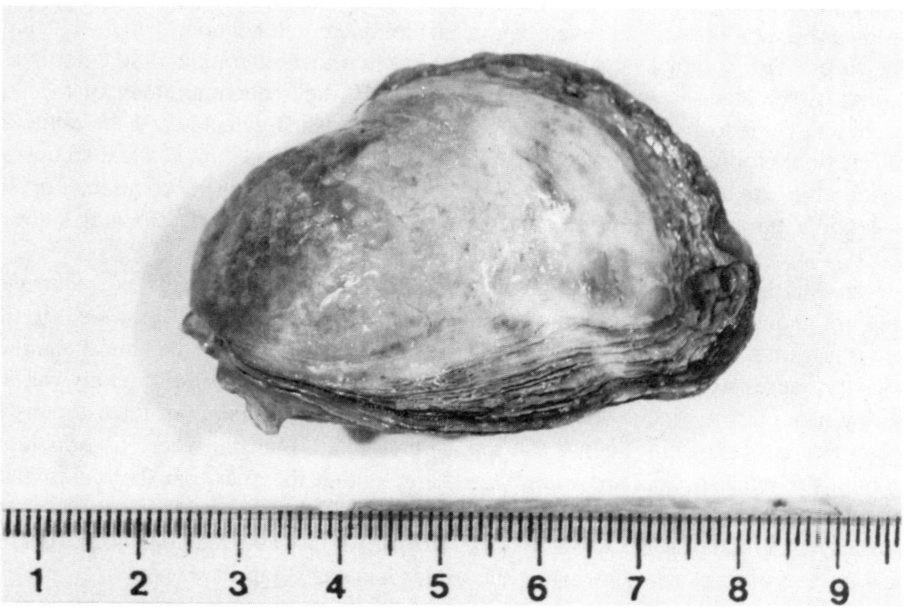

FIGURE 2. *Crassostrea rhizophorae* collected from natural populations at Jacuruna, Bahia, Brazil.

stratum. On most of the Brazilian coastline, the tidal range is about 2.5 m. Mangrove roots generally are found between 0.8 and 1.0 m from the 0.0 level of spring tides. It is nearly impossible for oysters to settle or survive below this level, because the substrate is extremely soft mud and anaerobic beneath the mud-water interface.[13] In addition, sublittoral survival is constantly threatened by severe pressure from predators (crabs, oyster drills, fish), and parasites, particularly the shell-inhabiting polychaete *Polydora*.[11] Thus, the narrow vertical band appropriate for oyster settlement and growth is between 1.0 and 1.5 m above the 0.0 level of spring tides. Zones above 1.5 m are inappropriate, because of long exposure time (more than 60% of each tide cycle). In this narrow band, the competition for settlement space is intense, and the oysters are found mainly grouped in clusters attached to mangrove roots and branches.

C. REPRODUCTION

For *C. rhizophorae* the primary bisexual gonads form from associations of germinal cells in the connective tissues anterior to the heart by the time the animal reaches the size of 0.7 cm, approximately 45 d after setting. The gonad contains the antecedent cells of both sexes, but its protandric nature becomes manifest with the proliferation of spermatogenesis cells in 90% of the animals, which are sexually mature before reaching 2.0 cm in size, about 120 d after setting.[14] In the same population, among older individuals (6 to 18 months, 4 to 6 cm in size), 83.5% were found to be females, which suggests that a high rate of sex change occurs in oysters between 2 and 4 cm in size. However, only 0.5% of the animals in this older population were found to be hermaphroditic.[15]

The anatomy of the reproductive organs in *C. rhizophorae* does not differ from that already described by Galtsoff[2] for *C. virginica*. However, the sexual cycle determined by Nascimento[16] and confirmed by morphological,[17] cytological,[18] and cytochemical[19] characterization of the sexual phases in *C. rhizophorae* is specific to this oyster. From an undifferentiated stage, the active gonad goes through a prematuration and a maturation stage before spawning. Thus, after a phase of partial spawning, which develops to an advanced spawning phase, the gonad may enter into a recuperation stage, when gametogenesis begins again toward a new maturation, even before the complete cytolization of the sex cells or a resting stage. The complete cytolysis of the gamete and obliteration of the follicles by connective tissue leads to an undifferentiated stage.[20] The sequential study of the sexual cycle phases seems to indicate that cytolization of the unshed gametes and the obliteration of the functional follicles by connective tissue is a slower process than gonadal recuperation and maturation. As a result, most adult oysters (4 to 6 cm in length) become mature again, without going through an undifferentiated stage, after the advanced spawning or resting stage.[20]

Because of the almost constant high water temperature, gametogenesis in *C. rhizophorae* occurs throughout the year, with two spawning peaks around March and October. Glycogen storage is less marked in *C. rhizophorae* than in other species, because the almost continuous gonad activity uses stored energy. The spawning peaks occur when drastic weekly variations in salinity take place, coincident with the onset of the rainy periods, but intense rains (150 mm/week) depress spawning.[12] In a subtropical area like Cananéia, where *C. brasiliana* is the dominant oyster species, spawning occurs throughout the year, but there are peaks in the periods of January to June and October to December.[10]

Although literature on the sexuality of *C. brasiliana* is scant, the gonads are functional about 6 months after attachment, when the oyster is about 3 cm in size.[4]

In the laboratory, embryonic development of *C. rhizophorae* can be achieved in 24 h. The highest proportion of normal D-larvae is obtained at a density of 10^4 to 4×10^4 ovocytes per liter when they are fertilized at concentrations of 500 to 5000 spermatozoans per ovocyte. The best range of salinities for embryonic development is 25 to 37‰. The best temperatures are around 25°C and below 30°C.[8]

Planktonic larval stages of *C. rhizophorae* and *C. brasiliana* have not been described, but they are easily recognized in their general aspects based on the description for *C. virginica* (Gmelin).[2] The umbo larvae are generally found 10 to 15 d from the beginning of larval life. Fixation occurs in the laboratory between 15 and 17 d, at 27°C.

The best period for larval collection of *C. brasiliana* is the fall (March to June), when the water temperature is around 25°C.[4] For *C. rhizophorae* the best periods correspond to the spawning peaks, around March (fall) and October (spring).

D. GROWTH

Because the temperature changes are so small, the growth rate of *C. rhizophorae* and *C. brasiliana* depends on the rainy season, which transports nutrients to the estuarine areas, promoting higher food supply. The older these oysters get, the slower they grow.[2]

Extensive studies have been made of the growth rate of *C. rhizophorae* and of the best size for meat yield. In the natural environment, this oyster grows to 4 to 7 cm the first year after attachment,[21] but about 18 months are required for the majority to reach the commercial size of 7 to 8 cm. Newly attached spat grow rapidly (1 cm/month) for the first 3 months, then more slowly for the next several months (0.78 cm/month); when they reach a size of 6.5 cm, the rate is 0.5 cm/month.[21] A comparison of shell cavity fluid, meat yield, and dry body weight showed that optimum size for harvesting is 7 cm.[22] The condition index varies with the reproductive cycle; it is highest during the peak of the spawning period (59.0 to 71.8) and lowest after spawning.[23,24] Based on the condition index of oysters between 2.1 and 10 cm, the size classes between 4.1 and 6.0 cm yield the most meat.[25]

Crassostrea brasiliana has only a small weight gain in meat during the first 8 months after attachment.[4] At the age of 1 year, the average weight is 2.7 g, but it averages 13 g at 2 years. Thus, it is recommended that this species be allowed to grow for 2 years before harvest.[4]

II. FISHERY

A. HISTORY AND METHODOLOGY

Crassostrea rhizophorae is found in the coastal waters of the various countries bordering the Caribbean and South Atlantic from the West Indies to Santa Catarina, Brazil. In all these countries it has represented an important fishery resource for over 100 years, especially for the maritime communities, which exploit it by using the simplest food-gathering techniques.[11] Small boats make the harvesting less difficult and more effective.[9] Production in excess of family food requirements is sold in the nearest market to provide other necessities. In some areas, especially those near townships and cities, exploitation and indiscriminate harvest have resulted in declining populations. Consequently, in some countries (especially in Cuba, Venezuela, and Brazil), farming of the mangrove oyster has been considered a means of sustaining, if not increasing, production. Such activity requires knowledge of the biology of the species and of the environmental conditions needed for efficient production.

First attempts to cultivate oysters were made in Cuba in 1951.[26] In Venezuela in the early 1960s[27] and in Brazil in 1972,[11] attempts were made to introduce foreign techniques, like the Spanish techniques of spat collection (Cuba) and oyster fattening (Venezuela) and the Australian technique of using trays placed in the intertidal zone (Brazil). Although preliminary results were promising in all these countries, later research has improved the techniques by adaptation to the different characteristics of the different areas, especially with *C. rhizophorae*. Thus, mangrove oyster culture systems have been developed that use inexpensive and easily accessible materials.

In Venezuela, the raft technique has been successful. The abundance of larvae insures a constant source of seed. Collectors (40 × 80 cm) are suspended below the water surface from wooden rafts that vary in size from 8 to 12 m^2. After setting, the spat are allowed to grow to about 2.5 cm in length, and only 200 to 300 are kept on each collector. From this point on, the juvenile oysters are treated at monthly intervals with a lime suspension for 15 min to eliminate competitors. The oysters reach marketable size in about 6 months. Between

FIGURE 3. View of an oyster farm in Bahia, Brazil.

210 and 300 kg of oysters per collector can be obtained. Based on this yield, a production of 200 to 300 t/ha/year can be projected, on the basis of one crop per year.[27] These figures are possible because the water depths in Venezuela permit the use of raft culture, which has the advantage that the entire vertical water column within the euphotic zone can be used.

The Cuban technology for growing *C. rhizophorae* involves the use of terminal branches of the red mangrove, without the leaves, as spat collectors. These collectors are suspended from longitudinal beams supported by crossed pillars implanted in the substrate, in such a way that their submerged portion corresponds to the 30- to 40-cm tidal band of the lower eulittoral zone. Each month, undesirable fouling organisms are cleaned from the collectors while the oysters continue to grow. The first harvest takes place 5 to 6 months after the initial placement of collectors; the largest oysters are collected, and the remainder are left attached to attain further growth. Subsequent harvest can occur each month. A farmer should be able to manage at least 7500 collectors.[9] Of these, about 3750 would be in production, yielding, after 8 months, approximately 26.2 ton of oysters in 600 to 700 m^2. This technique is more suitable for the shucked oyster than for the shell market, because the resulting product does not have a regular shape.

The practical technique for the cultivation of the Bahian mangrove oyster (in Brazil) involves the transfer of slow-growing spat from collectors high on the shore at downriver locations to predator-proof racks sited further from the shore at upriver locations (Figure 3). The optimum intertidal level for oyster production, in terms of productivity (survival × growth), lies at 1.5 m above low water mark of mean spring tides when the spring tide is

FIGURE 4. View of an oyster farm in Cananéia (SOSTRAMAR, S.A.).

2.5 m. Oysters located below this critical level succumb to parasite attack.[11] One hundred (1 × 20 m) racks supporting trays made of bamboo sticks would produce, in 1 year, 20 ton of well-shaped oysters, with a mean size of about 6 cm. The oysters are about 3 to 4 cm in size and of an age from 3 to 6 months when planted in the tray.[13]

All these techniques are artisanal and do not require mechanization even for harvesting. In northeastern Brazil, *C. rhizophorae* is marketed as boiled meat; fresh oysters in the shell are regarded as a delicacy to be consumed at one's own risk, in the absence of purification procedures. In the southeast the commercialized species is *C. brasiliana*. They are sold mostly as raw oysters in the shell.

Crassostrea brasiliana is cultivated in the lagoon-estuarine region of Cananéia. The cultivation technique includes spat collection from the wild on collectors each made of 100 large oyster shells placed on a bamboo frame at a depth of 7 to 8 m. After fixation, the spat are kept in the intertidal zone for a period of 4 months as a selection process. The remaining oysters are then transferred to growing lines suspended vertically from buoys or rafts.

In Cananéia, a private company is producing 300,000 dozen per year of well-shaped oysters by fattening them on racks (Figure 4). A hatchery has been built to prevent a shortage of natural seed in the area. The oysters reach commercial size (8 to 10 cm) in 2 years.

B. MARKET PROBLEMS AND POSSIBLE SOLUTIONS

Considering the potential of the Brazilian coastline for oyster production, the market is still limited. In Brazil, most of the marketed oysters (about 90%) come from wild sources. This production is around 900 ton, which corresponds to only 0.1% of world production per year (900,000 ton). Facilities for storage, transportation, and processing would surely enhance the harvest and improve the market. Oyster processing methods have been worked out, and the main problem would be to adapt them to artisanal use.

Only in the south of Brazil are oysters depurated before they are offered on the shell market. There is only one oyster purification plant (using ozone), at an oyster farm in Cananéia. Most of the oysters collected in other areas are sold only as boiled meat.

The relationship between culture methods and species is determined by the kind and form of oyster products to be marketed. In Brazil, there are two markets: fresh oysters in the shell and shucked oysters. For the latter, oyster size and shape are not important; *C. rhizophorae* is well adapted to this market. For the fresh shell market, the oyster must be well shaped and about 8 cm in size. *Crassostrea brasiliana* can be used for this market, although it is slower growing than the recently introduced Japanese oyster, *C. gigas* (Thunberg). Up to now, however, this species has not reproduced naturally, and hatcheries must produce seed.

C. CULTURE PROBLEMS

Depending on the technology used, problems develop. One is fouling, which may inhibit or reduce the growth rate of the oyster. Fouling is more common where the oyster grows permanently immersed (as on rafts). Methods of control are (1) desiccation, in which oysters are removed from the water for periods long enough to kill the fouling organisms, and (2) brief dipping of the oysters in freshwater or a concentrated saline solution. These methods are costly, especially in terms of labor. One suggestion has been to take advantage of the euryhalinity of *Crassostrea* species and the stenohalinity of most fouling organisms by locating the farm where the salinity is high enough to permit oyster growth, yet low enough to inhibit settlement of most fouling organisms, including oyster larvae.

Another problem is mortality due to low salinities during the rainy season. Alternatives are to lower the oysters below the halocline if the water is sufficiently deep or to move rafts of captive stocks to areas of higher salinity during the rains.[3]

Other culture problems are space and food competitors. In Bahia (Brazil), the use of bamboo-stick collectors for *C. rhizophorae* avoids space competition, because the main competitors (barnacles) prefer to settle on the upper surface of the bamboo sticks, whereas oysters settle on the more protected, concave, lower surface.[21] Spat collection of *C. brasiliana* is more effective at depths of 7 to 8 m. In more shallow waters (2 to 3 m), spat collection is not feasible, because of the high incidence of barnacle attachment.[10]

Epidemic oyster diseases, like those caused by *Haplosporidium* or *Perkinsus marinus* (Mackin, Owen, and Collier), have never been reported in *C. rhizophorae* or *C. brasiliana*. However, pathologies such as ceroidosis, reduction or loss of stored glycogen, decreased gametogenesis, edema, and metaplasia of the digestive diverticula were noticed in *C. rhizophorae* submitted to environmental stress.[28]

III. PROSPECTS

Oyster fattening technology should be improved. The development of hatchery technology will provide seed for the production of *C. gigas*, which has a higher growth rate than native oysters.[30] This species could be cultivated to fulfill the demands of the raw market, while the native oysters could be used for the shucked-oyster market.

The native oysters have an extraordinary potential, because their fecundity assures the almost continuous availability of seed, and the environment permits a high growth rate. However, the rapid-growth produces specimens characterized by very thin shells, needing careful handling. Additional research is required. The priorities should include improvement of culturing techniques, controlled production of seed to support further expansion of the industry, and even hybridization.[29]

In the last analysis, the future development of the oyster cultivation industry will be determined by the expansion of the market, which will depend on the increased and enhanced product offered. The prospects may be limited by the problems of transfer of culture technology for artisanal participation and the provision of financing for this group of producers.

REFERENCES

1. **Rios, E. C.,** *Coastal Brazilian Seashells,* Fundação Cidade do Rio de Janeiro, Museu Oceanográfico do Rio Grande, Rio Grande, R.S. XI, 255, 1970.
2. **Galtsoff, P. S.,** The American oyster *Crassostrea virginica Gmelin, Fish. Bull.,* 64, 425, 1964.
3. **Kamara, A. B., McNeil, K., and Quayle, D. B.,** Tropical mangrove oyster culture: problems and prospects, *FAO Tech. Conf. on Aquaculture,* FIR:AQ/Conf./76/E.58, 1976.
4. **Wakamatsu, T. A.,** Ostra de Cananéia e Seu Cultivo, S. Paulo Superintendência do Litoral Paulista, Instituto Oceanográfico da Universidade de São Paulo, 1974.
5. **Fernandes da Costa, A.,** Fisioecologia da Ostra da Mangue de Pernambuco *Crassostrea rhizophorae* (Guilding, 1828), tese de Doutoramento, Instituto de Biociências, Universidade da São Paulo, 1975.
6. **Lima, F. R. and Vazzoller, A. E. A. de M.,** Sobre o desenvolvimento das ostras e possiblidades da ostreicultura nos arredores de Santos, *Bol. Inst. Oceanogr. Univ. São Paulo,* 177, 1, 1963.
7. **Fernandes, L. B.,** Aspectos Fisio-Ecológicos do Cultivo de Ostra de Mangue *Crassostrea rhizophorae* (Guilding, 1828). Influência da Salinidade, tese de Doutoramento, Instituto de Biociências, Universidade de São Paulo, 1975.
8. **Santos, A. E. and Nascimento, I. A.,** Influence of gamete density, salinity and temperature on the normal embryonic development of the mangrove oyster *Crassostrea rhizophorae* Guilding 1828, *Aquaculture,* 47, 335, 1985.
9. **Nikolić, M., Bosch, A., and Alfonso, S.,** A system for farming the mangrove oyster (*Crassostrea rhizophorae* Guilding 1828), *Aquaculture,* 9, 1, 1976.
10. **Akaboshi, S. and Pereira, D. M.,** Ostreicultura na região lugunar-estuarina de Cananéia São Paulo, Brasil. I. Captação de larvas de ostras *Crassostrea brasiliana* (Lamarck, 1819), em ambiente natural, *Bol. Inst. Pesca,* 8, 87, 1981.
11. **Shelbourne, J. E.,** The mangrove oyster of Brazil, Newslett., Ministry of Agriculture, Fisheries, and Food, U.K., June 1977, 6.
12. **Nascimento, I. A., Santos, J. J., Andrade, J. R. C., and Silva, E. M.,** Influência de fatores ambientais na reprodução da ostra de mangue (*C. rhizophorae* Guilding, 1828), *Resumos do I Simpósio Brasileiro de Aquicultura,* Academia Brasileria de Ciências, 1978, 54.
13. **Nascimento, I. A.,** Cultivo de ostras no Brasil: problemas e perspectivas, *Cienc. Cult. (Maracaibo),* 35, 871, 1982.
14. **Nascimento, I. A., Silva, E. M., Ramos, M. I. S., and Santos, A. E.,** Desenvolvimento de gonada primária em ostras de mangue *C. rhizophorae*: idade e tamanho minimos de maturação sexual, *Cienc. Cult. (Maracaibo),* 32, 736, 1980.
15. **Nascimento, I. A., Ramos, M. I. S., and Santos, A. E.,** Sex-ratio e ocorrências de hermafroditismo em *Crassostrea rhizophorae* Guilding, 1828, *Resumos do I Simpósio Brasileiro de Aquicultura,* Academia Brasileira de Ciências, 1980, 385.
16. **Nascimento, I. A.,** Reprodução da ostra de mangue *Crassostrea rhizophorae* Guilding, 1828: um subsídio ao cultivo, Tese de Doutoramento, Instituto de Biociências, Universidade de São Paulo, 1978.
17. **Ramos, M. I. S. and Nascimento, I. A.,** Variações do indice gonadal na ostra de mangue *C. rhizophorae,* Guilding 1828, *Cienc. Cult. (Maracaibo),* 32, 1673, 1980.
18. **Santos, A. E. and Nascimento, I. A.,** Variação do diâmetro dos ovócitos em ostras (*C. rhizophorae*) em função do ciclo sexual, *Cienc. Cult. (Maracaibo),* 32, 1676, 1980.
19. **Nascimento, I. A. and Miraglia, T.,** Dados histoquímicos sobre as gonadas de ostras de mangue *Crassostrea rhizophorae,* Guilding, 1928, *Cienc. Cult. (Maracaibo),* 35, 1496, 1983.
20. **Nascimento, I. A. and Lunetta, J. E.,** Ciclo sexual da ostra de mangue e sua importância para o cultivo, *Bol. Fisiol. Anim. Univ. São Paulo,* 2, 63, 1978.
21. **Santos, J. J.,** Aspectos da Ecologia e Biologia da Ostra *C. rhizophorae* (Guilding, 1828) na Baía de Todos os Santos, tese de Doutoramento, Instituto de Biociências, Universidade de São Paulo, 1978.
22. **Nascimento, I. A., Pereira, S. A., and Costa e Souza, R.,** Determination of the optimum commercial size for the mangrove oyster (*C. rhizophorae*) in Todos os Santos Bay, Brazil, *Aquaculture,* 20, 1, 1980.
23. **Quayle, D. B.,** Pacific oyster culture in British Columbia, *Bull. Fish. Res. Board Can.,* 169, 193, 1969.
24. **Walne, P. R.,** The seasonal variation of meat and glycogen content of seven populations of oysters *Ostrea edulis* L. and a review of the literature, Fishery Investigations, Ministry of Agriculture, Fisheries and Food, U.K., Ser II, 1970, 26.
25. **Nascimento, I. A. and Pareira, S. A.,** Changes in the condition index for mangrove oysters (*Crassostrea rhizophorae*) from Todos os Santos Bay, Salvador, Brazil, *Aquaculture,* 20, 9, 1980.
26. **Perez-Farfante, I.,** El ostion de Cuba, Inst. Nac. Pesca Cuba, *Cent. Invest. Pesq. Contrib.,* 3, 1, 1954.
27. **Mandelli, E. F. and Acuna, A. C.,** The culture of the mussel *Perna perna* and the mangrove oyster *Crassostrea rhizophorae* in Venezuela, *Mar. Fish. Rev.,* 37, 15, 1973.

28. **Nascimento, I. A., Smith, D. H., Kern, F., II, and Pereira, S. A.,** Pathological findings in *Crassostrea rhizophorae* from Todos os Santos Bay, Bahia, Brazil, *J. Invertebr. Pathol.*, 47, 340, 1986.
29. **Menzel, R. W.,** Possibilities of molluscan cultivation in the Caribbean, *FAO (F. A. O. U.N.) Fish. Rep.*, 71, 183, 1971.
30. **Akaboshi, S.,** personal communication.

Chapter 11

CHILEAN MUSSEL CULTURE: *MYTILUS EDULIS CHILENSIS* (HUPE, 1854), *CHOROMYTILUS CHORUS* (MOLINA, 1782), *AULACOMYA ATER* (MOLINA, 1782)

Matoira H. Chanley and Paul Chanley

TABLE OF CONTENTS

I. BIOLOGY

A. MORPHOLOGY

Mytilus edulis chilensis Hupe, the common, edible, or blue mussel, is commonly called *chorito, dayes,* or *quilmahue* in Chile, and *mejillón chileno* or *mejillón del sur* in Argentina. This species is of typical mytilid shape with noticeable concentric growth lines.[1,2] The valves are covered with a black or blackish brown to purplish periostracum.[1] The umbones are pointed and slightly inclined. The dorsal shell margin is angular in the central portion, whereas the ventral margin is almost straight. The hinge has numerous small teeth.[1,2] According to Dell,[2] the taxonomic status of this subspecies requires further study.

Choromytilus chorus (Molina), commonly called *choro* or *choro zapato* in Chile, is similarly of a typical mytilid shape. It displays only concentric growth rings on the exterior valve surfaces.[1] The periostracum is coarse[13] and black or black-violet.[1,3] The umbo is pointed, but less so than in *Aulacomya ater*. The dorsal shell margin is somewhat angular along the central portion, while the ventral margin is slightly concave. There is one hinge tooth in the right valve, and two teeth are in the left valve.[1]

Aulacomya ater (Molina), the black-ribbed Mussel, commonly called *cholga* in Chile and *mejillón rayado* in Argentina, has a typical bivalve-mytilid shape. The valves display both concentric growth lines and strongly marked radial lines. The adult periostracum varies in color from brilliant black to purplish or bluish black, to polished dark brown.[1,3] The shell beneath is a purplish white streaked with purple radial lines. The shell interior is pearly white.[3] In young specimens, the color of the periostracum is a lighter or yellowish-brown. The umbones are pointed and angled anterioventrally. The dorsal shell margin is rounded and without prominent angles. The highest portion is at about the middle of the dorsal line. The ventral margin is, in general, slightly concave. The hinge possesses a single tooth that fits into a groove in the opposite valve.[1]

B. RANGE

Mytilus edulis is well known for its worldwide distribution.[4] In the area of current interest, *M. edulis chilensis* is found from Iquique (20°S)[1,2] to Isla Navarino (55°S),[3] the Strait of Magellan (53°S), and continuing up the Atlantic coast of Argentina to northern Brazil.[1] *M. chilensis* is found intertidally to a depth of 25 m.[2]

Choromytilus chorus has been reported from Pacasmayo, Peru (7°S)[1] to Tierra del Fuego (53 to 54°S) and Canal Beagle[2] and the Strait of Magellan,[1] continuing up the Atlantic coast of South America to the south of Brazil, including the Malvines (Falkland Islands).[1] It occurs intertidally to a depth of 20 m.[6]

Aulacomya ater has a wide range of distribution from Callao, Peru (12°S),[1-4] down the Pacific coast of South America, around the islands at the tip of the continent,[1,2,5] and up the Atlantic coast to southern Brazil,[1] including the Juan Fernandez archipelago[1,5] and the Malvines.[1,3,5] According to Olsson,[3] however, *A. ater* is plentiful only from Paracas, Peru (14°S) southward. The species is found intertidally to a depth of 40 m[2], but is mainly a lower littoral form. The related *A. ater maoriana* from New Zealand may be the same species.

C. SALINITY AND TEMPERATURE TOLERANCES

Except for the widely distributed *M. edulis*, salinity and temperature tolerance data, based on experimental evidence, for the Chilean mussels are unknown to the authors. However, based on the distributional pattern, it seems likely that these mussels can survive at least over the range of 10 to 18°C, and very likely in warmer conditions for those species extending into Peru and Brazil. Dell[2] has quoted some temperature data indicating a range of 10 to 13°C from the Strait of Magellan to Chiloé (42 to 43°S). He says that the Humboldt current impinges on the coast near Chiloé and then northward throughout the length of the coastline, maintaining a temperature of 18 to 21°C from Chiloé to Point Aguja, Peru (6°S). Data from other sources, which he also quotes, indicate surface temperatures of 12 to 15°C from Talcahuano (36°40′S) to Coquimbo (30°S) and 13 to 19°C from Coquimbo to Arica (18°30′S).[2] *M. edulis* reportedly dies when water temperatures are 26°C and above.[7]

Salinity data are not presented by Dell. However, many of the locations where the Chilean mussels are found are along the rocky coast with no map evidence of significant freshwater influence. Of the three species covered here, only *M. chilensis* has been reported by Dell[2] from locations at or near the mouths of rivers. In those areas with freshwater runoff, the quantity and resultant salinity drop are unknown to the authors. The minimum salinity for *M. edulis* growth and survival is over 15‰,[8] and they have been found at 40‰.[9] *C. chorus* reportedly grows better on the bottom, where salinity is high, than suspended in low salinity culture systems. Normally, growth is about twice as rapid in suspended mussels.[5]

D. TYPE OF SUBSTRATE

A. ater ater and *M. edulis chilensis* were the only two mussel species reported by Dell from the 1958 to 1959 Royal Society of London's expedition to Southern Chile.[2] They surveyed 52 stations between 42 and 55°S, which he reported. He included the type of habitat at each station as well as its location and the molluscan species collected. *M. edulis chilensis* was usually found attached to intertidal or sublittoral rocks (granite, gneiss, argillite, schist, basalt), on shingle or boulder beaches, or on the fronds and/or holdfasts of *Macrocystis*. Collection sites where *M. edulis chilensis* was found ranged from the very sheltered to the very exposed. A few of the stations were at or near the mouth of a river.[3]

Where found, usually in semiexposed locations, *A. ater* was attached to various types of intertidal rock (schist, gneiss, basalt, argillite, or granite), the fronds and holdfasts of *Macrocystis*, or the shells of other bivalves including *M. edulis chilensis*. Dredge stations from which he reported *A. ater* had stone and/or shell over a sandy mud or sandy substrate.[2]

E. SEXUAL DEVELOPMENT AND SPAWNING

Sexes of the Chilean mussels are normally separate. Gonads extend into the mantle, giving it the color of the gametes. At spawning, gametes are broadcast into the water, where fertilization and larval development occur.

M. edulis chilensis males have a creamy yellow to white mantle and the females one of creamy orange. Occasionally and exceptionally, functionally hermaphroditic specimens have been observed. In these cases, both mantle colorations are present, but in separate areas. At Ensenada de Nercón, spawning occurs between September and February, and there is a direct relationship between sexual maturity and temperature. Spawning begins when the temperature rises above 11°C. At Putemún (Chiloé), partial and successive spawning appears to occur throughout the year, and the greatest larval capture takes place during the fall months.[1]

Sexual maturity of *C. chorus* appears to occur between the 11th and 15th month and coincides with a marked slowdown in rate of growth.[1] The spawning season varies, depending on environmental factors. In Quele River in southern Chile, a major spawning season is in late spring, October to November. A minor season occurs in January to February. Further south, at Putemún, the spring season is longer, lasting from September through January.[6] Optimum collection of juveniles on artificial collectors is obtained during the summer and autumn months at the government mussel culture stations at Putemún and Talcan (Chiloé). Generally, commercial sets of appropriate magnitude are scarce and insufficient to supply demand.[1,6]

In *A. ater* there is marked internal sexual dimorphism. The males display a whitish-yellow mantle, while that of the females is light brown with darker spots. The spawning season, at Putemún, Estero de Castro (Chiloé, about 43°S), runs from the end of November to the end of February at water temperatures of 18 to 19°C. The smallest animal reported to spawn there was 65 mm long. At Bahia de Mejillones (near Antofagasta, about 23.5°S), animals are reportedly sexually mature in December and January. In the region of the Strait of Magellan (about 53°S), the period of reproduction begins with gonadal development in June and extends to February, when maximum spawning is observed.[1]

F. LARVAL DEVELOPMENT

The above-referenced three species of mytilids were raised by the authors and staff at the hatchery developed by Fundación, Chile at the Universidad del Norte marine center on Bahia Herradura, Coquimbo. However, to the best of our knowledge, the data have never been analyzed, written up, or published. Larvae of all species are similar to those of *M. edulis* described by Redfearn et al. in New Zealand.[10] The 63- to 71-μm eggs develop to 86- to 90-μm straight-hinge larvae in about 1 d. The hinge line is proportionately longer than in most bivalve veligers, and the round to knobby umbo develops late. The anterior end is much more pointed than the posterior in larger larvae. Metamorphosis typically occurs after about 2 weeks, when larvae are 260 to 350 μm in length.[10]

G. GROWTH

The minimum commercial size (5 cm in length) for *M. edulis chilensis* has been realized within 11 to 15 months at Putemún on collectors and rafts. The maximum reported size for the subspecies is 10.6 cm.[1]

C. chorus (choro zapato) is the largest mussel along the Chilean coast (*zapato* means shoe and describes a size once common) and is generally regarded as the best of the edible shellfish.[3] A maximum size of 26 cm has been reported.[6] This is not, however, a fast-growing species. It takes 2 years to reach a 6-cm size at Putemún, Chiloé from rafts. Growth is usually more rapid from rafts and long-lines than on natural banks. Growth slows between the 11th and 15th months of age, when spawning begins.[1]

A. ater grows to its largest size in the Magellanic region where a maximum size of 15.6 cm has been reported. At Mejillones, growth is rapid during the first 2 years, with the mussel gaining 25 to 60 mm; then growth slows. A maximum size of 9.8 cm has been reported from this location. In Isletilla (Estero de Castro, Chiloé), *cholga* farmed from long-line or raft reach 7 cm, the minimum legal harvest size, in 2 years.[1]

II. FISHERY

A. HISTORY

According to Osorio,[1] the mussel species harvested today in Chile are the same ones consumed by the coastal peoples of the region for over 6000 years. Methods for harvesting wild mussels have changed little. The fishery is largely artisanal, and fishermen frequently sell their catches locally. Consequently, production figures are low; a major portion of the legal and illegal catch is consumed without being reported.

B. IMPORTANCE

Castilla[5] has reported that the 1960 harvest of *M. chilensis* amounted to 2950 ton with a value of $189,000 in U.S. dollars. In the 1972—1973 season, more than 9000 ton was harvested, and the 1976 production was reported by Osorio[1] at 12,208 ton. This has been the record harvest, with production ranging between 5000 and 10,000 t through 1981.[5] Principal banks have been reported from Mehuin, Puerto Montt, Ancud, Puerto Aisen, and Punta Arenas.[1]

Mussel farming began in 1975, and in 1981, this species was cultivated in at least 50 sites, producing 582 t, or about 7.5% of the total harvest of this species. Aquaculture of this mussel accounted for 77.9% of the 747 t produced by all molluscan aquaculture in Chile in 1981.[6]

The 1938 harvest of *C. chorus* was 1739 ton,[1] and then in 1940, more than 3000 ton of this popular species was harvested.[5] By 1950, however, only 500 ton was found and harvested,[1] and in 1961, fishery extraction was prohibited.[5] Because of intense exploitation of this extremely popular product, many of the best-known banks, in and around Ancud, the only historic major center of the fishery, were completely exterminated.[1] In 1975, production was again resumed on a very small scale. This gradually increased to 110 t in 1981. Of this, 31.4 t was produced by aquaculture.[6]

An intense fishery for *A. ater* rose from something more than 7000 t harvested in 1960 to a peak of 23,564 ton in 1967. Harvests then decreased, and by 1972 totaled 15,376 ton, with a total crop value of $3,500,000.[5] The 1976 crop weighed in at 16,250 ton,[1] but since then the annual yield has been at about 8000 ton. This species usually accounts for the majority of the Chilean mussel harvest, but most of the harvest is from the fishery. Aquaculture production is low, amounting to only about 50 t in 1981, because natural banks of this species are still very productive in the south.[6]

Commercial fisheries for *A. Ater* have been located at one time or another at Antofagasta, Caldera, Puerto Montt, Ancud, Puerto Aisen, and Punta Arenas. During the 1950—1960 decade, the principal banks were located in the Gulf of Ancud (Chiloé). However, because of intense fishing pressure and exploitation, the major producing population has been gradually moving south. By 1969, the principal banks were located at Puerto Eden (49°S).[1]

C. RECRUITMENT AND METHODS OF CULTURE

Various artificial substrates from fiber or plastic rope and netting to tree branches and shells of other bivalves have been used successfully as collectors. Mussel seed collectors are placed in areas where banks of mussels are protected and maintained for spawning.

Collection of *M. edulis chilensis* seed, at Putemún and Talcan, was best at a depth of 0 to 2 m.[1] Sections of anchovy net 0.25 $\times$ 8.0 m can catch up to 2.2 million seed per square meter, but there is much variability in intensity of set from time to time and site to site.[6] During both 1972 and 1973, in April and in June, at Putemún a problem occurred in which the seed detached from the ropes.

Mussels can be grown right on the collectors, but more often are removed from them and restrung, according to either the "French" or the "Spanish" method, for further raft or long-line culture.[5] In the Spanish method, a cord is stretched out between stands, and little plastic pegs are inserted every 25 cm. The seed is scooped up from a mobile bin with the left hand while the right hand whips a band of rayon mesh around it to hold the seed in place. The operator moves the bin backward down the line as he attaches the seed. By the time the rayon disintegrates, in about a week, the seed has attached to the central cord and resumed its growth. The pegs spaced along the cord help distribute the weight of the column of growing mussels more evenly on the growing line. This method appears to be more popular than the French method, which does not use a central string. The seed mussels are placed inside a cotton or nylon mesh tube for continued growth. The mussels attach to and grow through the mesh which forms the growing line. The restrung mussels may be suspended from rafts (most popular for *choritos*) or from long-lines.[5] Rafts are commonly 10 $\times$ 10 m or 16 $\times$ 16 m and hold 300 to 500 culture ropes 3 to 12 m long, but typically 8 m in length. Production is usually 12 to 20 ton per year per raft. Age at harvest is usually 8 to 14 months.[6]

The demise of the popular *choro zapato* fishery led to governmental conservation measures as early as 1943, when the Ministry of Agriculture established its experimental mussel culture station at Putemún (Chiloé). Later, in 1965 and 1971, additional governmental agencies established more culture and experimental facilities in the area. These studies have shown, in part, that the best capture of seed of this species (*C. chorus*) on artificial collectors occurs between 0.5 and 3 m above the bottom and takes place between summer and autumn of each year. The poorest collectors, apparently, have been *M. edulis chilensis* shells.[1] Scarcity of seed severely limits farming production. This species frequently drops from growing lines when 5 cm in length, and consequently larger sizes are often cultured in trays or bags. *Choro* may be 3 to 6 years of age at harvest.[6]

Cholga seed are generally collected on branches in the Castro (Chiloé) area and later reattached to strings for further suspended culture from rafts or long-lines. In addition to various locations around Chiloé, cholga are now also farmed at Bahia de Mejillones (Antofagasta) in northern Chile.[1] Farmed *cholga* usually reach a marketable size of 70 mm in 1 or 2 years, depending on water temperature. On the natural bottom, 5 years may be required to reach a comparable size.

D. METHODS OF HARVESTING

Even at a commercial fishery, harvest has been by hand or hand-held tools. The most popular tool for mussel harvest (all species) was the "pangareta," a semicircular metal frame with a long handle attached to the middle of the upper arched half and teeth along the bottom straight diameter. A trailing collecting bag attached to the frame completed the tool.[1] With raft or long-line culture the strings are pulled up by hand, and the mussels are hand-stripped from their growing lines.

E. USES AND PROCESSING

All mussel species, when available, are widely marketed domestically live in the shell for both home and restaurant consumption. They are an important ingredient in the popular "sopa marinera," a thick soup featuring a variety of seafoods. *Cholgas* and *choritos* are also preserved by pickling, semidehydration, and smoking.[1] They make a very popular smoked mussel product.

F. CONSTRAINTS

The major constraint for mussel farming at the present time appears to be a lack of seed (except for *M. edulis*), which limits farmers' production and restricts the natural fishery. Lack of sufficient capital hampers increase of both seed production and farming efforts.

Metal ion pollution most probably precludes production from many otherwise potentially suitable bays in northern Chile. However, with few exceptions, these areas are few and far between and have not figured importantly in the historic fishery. Southern Chile has large areas that conceivably could be used for mussel farming.

G. MANAGEMENT

Minimum sizes and closed seasons have been established for the three important commercially harvested mussel species.

Because of intense harvesting pressure, a closed season for *M. edulis chilensis* was established on August 9, 1966 to run from August 15 to October 30 annually, with a minimum harvest size of 50 mm.[1]

Harvest of *choro zapato (C. chorus)* from Llanquihue Province south was prohibited indefinitely on August 24, 1971, at which time a 3-year harvesting ban was imposed on the rest of the country. In the same year, a minimum size of 120 mm was established for this species.[1] Harvesting was resumed in 1975.[6]

The minimum harvest size for *cholga (A. ater)* was established at 70 mm[5] in October 1969. When cultured from long lines or rafts, the minimum legal harvest size is commonly reached within 2 years. A partial closed season is observed annually from October 1 to January 31.[1]

FIGURE 1. Mussel-growing lines supported from mussel culture boat, Universidad de Concepcion, Dichato, Coliumo, Chile.

III. RESEARCH AND DEVELOPMENT

A. BIOLOGICAL TECHNIQUES

The government and university experimental stations have made a good beginning.[1] In 1943, the Ministry of Agriculture established its experimental mussel culture station at Putemún (Chiloé) in response to the need for conservation of *choro zapato* (*C. chorus*). Later, in 1965, the Instituto de Fomento Pesquero established complementary raft culture activities at Tubildad and Talcan (Chiloé). *Cholga (A. ater)* and *chorito (M. edulis chilensis)* culture programs were added at the Putemún station. Government activities have concentrated largely on cultivation technology, seed production, and subsidies to mussel farmers.

Among the university activities, the university at Concepcion has figured prominently in raft culture studies (Figure 1), the University of the North in hatchery production of many species (Figures 2 and 3), and the University of Chile in studying their basic biology and life histories at the Montemar Laboratory.

B. FUTURE DEVELOPMENT

The basic biology of these animals appears to be somewhat less well understood than many of the practical aspects of seed collection and culture. Continued experiments in selection of better materials and configurations for both collectors and growing ropes appear appropriate.

More sophisticated government and university hatcheries to deal with these and other molluscan species would appear to be a good long-term investment for this country, with its long coastline, spanning over 36° of latitude.

Farming activities in northern Chile will probably be limited because of the lack of suitable protected areas. Expansion should be possible in southern Chile, but markets are distant.

FIGURE 2. Fiberglass algal culture tanks at first shellfish hatchery in Chile at Universidad del Norte, Coquimbo. Identical tanks were used for larval culture.

FIGURE 3. Fiberglass nursery trays for culture of recently set mussels and other shellfish species; Universidad del Norte, Coquimbo.

REFERENCES

1. **Osorio R., C.,** Moluscos marinos de importancia economica en Chile, *Biol. Pesq. Chile,* 11, 3, 1979.
2. **Dell, R. K.,** The marine mollusca of the Royal Society expedition to southern Chile, 1958-59, *Rec. Dom. Mus. Wellington,* 7, 155, 1971.
3. **Olsson, A. A.,** *Mollusks of the Tropical Eastern Pacific. Panamic-Pacific Pelecypoda,* Paleontological Research Institution, Ithaca, 1961, 574.
4. **Abbott, R. T. and Dance, S. P.,** *Compendium of Seashells,* E. P. Dutton, New York, 1983, 410.
5. **Castilla, J. C. Z., Santelices, G. G., and Becerra, H. R.,** Guia para observacion e identification de mariscos y algas de Chile, *Expedicion a Chile,* Editora Nacional Gabriela Mistral, Santiago, 1976.
6. **Winter, J., Toto, J., Navaroo, J., Valenzuela, G., and Chapparo, R.,** Recent developments, status and prospects of molluscan aquaculture on the Pacific coast of South America, in *Recent Innovations in Cultivation of Pacific Mollusca. Developments in Aquaculture and Fisheries Science,* Morse, D., Chew, K., and Mann, R., Eds., 14, Aquaculture 39:95, Elsevier, Amsterdam, 1984.
7. **Wells, H. W. and Gray, I. E.,** Seasonal occurrence of *Mytilus edulis* on the Carolina coast as a result of transport around Cape Hatteras, *Biol. Bull.,* 119, 550, 1960.
8. **Castagna, M. and Chanley, P.,** Salinity tolerance of some marine bivalves from inshore and estuarine environments in Virginia waters on the western mid-Atlantic coast, *Malacologia,* 12, 47, 1973.
9. **Motwani, M. P.,** Experimental and ecological studies on the adaptation of *Mytilus edulis* L. to salinity fluctuations, *Proc. Natl. Inst. Sci. Biol. Sci.,* 21, 227, 1955.
10. **Redfearn, P., Chanley, P., and Chanley, M.,** Larval development of four species of New Zealand mussels: (Bivalvia, Mytilacea), *N. Z. J. Mar. Freshwater Res.,* 20, 157, 1986.

CHAPTER 12

CULTIVATION OF THE CHILEAN OYSTER, *TIOSTREA CHILENSIS* (PHILLIPI, 1845)

Matoira H. Chanley and Paul Chanley

TABLE OF CONTENTS

I. BIOLOGY

A. MORPHOLOGY

The shell of *Tiostrea chilensis* (Philippi, 1845) (family Ostreidae; common names: Chilean oyster, *ostra*) is oval or subcircular in outline, and the lower, or left, valve is cupped and attached to a substrate. The smaller right valve is nearly flat. Maximum size is about 8.7 cm. Irregular lamellae are found on the surface of the upper valve, especially near the margin. External shell color is gray to grayish-white. Internally the shell is usually white, but sometimes greenish. The pallial line is not conspicuous and disappears over the adductor muscle scar.[1] The Chilean oyster is nearly identical with the typical New Zealand *lutaria* ecomorph and is now considered to be the same species as the New Zealand dredge oyster, *Tiostrea lutaria,*[2] illustrated in Figure 1 of Chapter 36.

B. RANGE

The range of this species is usually listed as occurring from the Island of Chiloé in southern Chile (latitude 42°S) to the coast of Ecuador (latitude 3.5°S).[1,3] The main oyster beds, however, are confined to the Chiloé region. The species is scarce in central and northern Chile,[3] but experimental cultivation has been practiced in Tongoy, Coquimbo (latitude 30°S), Mejillones, Iquique (latitude 20°S), and probably other locations in northern Chile, using broodstock imported from Chiloé. The species is found intertidally to a depth of 11 m.[1]

C. SALINITY AND TEMPERATURE TOLERANCES

Salinity and temperature tolerances for the Chilean oyster have not been determined experimentally. The oyster is found naturally in areas of high, almost oceanic, salinity and is known to survive over a temperature range of at least 7 to 20°C. Larvae are liberated at water temperatures of 10.7 to 14.5°C.[1] Walne[4] reported that the Chilean oysters with which he was working in Wales during 1962-1963 did not survive the winter outdoors and could not withstand quite moderate frost.

D. SUBSTRATE

T. chilensis lives naturally attached to rocks or shells on a stony, sand, or hard mud bottom.[1,3] At times the oysters grow attached to each other, forming irregular clusters.[3] When cultivated, the spat are usually attached to other bivalve shells, particularly the shells of the large mussel "cholga" *Aulacomya ater* (Molina).

E. SEXUAL DEVELOPMENT AND SPAWNING

Chilean oysters are protandric, alternate hermaphrodites.[1] Sexual development is believed to be similar to that described for the New Zealand dredge oyster in Chapter 36. The smallest oysters found with sperm were 27 mm in length and were slightly more than 1 year old. The smallest oysters incubating larvae had shell lengths of 37 mm and were slightly more than 2 years old.[5] Spawning begins in September or October, when water temperature is 10.7 to 13.3°C, and continues through December (spring). This constitutes the major spawning season. There is a second, minor, spawning in March (fall).[1]

After spawning, eggs (264 to 323 μm) are fertilized and held within the mantle where fertilization occurs. Approximately 70,000 eggs are held by a large incubating adult.[1]

F. LARVAL DEVELOPMENT

Larval development is completed within the mantle cavity. Larvae are 416 to 514 μm in length when released,[6] after about 50 d of incubation.[5] They are roughly oval in shape, with the hinge area generally flatter than the rest of the shell margin. Most are capable of attaching to a substrate immediately after release, although some may not attach for 2 to 3 d. The minimum observed size at attachment was 426 μm. In Conwy, Walne[4] reported recently released larvae from imported adults, ranging in length from 390 to 520 μm. *Tiostrea* larvae differ from all other oyster larvae in having equal or nearly equal values and in lacking the posterior dorsal sulcus, umbones, and hinge teeth of other oyster larvae. They also grow to a size much larger than other oyster larvae before metamorphosis.[6]

G. GROWTH

This species is slow growing, increasing in size about 15 mm/year for the first 3 years and subsequently only about 10 mm/year. The minimum legal marketable size, with a 6-cm diameter, usually is reached during the fifth year of growth. However, growth rate has been increased 50 to 60% with off-bottom culture.[1] Nonetheless, 4 or 5 years typically is required to produce a marketable product. A recent publication[5] reports the harvesting of 5-cm-diameter oysters from long-lines after only 3 years, suggesting that the minimum size regulation has been changed or does not apply to cultured oysters.

II. FISHERY

A. HISTORY AND IMPORTANCE

Chilean oysters have been harvested from littoral areas for human consumption for at least 6000 years.[1] The fishery historically has been centered in the Chiloé region. Oysters generally were collected by hand from rocks exposed at low tide or by divers from submerged banks.[3] Production dropped to virtually nothing, from an era of reportedly high production during the 1950s, following a particularly destructive earthquake in 1960. Overfishing has also been cited as a factor contributory to the demise of the oyster fishery.[1] In 1963, only 50 t were harvested.[1] Recovery was slow through the 1970s, but oyster farming began in 1975 and production increased to 500 t from 1979 to 1981. At this time, oysters were being farmed in over 30 sites and about 1/4, or 120 t, of the total harvest was composed of farmed oysters.[5]

B. RECRUITMENT AND CULTURE

The very brief pelagic stage of this species greatly facilitates collection of oyster juveniles for cultivation. Adult oysters are collected and maintained in trays in a bay, usually at Pullinque, where water temperatures are slightly warmer than in the surrounding area. *Cholga* or other bivalve shells, strung on wire (Figure 1), are hung around the trays of adults to collect the spat. Collectors are monitored regularly and replaced when sufficient oysters have attached.

When oysters are about 2 cm in diameter, the shells are then restrung onto 2- to 3-m growing lines (Figure 2), allowing space for growth, and attached to long-lines, rafts, or staked-lines to continue growth.

Long-lines are the most popular, either as a single line or a pair of mother lines attached between a series of floats (Figure 3). Growing lines hang down from the mother lines. Floats may be metal drums or a variety of plastic materials that are available. Rafael Coronel, on Isla Chiloé, uses rectangular styrofoam blocks. These are wrapped tightly with strips of old tires, and it is to the tires that the long-lines are attached. They are regularly inspected and maintained. Anchorage at both ends may be to land across the bay or to cement on the bottom. Long-lines are most commonly used in the deep, narrow, fjord-like bays of southern Chile.

Rafts are typically about 16 × 16 m with 600 to 1000 strings of seed. String length may be up to 8 m, varying with water depth. Raft flotation may be provided by wood frames, metal drums, or even logs.[3] In southern Chile, rafts are generally wooden, but in

FIGURE 1. Strings of *cholga* shells with attached oyster spat ready to be restrung onto growing wires.

northern Chile, where wood is scarce, mariculture rafts may be made from a variety of materials. One fishing company, for example, ties the floats from worn-out fishing nets into a raft configuration and uses them to support the growing lines. Such a flexible "raft" will not support additional weight and must be serviced from a work boat. Staked-line culture is used in shallow-water areas, less than 4 m deep. Stout stakes are driven into the bottom in parallel rows, and polyethylene rope is attached to stake, forming a grid parallel to the bottom. The growing lines are hung from the polyethylene rope grid in a manner similar to those hung from long-line mother lines.[3]

C. HARVESTING AND PROCESSING

Cultured oysters are harvested manually by removing growing lines from the culture system and culling the oysters. However, a hand scrape or *pangareta* resembling an oyster dredge with a rigid handle, may be used in some shallow-water areas. In deeper water, a small dredge, without teeth, is sometimes towed from a boat powered by either motor or oars and hauled by hand.[1]

Culling and packing the cleaned shellstock are the only processing procedures. Chilean oysters are always eaten raw. They are opened at the time and place of consumption. (Interestingly, this species in New Zealand is almost always a fried product, the staple oyster of the popular fish and chips "take aways".)

D. CONSTRAINTS

The Chilean coast is longer than the combined length of the Atlantic, Gulf, and Pacific

FIGURE 2. String of *cholga* shells with attached oyster spat, restrung onto growing wires and ready to hang from long-line or raft.

coasts (excluding Alaska and Hawaii) of the U.S. However, there are few harbors, bays, or extensive estuarine systems in the warmer waters of northern Chile. Most of these are heavily polluted, largely as a result of mining activities, and are unsuitable for mariculture. Furthermore, the bottom drops quickly from shore to oceanic depths. Although this configuration helps produce the famous upwelling that provides nutrients for major fisheries, it leaves surprisingly little area for molluscan fisheries or mariculture. Nevertheless, oysters, scallops, and several species of mussels are being farmed, especially in, or near, Mejillones, Tongoy, and Coquimbo.

In southern Chile, bays and estuaries are abundant, but water temperatures are colder, and the distance to significant markets can be considerable.

Another problem is the low growth rate of the Chilean oyster. It is difficult to farm profitably any crop that requires 4 to 5 years of growth before harvest. The Pacific oyster, *Crassostrea gigas* (Thunberg), has been introduced into Chile and cultured experimentally, mainly in the north-central part of the country. It grows much more rapidly than the Chilean oyster. However, it is generally considered inferior as a raw oyster, which is the only existing domestic market. It is unlikely that Chilean production of the Pacific oyster could compete in export markets with other sources of supply.

E. MANAGEMENT, RESEARCH, AND DEVELOPMENT

Chilean oysters may not be harvested until they are a minimum of 6 cm in diameter, except as noted previously.[5]

Government-operated experimental stations at Pullinque, Hueihue, Butachaques, Pumalin, and the islands of Teresa and Apiao have worked with oysters. These stations, especially

FIGURE 3. Long-lines at Ancud on Chiloé showing variety of flotation materials used.

at Pullinque, produce seed oysters for sale to oyster farmers and to "increase the natural banks," thereby increasing the commercial production of the species.[1] There are also several university experimental stations that have worked with oysters.

The technique of refrigerating eyed *Crassostrea* oyster larvae and shipping them to convenient locations, where they are placed with cultch in setting tanks, has recently been used successfully with Chilean oysters at Quempillēn, Isla Chiloé. A lower percentage of larvae are reportedly lost than when setting occurs in bays, as described previously.[5]

F. FUTURE DEVELOPMENT

Except for their slow growth, *Tiostrea* oysters are well suited for mariculture. Because larvae are incubated throughout development, the adults serve as a hatchery, thus obviating the expense of hatchery production. Juveniles can be collected by abundant cultch placed around trays of selected (fast-growing, disease-resistant, etc.) adults. Alternatively, selected adults can be maintained in troughs or tanks and the larvae collected twice a day from a screen placed over the overflow. Collected larvae can be permitted to attach to a cultch or can be raised cultch-free.

Decreasing the time to harvest would greatly improve the desirability of this species for mariculture. It seems likely that harvest time could be at least halved through selective breeding and manipulation of culture conditions. Use of off-bottom culture, protection from predation, and culture in areas with higher water temperatures should be particularly productive. Selective breeding could be accomplished readily by maintenance populations of fast-growing (or otherwise desirable) oysters maintained in breeding troughs or trays to produce larvae.

It is interesting to note that in New Zealand, where there are three species of commercial oysters, this species is the only one that has not been farmed. In Chile, except for the

FIGURE 4. String of harvested oysters ready to be removed from the growing line, culled, cleaned, and sold.

experimental introduction of the Pacific oyster, this was the only species of commercial oyster, and it is farmed.

If this species can be reared to a marketable size in a shorter period of time, it could become important on world markets. Presently, production could be increased considerably for domestic markets. The local demand for Chilean oysters exceeds the supply.

REFERENCES

1. **Osorio, R. C.,** Moluscos marinos de importancia econõmica en Chile, *Biol. Pesq. Chile,* 11, 3, 1979.
2. **Buroker, N. E., Chanley, P., Cranfield, H. J., and Dinamani, P.,** Systematic status of two oyster populations of the genus *Tiostrea* from New Zealand and Chile, *Mar. Biol.,* 77, 191, 1983.
3. **Castilla, J. C. Z., Santellices, G. B., and Becerra, H. R.,** Guia para la observacion e identification de mariscos y algas commerciales de Chile, *Expedicion de Chile,* Editora Nacional Bagriela Mistral, Santiago, 1976, 120.
4. **Walne, P. R.,** *Culture of Bivalve Molluscs: 50 Years Experience at Conwy,* Fishing News Books, Farnham, Surrey, England, 1974, 189.
5. **Winter, J., Toro, J., Navarro, J., Valenzuela, G., and Chaprro, O.,** Recent developments, status, and prospects of molluscan aquaculture on the Pacific coast of South America, in *Recent Innovations in Cultivation of Pacific Molluscs. Developments in Aquaculture and Fisheries Science,* Morse, D., Chew, K., and Mann, R., Eds., 14, Elsevier, Amsterdam; *Aquaculture,* 39, 95, 1984.
6. **Solis, U. I.,** Observaciones biolõgicas en ostras (*Ostrea chilensis* Philippi) en Pullinque, *Biol. Pesq. Chile,* 2, 51, 1967.
7. **Chanley, P. and Dinamani, P.,** Comparative descriptions of some oyster larvae from New Zealand and Chile, and a description of a new genus of oyster, *Tiostrea, N. Z. J. Mar. Freshwater Res.,* 14, 103, 1980.

Chapter 13

OYSTER CULTURE IN EUROPEAN COUNTRIES

M. Héral and J. M. Deslous-Paoli

TABLE OF CONTENTS

I. INTRODUCTION

Three species of oysters have been or are cultivated in Europe: *Crassostrea angulata, C. gigas,* and *Ostrea edulis.* If we follow the taxonomic critera of Grassé,[1] the European oysters belong to the group of Mollusca, class of Lamellibranchia or Bivalves, order of Filibranchia, family Ostreidae with two genera: *C. angulata* (Lamarck), Portuguese, cupped oyster; *C. gigas,* (Thunberg), Japanese or Pacific, cupped oyster; *Ostrea edulis* (Linné), flat European oyster.

Some authors think that *C. gigas* and *C. angulata* belong to the same species. Indeed, Ranson[2,3] stated that the characteristics of the larvae are the same, and Menzel[4] obtained viable hybrids (F_2) between these two oysters. Also, Buroker et al.,[5] in studying the genetic variations of proteins and enzymes of the flesh, revealed a genetic similarity of 99% between *C. gigas* and *C. angulata* on 24 loci. These authors proposed the hypothesis that Japanese oysters would have been imported from Japan to Portugal by boats in the 16th Century or vice versa. However, these two oysters obviously show different characteristics in the metabolic rate,[6] the filtration rate,[7] growth performance,[6-9] reproduction mode,[9] and resistance to disease.[10] All these latter elements combine to affirm that the Portuguese and the Japanese oysters are two species with well-defined physiological characteristics, particularly with regard to oyster culture.

II. BIOLOGY

A. MORPHOLOGY

The cupped oysters *(Crassostrea)* have a shell longer than it is broad with chalky deposits divided into sheets. The left valve is cupped (Figure 1). The right valve for the *gigas* oyster is flat, more irregular, and can sometimes be curled or scaly. The flat oyster *(Ostrea)* has a rounded or oval shell, cemented on the left valve, which is more curved (Figure 2). In both oysters, the flesh is covered by a tegument, the mantle (Figures 1 and 2). The free space between the two lobes of the mantle is the pallial cavity, which the gills divide into an inhalant and an exhalant part. The mouth, surrounded by the labial palps, is near the hinge, whereas the anus is just above the adductor muscle (Figures 1 and 2).

B. RANGE

Before diseases, which comprised the range of the flat species particularly on the Atlantic coast, *O. edulis,* the native oyster, was abundant in several European countries: from Norway, Denmark, the Netherlands, Germany, Belgium, France, Great Britain, and Ireland in the north, to the south Atlantic coast in France, Spain, Portugal, and Morocco, and the Mediterranean and Adriatic coasts of France, Italy, Yugoslavia, and Greece. It is or was cultivated mainly in France, the Netherlands, Spain, and to a lesser degree in the U.K. and Ireland.

C. angulata is still present in Portugal and the south of Spain. Introduced in France in 1866, this oyster was cultivated largely in this country until 1966 to 1970, when two diseases stopped production. This species seems to be still present on the Adriatic coast of Italy.

C. gigas has taken the place of the Portuguese oyster in European culture since 1970. It was cultivated first in France with adults imported from British Columbia (Canada) and spat from Japan. Nowadays, the only cultivated cupped oysters produced in European countries are *C. gigas.* There is a natural breeding ground only along the southwest part of the French Atlantic coast. This species is beginning to be cultivated in several other countries, such as Great Britain, Ireland, Denmark, Germany, Norway, Spain, and Italy, that have hatcheries to produce seed.

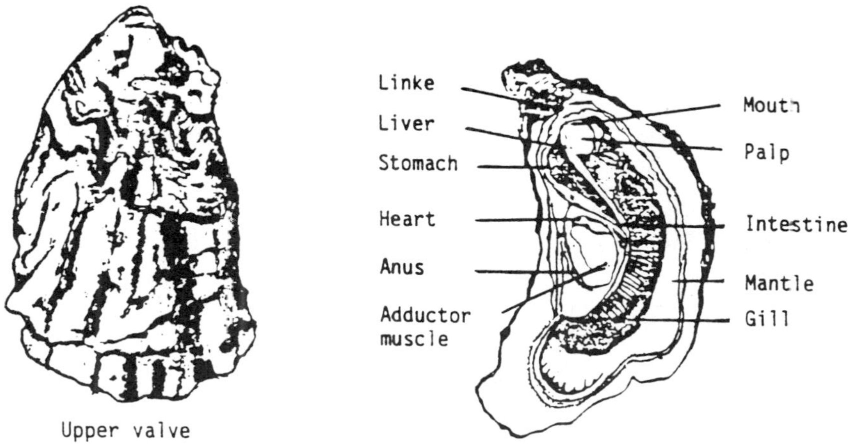

FIGURE 1. Anatomy of the cupped oyster, *Crassostrea gigas*.

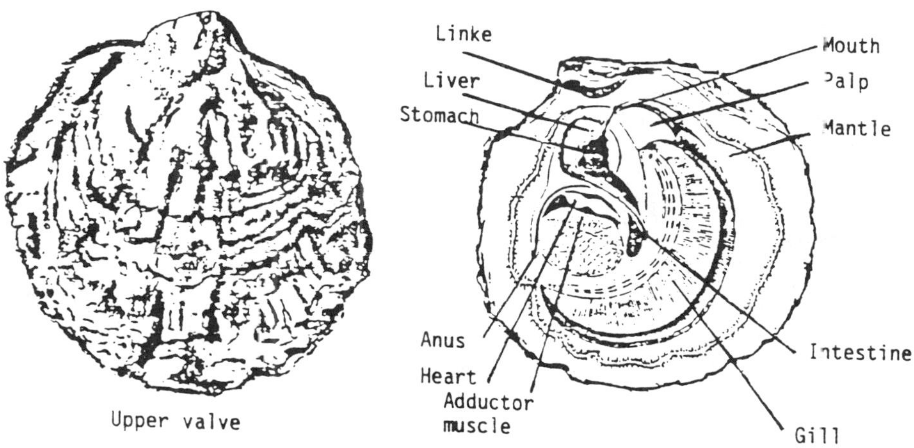

FIGURE 2. Anatomy of the flat oyster, *Ostrea edulis*.

C. SALINITY AND TEMPERATURE TOLERANCES

1. Larvae

On the Atlantic coast, the reproduction of *C. gigas* occurs in July and August in European countries. The temperature must be above 18°C with an optimum at 21 to 22°C, and the salinity can vary between 25 and 35‰. The higher the salinity is, the higher the temperature needs to be. Larval production is compromised when temperature is under 18°C and salinity above 34‰ (Figure 3).

In the lagoons on the Mediterranean coast, when in summer the temperature reaches 24 to 25°C and the salinity 37 to 38‰, *C. gigas* does not reproduce; nor does it reproduce in the open Mediterranean Sea where salinity is nearly 37‰ and temperature in the range of 19 to 22°C, although in the Adriatic Sea, where the salinity is lower, reproduction occurs. *O. edulis* spawns in July, with larvae incubated in the mantle cavity. The salinity should be in the range of 25 to 36‰,[11] but otherwise does not interact with the temperature, which can vary from 14 to 22°C; there is no larval development under 14°C.[9] *O. edulis* reproduces regularly in the Mediterranean gulfs, which permits a new oyster culture in open waters.

2. Juveniles and Adults

Temperature and salinity conditions determine species distribution not only for breeding,

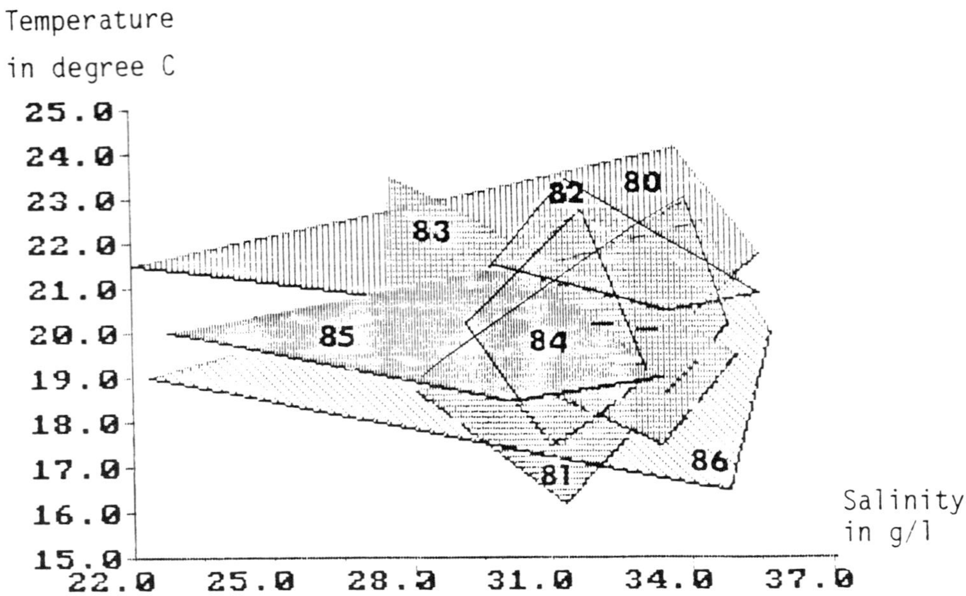

FIGURE 3. Temperature, salinity diagram of the bay of Marénnes-Oléron during the evolution of the larvae of the Japanese oyster (1980 to 1986). (From Héral, M., *Aquaculture*, Lavoisier, Paris, 1986, 346. With permission.)

but also for growing. *C. gigas* is a very euryhaline species that can be cultivated not only in oceanic waters, but also in upper estuaries with a mean salinity of 15‰ with variations between winter- and summertime of 2 to 25‰ (the Gironde estuary on the Charente Riber, for example). It can even be cultivated in salt marsh ponds or in "claires" in salinities that can reach 45 to 50‰. Above 50‰ of salinity, mortalities are observed. On the other hand, for the Japanese oyster, growth and fattening is affected by a mean salinity lower than 15‰. The oyster can persist for several weeks in wintertime when rivers are in spate, even in freshwater, by closing its valves. On a tidal cycle, it has an active metabolism during high tide and stops its activity at low tide with the freshwater input. For *O. edulis*, the optimal salinity is between 32 and 37‰.[9] At 25‰, the species grows well. Under 20‰ and above 40‰, it becomes feeble.

As temperature is an important parameter which controls all phenomena in mollusk physiology—filtering activity, metabolism, and, thus, respiration and excretion—Héral et al.[12] demonstrated that if the egg-laying period is excluded, temperature is the primary explanatory factor for shell growth and the third explanatory factor for meat production in *C. gigas*. This oyster can have quite a high level of filtration even when temperature is at 5°C,[13] whereas the filtration activity for the flat oyster is reduced at 8 to 10°C. In wintertime, the indirect effect of ice on the oyster beds can cause high levels of mortality as the mechanical action of the ice breaks the young oysters off the collectors and carries them away with the floating ice. Freezes can cause direct mortality by tearing up the muscle with ice crystals or by bursting the shell.

High temperature in summertime, during reproduction, can bring about physiological disorders which cause mortality. In one sample, on the south Atlantic coast of France, the summer mortality rate for Pacific oysters was 20%.[14] Death occurred mostly among 1-year-old oysters, before spawning, with the temperature higher than 21°C. A thinning of the digestive tubule epithelium was observed, but without evidence of infectious disease. Contrary to the observations of Mori[15] for specimens in Japan and Perdue et al.[16] for ones in America, the mortality did not appear to be related to the eutrophication of water and a high

TABLE 1
Quantities (Q) and Percentage (%) of the Dry Flesh (mg), the Energy (kJ) and the Proximate Constitution of the Flesh Lost by Males (♂) and Females (♀) during the Emission of Gametes

Oysters Collected during 1978

Year	1979				1980		1981			
	♂		♀		ND[a]		♂		♀	
Sex	Q	%	Q	%	Q	%	Q	%	Q	%
Dry weight (mg)	128.2	17.8	447.4	43.1	810	42.9	1510	55.9	1930	61
Energy (kJ)	—	—	—	—	19.48	52.7	30.98	61.1	45.3	69
Proteins	86.7	27.7	236.3	51.1	435.7	51.0	502.6	55.2	663.7	61
Lipids	11.6	18.7	88.6	63.1	257.0	67.3	89.5	49.1	334.2	78
Free carbohydrates	5.7	33.7	16.6	59.8	9.46	30.1	30.0	57.9	22.9	51
Glycogen	14.3	44.3	11.5	39.0	—	—	—	—	—	
Ash	4.1	4.4	42.6	32.3	66.1	27.7	151.8	40.5	125.5	36
	Calculated from 24/7/79 to 21/8/79				Calculated from 15/7/80 to 25/8/80		Calculated from 20/7/81 to 28/10/81			

[a] ND: nondetermined sex.

level of lipids bound to an overmaturation. Actually, it seemed to be a physiological disorder in relation to a deficiency of the energy balance of the oyster, limited to heat stress.

D. TYPES OF SUBSTRATE

At the end of the planktonic stage, cupped and flat oysters attach to hard natural substrates such as rocks, stones, gravels, and living or dead shells, or artificial substrates such as dikes, artificial reefs, etc. For cultivation, the spat are fixed on cultch, which can be limed or unlimed tiles, slates, oyster or scallop shell, plastic tubes, etc. The adult oysters, if they are cultivated in suspension, can stay on the collector until marketable size or can be detached and attached to wood or rope. For bottom culture the oysters are either scraped from the collector and spread directly on sandy or muddy grounds or cultivated on tables and racks in plastic, iron, or wood bags or pockets.

E. SEXUAL DEVELOPMENT AND SPAWNING

C. gigas is an oviparous oyster with alternative protandry and a high level of fecundity. The oyster is either male or female during a particular breeding season but can change sex the following year. A small percentage can remain hermaphroditic.[17] The environment (temperature and nutrition) and hormonal factors seem to determine the percentage of sexual change. The Japanese oysters mature during the first year if they have reached a minimum size of 50 mm. It appears (Table 1) that reproduction is a function of age.

O. edulis has a restricted fecundity and is larviparous. It is hermaphroditic, with a consecutive rhythmic sexuality. Male and female gametes are not ripe at the same time. If we follow Marteil,[9] the flat oyster is male in autumn; after spawning, remaining spermatozoids are lysed and the ovogonies develop for the next breeding season, when it becomes a female oyster. The sexual inversion is influenced by temperature and available food. Yonge[18] showed that in Scandinavia the flat oyster changed sex once a year, while in Great Britain and in France each oyster could be male or female several times the same summer. Fertilization in *O. edulis* occurs in the pallial cavity, where the larvae are incubated to an advanced stage before release ("swarming").

FIGURE 4. The four stages in development of cupped oyster larvae.[27] French oyster biologists divide stage 1 into two: the straight hinged larva (24-h-old larva) and the very early umbo stage. (From His, E. and Robert, R., *Rev. Trav. Inst. Peches Marit.*, 47 (1 and 2), 63, 1985. With permission.)

The duration of sexual maturation is a function of the quantity of heat that the reproductive oyster receives. Muranata and Lannan[19] showed that the mean value of day degree is nearly 2300 degree-days for *C. gigas* maintained experimentally at a temperature of 18 to 22 °C and salinities of 20 to 30‰. Mann[20] demonstrated that gametogenesis began only above 10.55°C. In addition, Héral et al.[21] described how, for the Bay of Marénnes-Oléron, which is the main sector of reproduction of cupped oysters in the European Atlantic coast, *C. gigas* bred after 2287 degree-days: the date of spawning was dependent each year on temperature fluctuation, with 95% of spawning occurring between July 12th and August 21st. For the flat oyster, Marteil[9] found that the minimal temperature necessary for the beginning of gametogenesis is 10°C, but only after a period of sexual rest, which is not required for the Japanese oyster.

The effect of temperature on the release of the gametes is also important. Lubet[22] showed that there was a minimal critical temperature under which the emission of sexual products could not take place. For *C. gigas* it is 17 to 18°C and for *O. edulis* 14 to 16°C.

Marteil[9] reported that *O. edulis* delayed its gametogenesis in estuaries when the salinity was under 20‰, but variations between 30 and 36‰ did not seem to play a role in the date of spawning and on the quantity of the gametes. For the Japanese oyster it seems that gametogenesis takes place sooner when the salinity is between 20 and 35‰.

C. gigas has complete spawning. In France, this oyster spawns mainly once in Marénnes-Oléron Bay, but in Arcachon, successive rebuildings of the gonads occur, with two or three spawnings depending on the heat of the summer. Provided the temperature is high enough, *O. edulis* can possibly have a phase of reconstitution of the gonads after the first estival laying, which permits another spawning in autumn. On the Adriatic coast, the spawning season is in March and April, but for the same species in Norway it occurs in August. These differences show the role of climatic conditions on the rapidity of gametogenesis and, consequently, the time of spawning.

When the sexual products are ripe, adult *C. gigas* eject their gametes (20 to 100 million eggs) in the water. After external fertilization by the spermatozoids, the ova develop into the trocophore larvae.

F. LARVAL DEVELOPMENT

For *C. gigas* the length of the youngest well-formed larva is nearly 60 μm, and the size at metamorphosis is circa 300 μm (Figure 4).[23-25] The duration of the larval life is in relation to the temperature. The time varies on the French Atlantic coast from 15 to 28 d at temperatures 20 to 26°C. The optimal temperature for the shorter larval development was 28°C for a salinity of 30 to 32‰.[26] At lower temperatures, the pelagic phase is lengthened. For

TABLE 2
The Stages in the Development of the Larva of the Japanese Oyster *Crassostrea gigas* as Described by Oyster Biologists and Malacologists

Days after the fertilization	Names used by malacologist	Height in μm
1—6	Veligerous larva D	57—105
6—18	Veligerous umbonated larva	105—260
18—22	Veligerous eyed larva	260—280
22—24	Pediveligerous larva	280—300
24	Plantigrade larva	300

the flat oyster, the planktonic stage is from 5 to 14 d at temperatures from 17 to 26°C. The size of the released larvae (mean 180 mm) depends on the length of time the mother oyster retains them in the mantle cavity. Setting occurs when the size is between 280 and 300 μm.[24] All the spawnings do not result in settlements because for *C. gigas* the temperature must be higher than 18°C (Figure 3) and for *O. edulis* higher than 15 to 16°C

Helm and Millican[28] showed that the optimal salinity for *C. gigas* is 25‰ for larval development. However, recruitment can be as good at 20‰ as at 35‰ and is more dependent on the temperature. The higher the salinity is, the more necessary it is for the temperature to be above 22°C (Figure 3). So, to aid in cultivation, twice a week each summer since 1930 biologists from fishery institutes such as IFREMER have analyzed the numbers and development of the larvae in places were oysters breed to predict the optimal times for the oystermen to immerse the collectors.

G. GROWTH

Substantial differences in the amount of growth are recorded not only from different sites in different countries, but also in the same site at different levels in exploitation of a bay (Tables 3 and 4). Thus, growth appears to be a function of temperature. In temperate regions, growth in wintertime is reduced, particularly for *O. edulis* under 8 to 10°C. *C. gigas* is more tolerant of low temperatures, and when the level of food is high enough, winter growth occurs.

The main factor which increases or decreases the growth rate is the available food. This, in turn, depends on two factors. The first one is the nutritional value of the bay for the oyster, which is a function of the current velocity, multiplied by the quantity of food in the water and the time of immersion. Different bays, lochs, and estuaries were tested in the U.K.[35] and Ireland[36] for *C. gigas* growth, and some sites showed growth ten times that of the others. In experiments such as these where the quantities of cultivated oysters are little and where there is no other large cultivation or wild population of mollusks in the area, the growth results indicate the level of the trophic capacity of each sector.

The second factor affecting growth is the impact of breeding density on the available food at two scales: the local density, which is the density of the unit of culture, and the total biomass of cultivated and wild population of mollusks in the area. It is well known that oyster growth relates to settlement surface.[32] For this reason, spat produced free of support (in hatcheries) has a better growth than natural settlement unless the young oysters are scraped early from the collectors. This observation for juveniles is also valid for adults. Several authors have shown that growth was a function of the density on the ground, in baskets, or in racks. The impact in the rural biomass on the growth rate has been studied by Héral et al.,[6,38] who found that for the Bay of Marénnes-Oléron there is a slight correlation (Figure 5) between the decrease of the growth rate and the development of the total cultivated biomass of Pacific oyster for the last 15 years.

The maximal growth of *C. gigas* occurred in Israel in subtropical fish ponds,[39] with

TABLE 3
Growth Performance of the Cupped Oyster *Crassostrea gigas* on the European and Mediterranean Coast

Locality	Total weight after 12 months (g)	Total weight after 24 months (g)	Ref.
Emsworth Harbour (U.K.)	34	100	41
	(5) 60		
Newton Bay (U.K.)	52		29
Linne Mhiurich (U.K.)	14	58	
Emsworth Harbour (U.K.)	20	70	31
		(16 months)	
Menai Straits (U.K.)	50	130	35
Rossmore (Ireland)	46	—	36
Carlinglard	16	—	
Bullinakill	6	—	
Carna	3	—	
Flensburg		60	42
Baltic North Sea	8		43
(Germany)	(12)60		
Flensburg Fjord (Germany)	24	—	44
Oualidia (Maroc)	(6) 43	120	37
Etel (France)	12—29	—	45
Marénnes-Oléron (France)	48	60	46
Marennes-Oléron (France)	35	60	47
Thau (France)	35—50	68	48
		116 (20 months)	
Corse (France)	—	100 (17 months)	49
Arcachon (France)	15	58 (30 months)	50
	(18 months)		
Marénnes-Oléron (France)	8	27 (30 months)	
	(18 months)		
Bretagne Sud (France)	28	99 (30 months)	
	(18 months)		
Marénnes-Oléron (France)	1970 50	100	6
	1972 33	75	
	1974 20	50	
	1975—1981	40	
	20		
	1984 15	30	
Fish pond (Israel)	(4) 79	—	39
	92		
Lim Canal (Yugoslavia)	26	103	34

Note: Number in parenthesis represents weight at the beginning of the culture.

high temperatures and food levels, where young oysters of 4 g reached 80 to 90 g in 1 year. The better growth in European countries is 40 to 50 g after 1 year and 100 to 130 g at the end of the second year of growth. This growth rate was found in England,[29,35] in France when the stock density was low,[36] and in Corsica and Morocco.[37] For *O. edulis*, the better growth is observed in France on the Mediterranean coast with a weight of 60 g after 2 years of cultivation. This performance could be achieved also in French Brittany but with a seed density of 1 ton/ha, which is five times less than the traditional density.

The growth of the flesh expressed in dry weight of the meat in wintertime represents about 1% of the total weight and in summertime about 2.5 to 4.5%. The dry weight of the flesh shows great variation (Figure 6) in relation to the reproduction and lack of food during wintertime.[40]

TABLE 4
Growth Performance of the Flat Oyster *Ostrea edulis* on the European and Mediterranean Coast

Locality	Total weight after 12 months (g)	Total weight after 24 months (g)	Total weight after 36 months (g)	Ref.
Emswork (U.K.)	(5) 30			29
Harbour (U.K.)	(5) 30			
Menai Straits (U.K.)	15—20	30—35	40—50	30
Emswork Harbour (U.K.)	10	28	50	31
Golfe du Morbihan (France)	4	15		9
Cancale	—	62		32
Bretagne (France)	—	47		
1983	—			
1984	—			
Thau	23—35	30—60 (26 months)		33
Agde (France)		30 (16 months)		
Lim Canal (Yugoslavia)	—	28	50	34

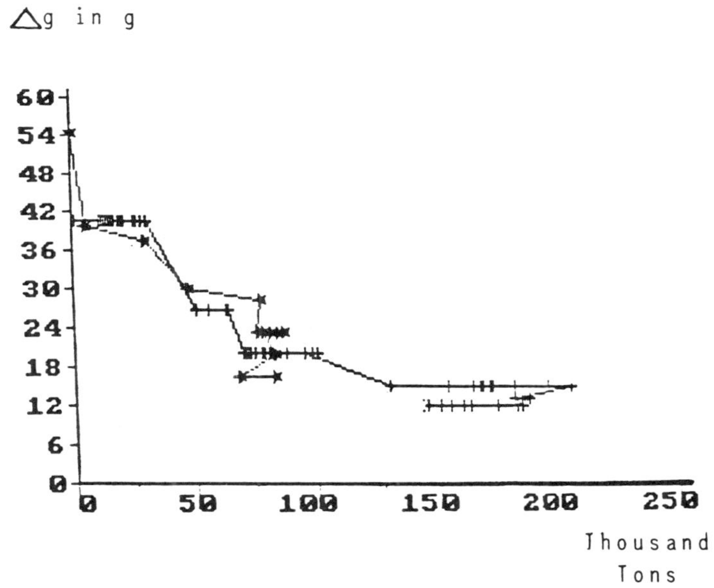

FIGURE 5. Evolution of the annual growth rate of Portuguese oyster (+) and Japanese oyster (*) in relation to the total biomass of oysters in the bay of Marénnes-Oléron. (From Héral, M., Deslous-Paoli, J. M., and Prou, J., *ICES, Mariculture Committee,* F, 41, 22, 1986. With permission.)

III. FISHERY

A. HISTORY

The flat oyster *O. edulis* is the native European species, which was very common until the last century. Shallow bays and estuaries around the coasts of Europe, including Ireland and the Mediterranean, abounded in centuries past with oysters that could be picked by hand in many places at low water or dredged by boats, resulting in a prosperous oyster fishery. In ancient times, Romans built some ponds to stockpile the harvest and tried to collect the spat on wood fascines. During the Middle Ages and Renaissance, fisheries on the oyster

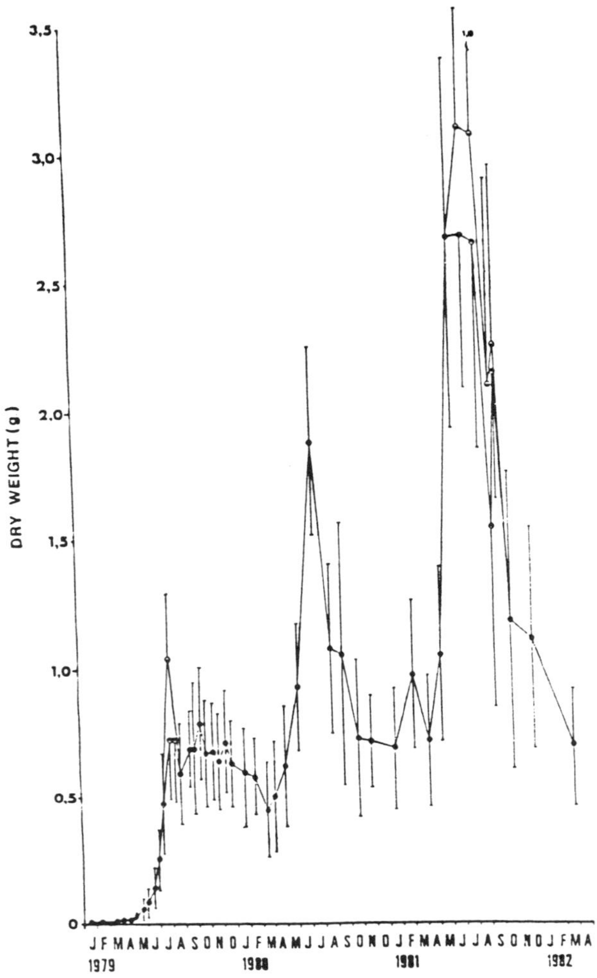

FIGURE 6. Evolution of the dry weight of *Crassostrea gigas* in the bay of Marénnes-Oléron. (From Héral, M., *Aquaculture*, Lavoisier, Paris, 1986, 346. With permission.)

beds went on. In the 18th Century, the increased population induced an increase in fishing efforts causing overfishing in the beds, whose production in preceding years had varied depending on the intensity of settlement. By the end of this century the oyster beds of France, Ireland, and England were largely overfished, and regulations were made to limit the fishing season and to suspend it altogether during the reproductive period.

Despite these more and more constraining regulations, the fishing effort grew. Increased consumption was encouraged by the development of transportation of goods and by the beginning of oyster culture. At this period there was no real culture; the oysters were only kept in stocks in ponds before the sale. On the Atlantic coast, near La Rochelle, old unused salt marshes were turned into "claires" for finishing the growth and fattening the flat oysters. It was only after 1860 that in France the technique of cultches was begun by enlarging the Italian wooden fascines used by the Romans. Culture techniques advanced very quickly with new materials—slates and tiles with lime for settlement—and new culture methods, bottom and rack. In 1920 to 1921, a disease decimated the beds of flat oysters throughout Europe. It was only after 1928 that settlement began to be abundant again, particularly in French Brittany. In several localities (Marénnes-Oléron, Arcachon), the wild flat oyster beds had definitively disappeared. In Ireland, Great Britain, the Netherlands, and the Mediterranean coast, some oyster beds persisted, and their exploitation by fishermen continued. In 1950 to 1951, a new disease appeared and destroyed the flat cultivated oysters of the Thau lagoon. The breeding of *O. edulis* continued in French Brittany, which sold juveniles to Spain (Galicia) and to the Netherlands. After a phase of large expansion (annual French production of 30,000 ton; Figure 7), flat oysters were attacked again by two parasites—*Marteilia refringens* Combs (since 1974)[51] and *Bonamia ostreae* Pichst and Coll (since 1979)[52]—that spread to other European countries, reducing the breeding in France, Holland, and Spain.

Since the beginning of European oyster culture in 1860, cupped oysters had been imported to France from the Portuguese wild oyster beds to palliate the lack of flat oysters. During one of these trips a boat threw its cargo of oysters out in the estuary of Gironde, and the species *C. angulata* spread quickly on the southwest Atlantic coast of France. After 1920, yield from cupped oysters was almost equal to that of the flat oyster (annual French production of cupped oysters more than 85,000 ton; Figure 7). From 1966 to 1969, throughout the French coasts the Portuguese oysters were infected by a gill disease and by an iridovirus that destroyed the species *C. angulata* in Europe. At the present time, only some oyster beds in Portugal are still active.

French importation of *C. gigas*, with spat from Japan and adults from British Columbia, occurred from 1968 to 1975 (Figure 8). Reproduction of Japanese oysters has been well established since 1971 on the southwest Atlantic coasts of France, and the level of production in this country has reached 100,000 ton since 1980. Other countries such as Great Britain, Ireland, Germany, Norway, Denmark, Spain, Italy, and Morocco have started very recently to cultivate *C. gigas* with spat produced mainly in hatcheries from Great Britain, France, or Ireland.

This brief historical study shows that during the last century, a constant series of events, particularly evident in France, manifested itself: first, overfishing of natural flat oyster beds followed by disease; second, importation of a foreign species *(C. angulata)*, bred on a large scale with high density and disease; third, importation of a new exotic species *(C. gigas)* and cultivation at a high level.

Each time the response to mortality caused by disease has been the importation of the next species. It would be better to understand the disease and to practice management of oyster bays. Thus, it appears that diseases occurred after a period of high level of production, and we suggest the hypothesis that when the cultivation densities in a bay are too high, the trophic equilibrium is surpassed, growth and fattening decrease, and the defense mechanisms of the oysters are lowered. By controlling the stock in culture, the risk of disease could be lessened.

B. IMPORTANCE

Statistics to estimate the level of production in different countries are very often not realistic and cannot be compared with one another (see, for example, Héral et al.,[38] for validation of the French statistics on oysters).

The data from the last yearbook of Fishery Statistics published by the Food and Agriculture Organization (FAO) of the United Nations can give bases for comparison. For the cupped oyster, which is only the Japanese one, the European production from 1981 to 1984 varied between 78,000 and 105,000 ton of total living weight including the shell. The French production represents 99% of the total production. For the flat oyster, the same source of data shows that the production is at a low level, around 4000 ton which represents only 4% of the quantity of the cupped oyster (Table 6). The French production (40% of total), which was nearly 15,000 ton, is now 1800 ton because of the two diseases, *Marteilia* and *Bonamia*. The harvest of *O. edulis* in the Netherlands has increased to 20% of the total by the use of native oysters that are uncontaminated by diseases and have tolerance for low temperatures, instead of importing seed from France.[29]

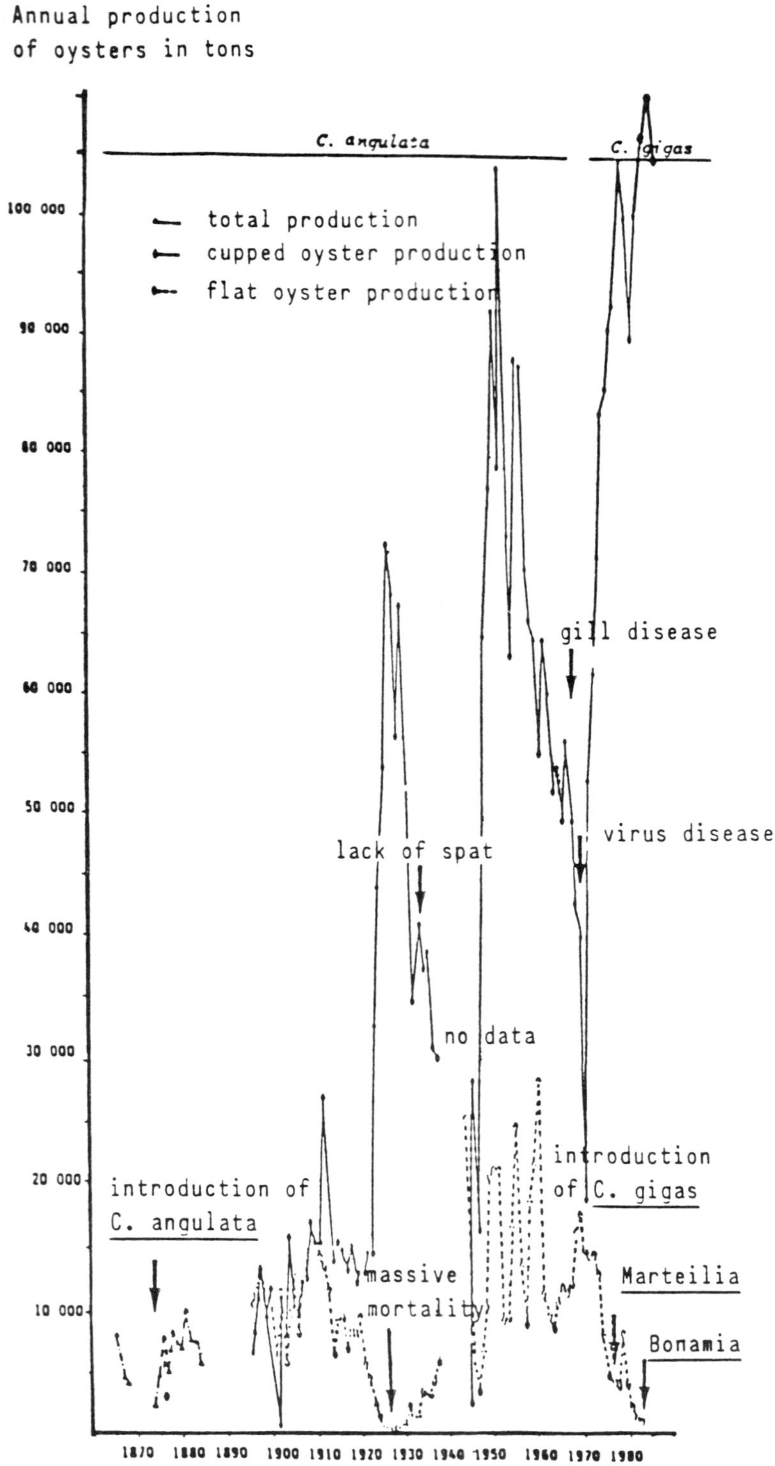

FIGURE 7. Evolution of the production of oysters in France from 1865 to 1985. (source: official statistics from Fisheries Department). (From Héral, M., *Aquaculture,* Lavoisier, Paris, 1986, 346. With permission.)

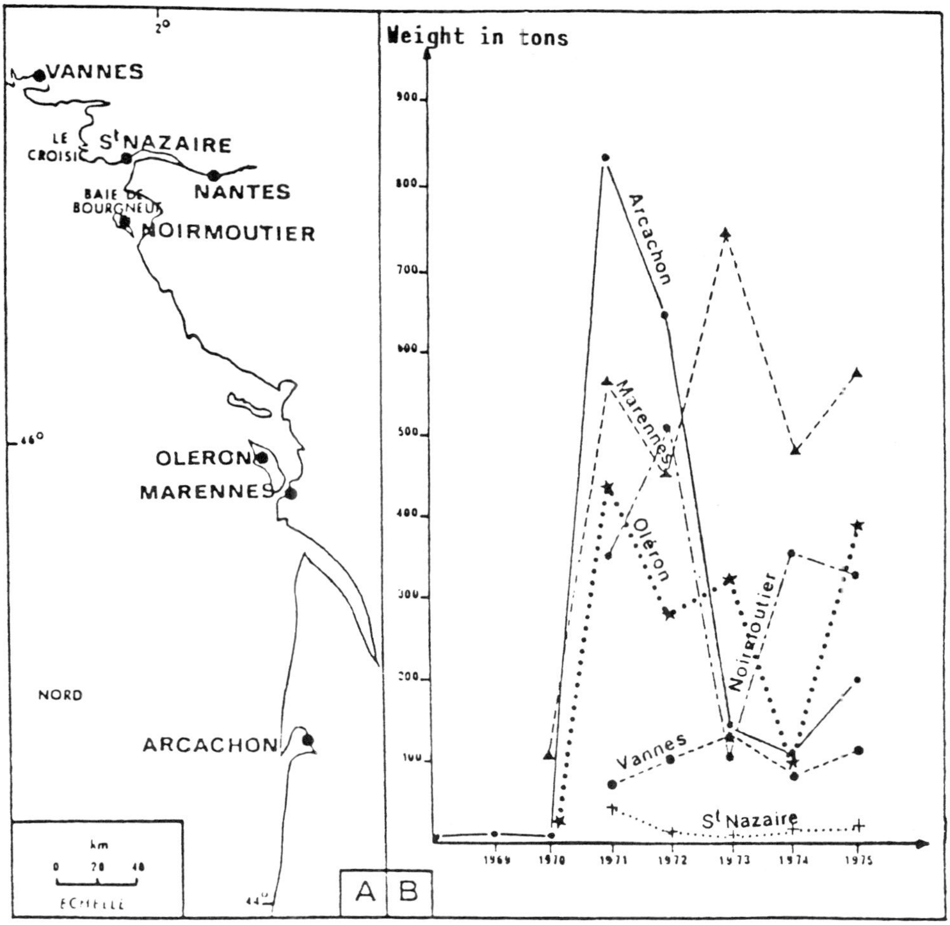

FIGURE 8. Quantities of spat *(C. gigas)* imported from Japan on the French Atlantic coast. (From Gruet, Y., Héral, M., and Robert, J. M., *Cah. Biol. Mar.*, 17, 173, 1976. With permission.)

TABLE 5
Nominal Catches for *Crassostrea gigas* by Countries in Tons of Total Weight

	1981	1982	1983	1984
France	90,242	80,082	104,849	77,755
United Kingdom	61	58	70	144
Germany	1	1	1	1
Total	90,304	80,141	104,920	77,900

From *Yearbook of Fishery Statistics,* FAO Statistics Series, Vol. 58, Food and Agriculture Organization, Rome, 1986.

In a recent FAO paper on development of aquaculture in Europe (Table 7), it is predicted that by 1990 oyster production in France will stabilize; Italy, Spain, and the U.K. will increase; and Denmark, the Federal Republic of Germany, and Yugoslavia are expected to develop a culture.

It is mainly in France that oyster culture plays a large economic role, with a production of oysters above 100,000 ton and with a value of nearly 1 billion francs ($150 million US).

TABLE 6
Nominal Catches for *Ostrea edulis* by Countries in Tons of Total Weight

	1981	1982	1983	1984
France	3.242	7.059	1.369	1.773
Ireland	400	861	316	371
Netherlands	572	560	680	843
Spain	280	539	549	560
Sweden	25	13	13	11
United Kingdom	560	399	269	360
Turkey	1	0	0	9
Yugoslavia	56	82	57	181
Total	5.136	9.513	3.253	4.108

From *Yearbook of Fishery Statistics,* FAO Statistics Series, Vol. 58, Food and Agriculture Organization, Rome, 1986.

TABLE 7
The Aquaculture Production of Oysters in Tons of Total Weight and Estimates of Future Production

	1980	1983	1985	1990
Denmark	0	15	30	60—100
Federal Republic of Germany, Japanese oyster	0	0	10—20	50
France, flat oyster	4,170	1,288	2,300	4,000
Japanese oyster	95,165	107,779	105,000	110,000
Ireland, flat oyster	60	35	—	—
Italy, flat and Japanese oyster	—	—	1,000	3,000
Netherlands, flat oyster	—	1,000	1,500	—
Norway, Japanese oyster	0	30	50	1,000
Spain, flat oyster	1,700	2,350	3,000	4,000
United Kingdom, flat and Japanese oyster	800	800	800	1,200
Yugoslavia, flat oyster	—	30	40	100
Total	101,895	113,312	113,735	123,430

From Ackefors, H., Mumo, A., Müller, F., and Querellou, J., Development of Aquaculture in Europe, FAO ERC, 4, Food and Agriculture Organization, Rome, 1986, 31.

This represents one fifth of the total French marine production value. The permanent employment for oyster culture is about 10,000, and more than 30,000 seasonal jobs are created in 5000 industries.[58] The number of permanent jobs represents one third of the fisheries employment. This activity, when concentrated in some bays or estuaries, plays a high social importance and is very often the primary activity in these places. There are many satellite jobs in the plastic industry—factories for crates, for boats, for tools, for special machines, and so on. The evolution of the costs of production since 1970 in constant francs shows a diminution or stagnation in relation to the increase in cupped oyster production, caused by the allocation of new areas for breeding and the conversion of flat oyster to Japanese oyster cultivation. Because of mass production of *C. gigas*, there is an oversupply, and new markets must be developed.

C. RECRUITMENT AND METHODS OF CULTURE

A steady supply of seed oysters can be realized in three ways: regulated fisheries of juveniles on natural oyster beds, settlement of the spat on cultch, and production of cultchless seed by hatcheries and nurseries.

TABLE 8
Main Hatcheries Producing Spat from *Crassostrea gigas* and *Ostrea edulis* in European Countries

Name	Address	Level of production
SATMAR	La Salive Gatteville-phare 50 760 BARFLEUR (France)	Commercialized 80 million in 1983
HEPC	Le Varquez 29 226 Carantec	Commercialized 18 million in 1986
Aquamare	Route diette 17590 Ars en Ré (France)	Commercialized 1—5 million
Tina Menor	Pesues Cantabria (Spain)	Commercialized 15 million
Planta de cultivos marinos	Muelle de Parcillan Ribadeo galicia (Spain)	Experimental
Conwy laboratory	Fisheries Experiment Station Benorth Road Conwy, Caernarvonshire (U.K.)	Experimental
Seasalter shellfish	Ltd. The Harbour Whistable Kent	Commercialized
Boet-mor/seafood	Cus-Hastrough Gladdaghduff Co Galway (Ireland)	Commercialized

The controlled population of oyster spat in experimental hatcheries has been well known since the work of Loosanof and Davis[59] and Walne.[60] The hatcheries have three main objectives: to produce seed when techniques for settlement are not known, to produce seed when the species do not reproduce in the area or reproduce only sporadically, and to practice genetic selection.

In Europe, as the natural spatfall is abundant for the cupped oyster and the flat oyster, the hatchery production remains small. Lucas[61] gives (Table 8) a list of hatcheries. These hatcheries, in the future, could play a large role for production if researchers of genetic selection are successful in obtaining new races (triploids, for example) with better growth. Cultchless oysters present technical advantages for better growth, as the density can be controlled. The higher price of this seed, ten times more than that from natural settlement, allows a large development only for countries without natural recruitment (U.K., Spain, etc.) Occasionally in France, when there is a lack of spat due to abnormal weather conditions (1 to 2 years out of 10), the hatcheries can partially make up for the deficient seed.

C. gigas is indifferent to the type of substrate to which it attaches. This explains the large diversity of the cultches. The only requirement is that the collectors must be clean, without fouling or silt, so they must be placed at just the right time before settlement. For cupped oysters the old use of stones is practiced less and less, and the wood stakes and fascines are obsolete. Limed tiles are used particularly in the Arcachon Bay, allowing an early removal of juveniles. Seed-collecting techniques have been developed over the years involving different kinds of collecting devices; metallic bars, slate, and oyster and scallop shells are often used. The shells are either placed in bags or strung on iron rods or wires. For the last 10 years plastic collectors have taken the place of the traditional collectors, especially with tubes used in packets of seven (Figure 9). Machines for removing seed have been used with tubes for the last 3 years.

Flat oysters are more selective of the cultch to which they attach. Spat collectors are usually tiles coated with a layer of lime mixed with sand. The composition of the lime is different for each bay and is described more precisely by Marteil.[9] The main advantage of using limed collectors is that oysters can be removed from the collectors after 6 to 10 months. In French Brittany, a new collector made of mussel shells stuffed in bags or tubular nets is nearly as efficient for cultivation in open waters,[62] with a very low cost of under 0.01 francs

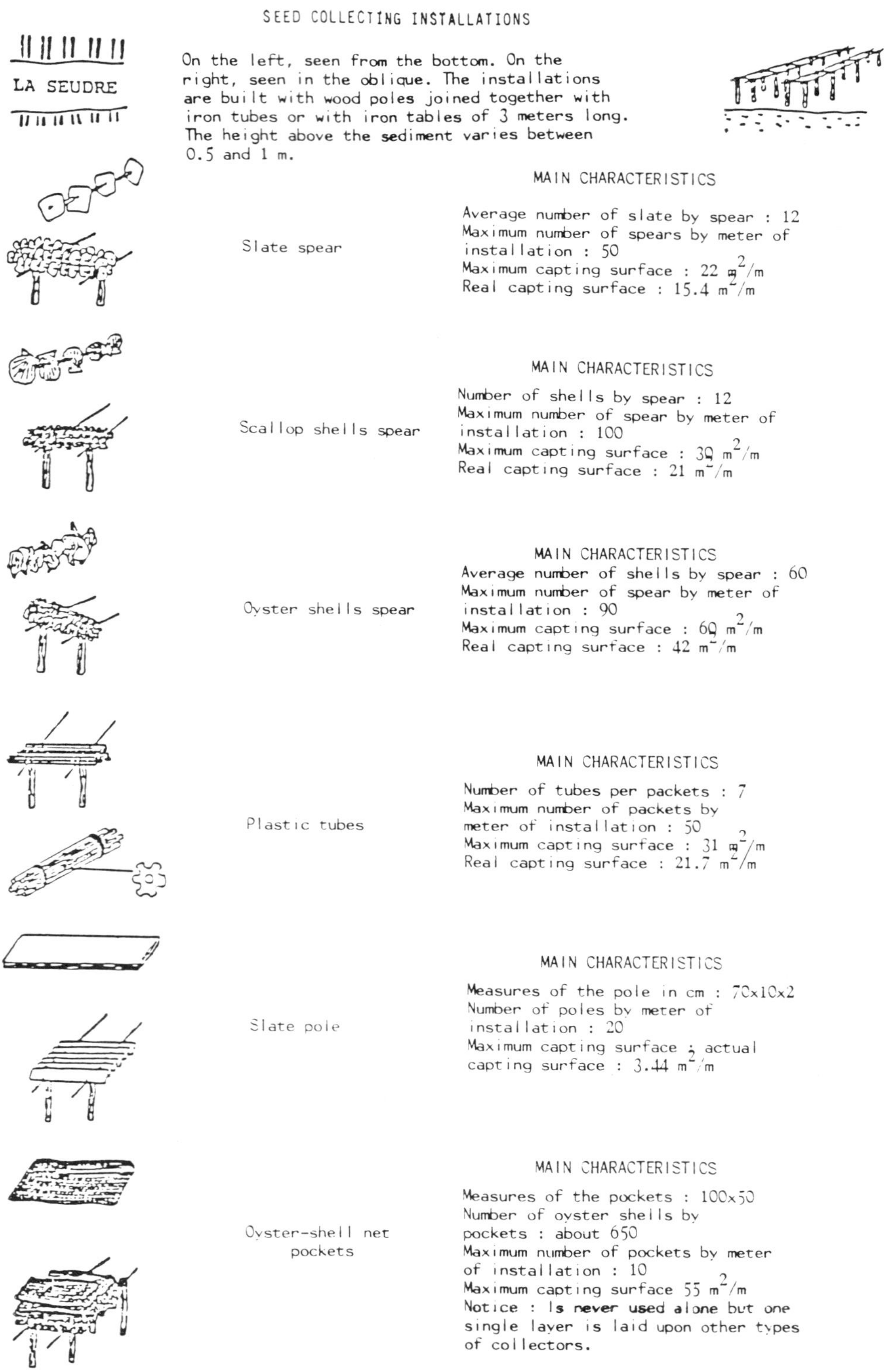

FIGURE 9. Different types of collectors used in the basin of Marénnes-Oléron. (From Berthomé, J. P., Prou, J., Razet, D., and Garnier, J., *Haliotis,* 14, 39, 1984. With permission.)

by unit in comparison with 0.06 francs per unit with tiles. Sampling has been done on the number of collectors used in the main bays to estimate the recruitments and predict the potential production that can be obtained in years to come. In the Arcachon Bay, nearly 5 billion young oysters settle each year, while an average of 10 billion spat are produced in the Bay of Marénnes-Oléron, furnishing the entire French production of cupped seed oysters.

After the seed collection, oyster culture is carried out in two phases: pregrowing and maturing phase. A variety of techniques are used, with the duration of each phase depending on the growth rate and on the cultural peculiarities of each bay.[9] The three main oyster culture methods used in Europe are bottom culture, rack culture, and hanging-rope culture.

Bottom culture is carried out in intertidal zones as well as in deep water. The ground is hardened, if necessary, and the young oysters are sown on the ground attached or not to the collectors. In the intertidal zones, to protect the oysters against crabs, plastic wire fences with or without wooden planks are placed around the beds. This pregrowing can last 1 or 2 years. During the maturing period all the oysters are scraped from the cultch, sorted by weight, and sown again on maturing beds, which can be protected against the storms by fences made with wood sticks 5 m high (in Arcachon). The mean densities for the cupped oyster are 5 kg (total weight) on a square meter for the pregrowing and 7 kg (total weight) for the maturing phase. For the flat oysters before the diseases, Marteil[9] indicated densities of 0.5 kg for the first year, 1 kg the second year, and 3 to 5 kg the third and fourth years. During growth, oysters are regularly harrowed or turned over with oyster forks. The average yield for cupped oysters is estimated as 1 ton of spat producing 20 ton; for flat oysters 1 ton of spat generally produces 12 to 15 ton after 3 years of cultivation.

In deep water along the coast of Brittany and Holland, the flat oysters are planted at a depth ranging from 3 to 10 m after dredging for stones, predators, and competitors. Before the diseases, the density was 0.5 kg for 1-year-old oysters and 0.7 to 0.9 kg/m^2 for 2-year-old oysters. Now the densities are five times less (1 kg/m^2), allowing optimal growth so as to avoid the large mortality rate caused by *Bonamia* that occurs in the third and fourth year. Cupped oysters are also cultivated in deep water, particularly since the decline of the production of flat oysters, at a density of 2 to 3 kg/m^2. Even with high mortality (due to storms and predation), the mean yields are better than in the intertidal oyster beds and with a lower cost, but a higher investment (dredging boat).

The most common culture used at the present time along the French Atlantic littoral is the rack culture on iron tables. The tables are generally 3 m long and 0.5 m above the ground. In the intertidal oyster beds, the collectors are fixed to a row of tables at a high density (50 to 100 collectors per meter) (Figure 9). A year later, they are thinned out to a density of 8 to 10 collectors per meter. After scraping, the juveniles are sorted and put in containers. The most common are plastic packets with a standard size of 1 m long, 0.5 m wide, and 0.1 m deep. The mesh size depends on the size of the seed. The weight per bag is 5 kg for flat oysters and on average 10 kg for cupped oysters, but varies with the age of the oyster: 5 kg per bag for the youngest and 18 kg for the oldest. Rows of tables (30 to 100 m) are placed parallel to each other. To avoid too great a density of oysters, the cultured tables should not total more than 1/3 of the leased area. One hectare should not have more than 6000 to 7000 oyster packets.

In the hanging method, the oysters are fixed on different supports: ropes, nets, and iron or wooden bars. These are hung from fixed tables or floating rafts. Fixed installations are mainly used in Mediterranean lagoons where there is no tide and quite a large depth (10 m). The metallic tables are 50 m long and 10 to 12 m wide. Each table has about 50 wooden or metallic bars which carry about 1000 supports.[64] The collectors of *C. gigas* coming from the Atlantic are directly hung under the table. Part of the oysters are harvested after 12 to 18 months; the rest may be stuck on wooden bars with quick cement and harvested 1 year later, giving large and fat oysters. The average yield is 5 to 7 ton per table. The oysters can also be cultivated on floating rafts (like those used in Spain for mussels). In Galicia, the "batea" is also used for flat oyster production. In Corsica, the same rafts are used for the cultivation of Japanese oysters. There have been some experimental cultures of flat oysters in suspension in Brittany and on the Mediterranean coast in places where the diseases *Marteilia* and *Bonamia* are not present.

D. METHODS OF HARVESTING

The harvest in the intertidal area is done by hand. For bottom cultivation, oysters are fished with oyster forks, put into baskets at the low tide, and at flood tide loaded onto flatboats. The yield is not more than 1.5 ton per man per tide. For rack culture on tables, the harvest is not mechanized, but the work is easier and faster. It consists of detaching the pockets from the table and loading them on flatboats. The yield depends on the size of the oyster beds. The average yield can be estimated at 200 pockets a day per tide, or nearly 3 ton.

In the subtidal areas, the oysters are dredged with special boats.[9] The yield is very high—10 ton/h—but the daily yield is not more than 15 ton after taking the access time and the unloading time into consideration. After harvest, oysters are washed with machines and sorted by weight, by hand or mechanically. The oysters will either be prepared for the market or cultivation will be finished by a fattening phase. Fattening developed particularly in the bay of Marénnes-Oléron in old salt marshes converted since the 17th Century, to oyster ponds called "claires". In these shallow (0.4 m) earth ponds, the seawater comes in by gravity during high tide only during the spring. In these "claires", there is a high level of phytoplankton productivity related to the input of a high level of nutrients in the estuarine waters and a low level of turbidity. However, to obtain fattening, the densities of oysters must be under 10 oysters per square meter, which represents a biomass of 0.8 to 1 kg/m^2. As there is no water change during the neap tide (10 d), the phytoplankton bloom is very quickly grazed if the biomass of mollusks is too high. At a density of 20 oysters per square meter, there was no fattening and, indeed, a decrease in the weight of the flesh.[65]

In these claires, a phytoplanktonic algae bloom, *Navicula ostrearia* Bary, occurs. At the death of the cell, the green pigment diffuses in the water[66] and is absorbed by the gills of the oyster, giving them a special dark-green color that is in demand. The blooms of *N. ostrearia* are attributed to the excretion of organic nitrogen by the oysters.[67]

E. USES AND PROCESSING

In European countries, oysters are sold fresh in their shells. They are very often eaten raw, requiring rapid turnover so that quality, especially bacterial quality, remains high. Some are cooked in apple fritters or as stuffed oysters, but this is the exception (mainly in restaurants), and cooked specialties are not very popular. Nearly half of the production is sold during the Christmas holidays and for New Year's. Mechanized structures are used, and larger firms have high daily preparations (20 ton/d) and sale.

F. CONSTRAINTS

1. Diseases and Parasites

From 1966 to 1969, the Portuguese oysters *C. angulata* were damaged by the "gill disease". The local lesions were first attributed to a protist. Research on viruses in bivalves is recent. It was only in 1972 that Farley et al.[68] found a virus in *C. virginica* (Gmelin). Comps and Duthoit[69] isolated viral particles in gill lesions that caused a mortality rate of nearly 40% for the Portuguese oyster in France. The description of this virus placed it in the group of iridoviruses.

The second disease which affected the cupped oyster from 1970 to 1973, destroying the culture of Portuguese oyster in Portugal, Spain, France, and Great Britain,[70] was due to another iridovirus (photo) causing a viral hemocytic infection.

The introduced *C. gigas* was resistant to the two iridoviruses that had destroyed the Portuguese oyster, signifying the specificity of the agent on the host. Meanwhile, during some summer mortalities in 1977 in Arcachon Bay, the same virus as that of the *angulata* oyster was found on the Japanese oyster,[71] which showed that the resistance of *gigas* oyster to iridovirus is not perfect.

The contamination of *C. gigas* by *Mytilicola orientalis* is recent. His[72] found that this copepod was present in the digestive tract of 10 to 40% of the oysters, varying with the season. It can proliferate in the intestine (up to 40 individuals), building an intestinal occlusion and damaging cellular walls of the digestive tube.[73] An infection of more than three female oysters caused a significant reduction in the concentration of glycogen and carbohydrates of the total flesh.[74]

The oldest epizooite which affected the flat oyster in France, the Netherlands, England, and Ireland occurred from 1920 to 1927.[4] A noninfectious agent had been described, but Orton[75] found abnormal cellular figures which Grizel[76] thinks was due to an intracellular parasite.

The largest mortalities which disrupted the production of the flat oyster in Europe began in 1968 in French Brittany. The disease spread progressively to different farming areas from 1973 to 1975 on the French Atlantic coast[77] (except in the open seawater), and later, with transportation of parasitized French flat oysters, to Spain and the Netherlands. *Marteilia refringens* is a protozoan considered to be related to the Paramyxea and not to the Haplosparidia.[78] This parasite and its cycle was first described by Comps[79] and confirmed by Grizel et al.[51] and Perkins.[80] It develops in the digestive epithelium with an annual cycle. The infestation period occurs in summer, when the temperature of water is above 17°C, but the development cycle of the parasite occurs when the temperature is above 12°C (Figure 10).[76] Franc[81] showed that the parasite could have a sexual phase. The pathogenic action of *Marteilia* could be due to the disaggregation of epithelial cells of the digestive diverticulum. The parasite has been found in *O. edulis* on the Mediterranean coast in the lagoon of Thau and in Morocco (Nador), in *C. gigas* (but only in young stages),[82] and in *Mytilus edulis* Linnaeus and *Mytilus galloprovincialis* Lamarck, where it can have a complete cycle.[83] *Marteilia refringens* is not associated with mortalities for these species.

When *Marteilia* began to decrease, new abnormal mortalities occurred in flat oysters in 1979 due to *Bonamia ostreae*, a new protozoan parasite of blood cells.[52,84] It seemed to be the same parasite, often called "microcell" disease because of the size of the cells (2 to 3 μm). The cycle of infestation has not yet been elucidated. Unlike *Marteilia, Bonamia* infestation can occur during the whole year.

This hemocytopathogenic parasite induces gill ulceration with perforations, indentations, and injuries to the conjuctive tissue. Severe infections spread to the different centers of North and South Brittany oyster production in France.[85] As a result of commercial transport activities, this disease now occurs in the entire flat oyster culture areas in Europe. It is present in Denmark, the Netherlands,[86] England, Ireland, and Spain.[87] The flat oysters of the Mediterranean coast are not contaminated, but in 1988, *Bonamia*-contaminated flat oysters were found on the French Mediterranean coast. If they are transported to the Atlantic coast, they become infected. Everywhere *Bonamia* is present, it causes heavy mortalities in *O. edulis*. As the spat is very little contaminated by *Bonamia*, and as the level of contamination is a function of age, in Brittany as in the Netherlands, cultivation is possible at low density in open waters, with fast growth (2 or 3 years) after an eradication of all the old oyster beds. Despite these precautions, Von Banning[86] found that *B. ostreae* has the ability to survive in very low density oyster stocks.

The lesson which can be learned with the severe diseases that appeared during the last 20 years of oyster culture in European countries, mainly in France, is that it is necessary to limit the transport of living populations, particularly of mollusks between different coun-

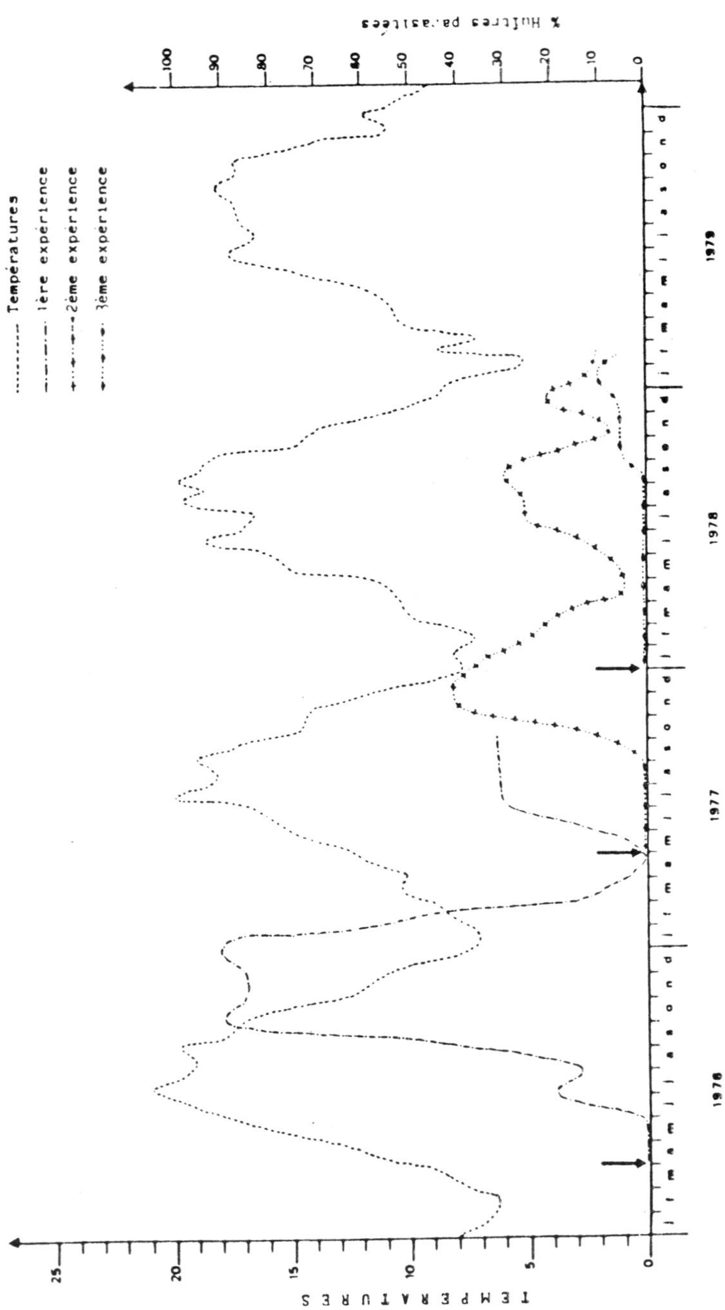

FIGURE 10. Evolution of the occurrence of *Marteilia refringens* in *Ostrea edulis* in relation to the temperature of the water. (From Grizel, H., Thése d'Etat, Université des Sciences et Techniques du Languedoc, Montpellier, France, 1985, 145.)

tries and in the same country between different bays. When for economic reasons (lack of spat, importation of new races with better growth and resistance to disease) importation must take place, the risk of spreading diseases and parasites is very high. For these reasons, international organizations have made some recommendations.[88] For example, ICES recommends that during introduction, broodstock should be kept in a closed system or quarantine with sterilized effluents. Only the progeny (F_1 or F_2) can be introduced in the natural water. The ICES council also recommended procedures for current commercial practice involving quarantine and disinfection to confirm freedom from pests and diseases.

2. Predators and Competitors

The predators are not very numerous; the most active is the green crab *Carcinus maenas* Linnaeus, which eats large quantities of young spat on the collectors. Birds can cause serious predation (as ducks do for mussels in Wadden sea), but for oysters the populations of "huitres pies" (oyster pies) are not large. Some 10 years ago, the oystermen protected the oysters planted on the bottom by erecting pieces of iron to avoid the large flatfish (teyre), but these have actually become very rare. In the open sea, the density of starfish can be controlled by dredging.

Competitors can be very numerous, particularly if we speak about trophic competition. It is thought that large populations of cockles, clams, and mussels, consuming the same type of food, can affect the growth of the oysters. *Crepidula fornicata* (Linnaeus), imported in Europe from the U.S. during World War II, is the more active trophic competitor. Even though the energy budget of this species[89,90] is four times less than that of the Japanese oyster, the very large biomass obliges the oystermen to dredge and destroy them.

The competition for space, particularly on the collectors for cultivated populations and on the rocks for natural oyster beds, is mainly between cirripeds and postlarva oysters during the month before and after settlement. During later stages of the culture, it is mussels and, again, *Crepidula* that attach to the shells of the oysters, sometimes completely covering them, forcing the oystermen to clean the oysters and the structures. Fouling is abundant on the Mediterranean coast where suspension culture techniques are employed. The main active species involved in this fouling are the Ascidians with active metabolisms.[91] They are competitors both for a place on the substrate and from a nutritional point of view. The only remedy available is to raise the oyster structures out of the water and allow them to dry for several days. The constraints caused by predators and competitors are dealt with quite well by the oystermen and are not a limit to the development of the culture as diseases or pollution can be.

3. Pollution

The sedentary character of bivalves makes them sensitive to natural or induced disturbances of the environment. Their filter-feeding characteristics make shellfish vulnerable to bacteriological and chemical pollution. As oysters are cultivated on the coast and particularly near estuaries, they are sensitive to any modification of the ecosystems by pollution. Culture areas are in conflict because of different uses for the water in the region. Conflicts over the use and the quality of the input of freshwater will perhaps be the main problems in European countries in the future. One difficulty in maintaining the trophic level of cultivated bays is the changing quantity and quality of the estuarine water with the use of fertilization and development of irrigation. On the seashore, conflicts for space between tourism and oyster culture occur. Tourism also brings pollution problems (bacterial quality of the water, products used by pleasure craft, and so on). Antifouling paints, used on sailboats for destroying fouling species which grow on the boat bottoms, use copper oxide, arsenic, and mercury salts. For the last 10 years, tributyltin (TBT) has been used. His and Robert[92] and Robert and His[93] have shown that the paints are severely toxic, particularly for the larval development

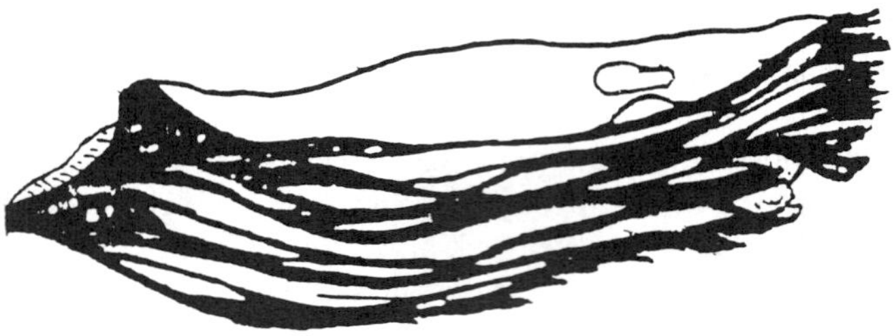

FIGURE 11. Transverse section of the upper shell of *Crassostrea gigas* showing abnormal growth in thickening with inhibition in length caused by TBT. (From Alzieu, C. and Héral, M., *Ecotoxicological Testing for the Marine Environment,* State University of Ghent and Inst. Mar. Sci. Res., Bredene, Belgium, 1983, 187.)

of the Japanese oyster. Also, this product at very low levels stopped the growth of certain species of phytoplankton (*Chaetoceros calcitrans* Takano, *Isochrysis galbana* [Parke]) which can be used by oyster larvae as food.[25] During the years 1977 to 1981, no settlement occurred in the Bay of Arcachon with the development of pleasure crafts. When the use of TBT was regulated (1982), high levels of spat attachment occurred again. It has been shown that this product causes a malformation of the shell of the Japanese oyster, modifying calcification, inducing growth in thickness, and inhibiting the growth of the shell in length.[94-98] Subsequently, the use of this type of paint was prohibited in 1982 on boats less than 25 m in France, and some years after in England and Germany. Thus, a product used for a concurrent activity can cause a severe disturbance of the production of oysters by acting on the larva, on its food, and on the growth of the shell.[99]

Another pollutant that cost quite of lot of money for the flat oyster culture in Brittany was the stranding of oil tankers near bays where cultivation of flat oysters was the main activity. Destruction of the oysters on the bottom occurred, and the remaining oysters could not be sold because they had directly absorbed the dissolved hydrocarbon, giving a horrible taste to the oysters. They could not be used for human consumption because of the toxic effect of these compounds. An evaluation of the consequences of the Amoco Cadiz stranding[100] for oyster cultures has been estimated in 1980 to be 114 million francs. It is important but less so than the two diseases (*Marteilia* and *Bonamia*) which caused a loss of 1.6 billion francs in turnover and 1.3 billion francs of added value.[101]

4. Other Constraints

For these euryhaline species, the freshets are not a large constraint. From the market point of view, the taste of the oysters changes with freshwater, causing a decrease in sales. Until now, the lack of seed has not been a problem for the oyster culture. When there is no pollution, recruitment remains high. Héral et al.[21] showed that the Portuguese oyster has been without settlement only 2 years during the last century. For the Japanese oyster, 3 years without high settlement occurred during the last 15 years. It seemed to be due to the higher temperature requirements of these oysters and their need for a temperature above 18°C for larval development. If climate conditions remain cold, European countries could find hatcheries useful to remedy the deficiency of natural settlement.

A new constraint has been developing for European countries during the last 10 years in relation to summer phytoplankton blooms producing toxins. Paralytic shellfish poisoning (PSP) is associated mainly with the species *Gonyaulax tamarensis* Lebour or *G. excavata* Bolech. It has been described in the U.K.,[102] Norway,[103] Spain,[104] and Portugal.[105] Diarrhetic shellfish poisoning (DSP) is found in association with *Dinophysis acuminata* Claparède and Lachman and perhaps with some other species of *Dinophysis* and *Prorocentrum*.[106] PSP is present in France,[107] the Netherlands,[108] Ireland,[109] and Spain.[104] Oysters are less sensitive to toxin than mussels and clams. For these reasons, monitoring, based on "mussel watch", is followed in several countries.

High mortalities, with no toxicity for human consumption, have occurred on the Mediterranean coast due to an anoxic environment. This happened in the Thau lagoon in 1975 and 1986, in conjunction with several phenomena such as high temperature, no wind, stratification of the water, and high density of organic matter, causing dystrophic crises with red water. A bloom of photosynthetic sulfide-oxidizing bacteria produced hydrogen sulfide in anaerobiosis.[101]

In European countries, the technical problems do not present as major constraints, such as lack of facilities and transportation. Also, economic problems arise in supply and demand and the cost of production.

G. MANAGEMENT

In France, permits to harvest are given by the public authority that leases the culture grounds in the coastal areas. It is for a period of 30 years and varies in other countries. The concession is transmissible to the same familial enterprise. The yearly cost is about 200 French francs per hectare. The regulations for exploitation of the grounds are fixed by the state for the type of culture and the nature of the installation, as well as densities and the time of harvesting during the year. The state and the local agencies manage the bays by dredging (mud, sand, or parasites). Each oyster farmer is responsible for the management of his own grants.

The organization of the sanitary and quality control is very different and not yet unified between the different European countries. Sanitary control of mollusks in France is based on the assumption that it is more effective to prevent contamination than to try to remedy its effect. It is the reason why emphasis is placed on monitoring the water quality in cultivated zones. The coast line is classified into sanitary and unsanitary sectors, based on criteria of the number of fecal coliforms in oysters, mussels, or clams (Figure 12). In cases where an area is classified as unsanitary, harvest is, in general, forbidden. It may be allowed in certain cases, but the mollusks must be relaid in clean waters for at least one month or depurated by chlorine or ozone. In cases where the water is of good bacterial quality before sale, the cultivated mollusk must stay for 2 to 3 d before shipment in a "degorgeoir". It is a large tank (50 to 100 m^2 or larger) where oysters or mussels stay in clean controlled seawater for self-depuration.[111]

In Holland and England, the sanitary regulation is based on the same regulations as in France. In Spain and Italy, all harvested mollusks are treated in depuration stations where water is disinfected with chlorine or ozone. The shellfish stay between 1 and 4 d in the depuration system in depending on the level of contamination. The European countries followed the recommendations of the International Code of Practice for Molluscan Shellfish by FAO/OMS.[112] The sanitary laws of each country are evolving quickly, taking into account the quality of the waters, dinoflagellates and their toxins, levels of chemical pollutants in the water, hydrocarbons, organic-halogenous substances, heavy metals, following the European Economic Community (EEC) recommendations.[113] As for the water, the different countries belonging to the European community will have to unify their regulations for the sanitary quality of the products before 1992.

IV. RESEARCH AND DEVELOPMENT

A. INTRODUCTION

Past success of molluscan culture has been due to natural spat settlement, the use of a natural phytoplanktonic free food, the large adaptation of sedentary species to the variation of the environment, and acknowledgment of cultivation techniques.

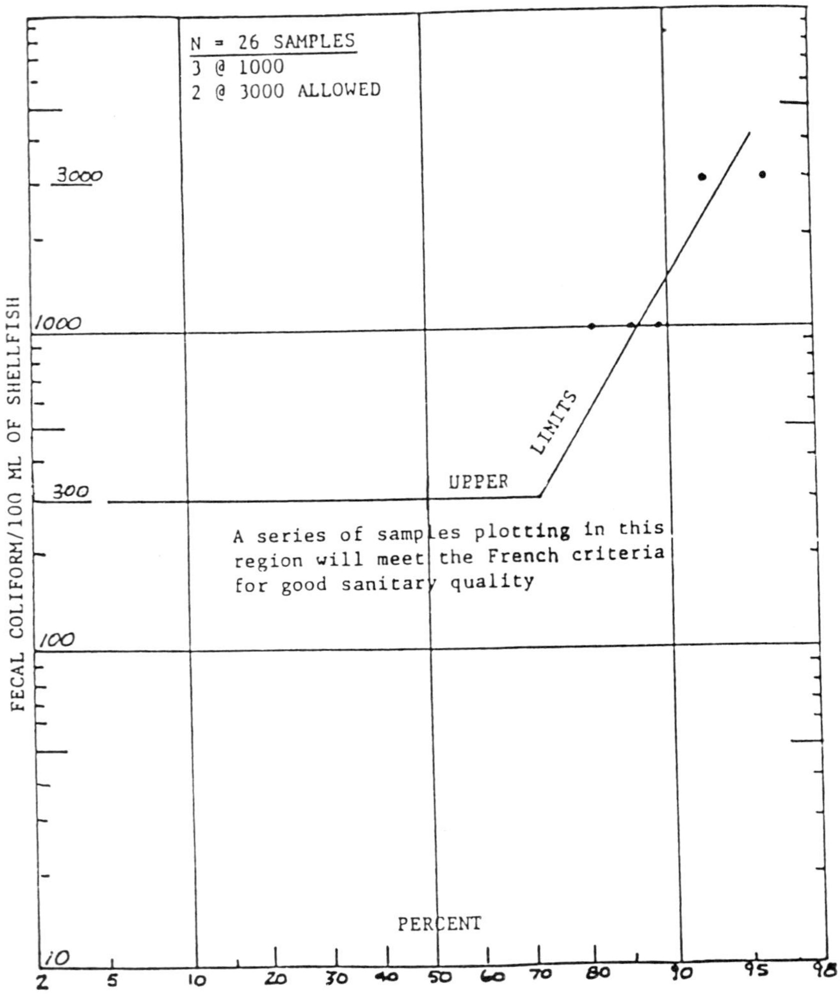

FIGURE 12. Criteria for classifying a culture site as sanitary. (From U.S. Food and Drug Administration and Sanitary Control of Shellfish in France, Washington, D.C.)

In spite of these opportunities and economic successes, there remain problems in maintaining the same level of productivity for countries where the cultures have reached a high level of production and in developing the potential of culture for other countries. Scientific research can resolve some of these problems, particularly new techniques of open-sea culture, prevention and determination of epizootic diseases, determination of the carrying capacity of the environment for cultivation of mollusks, prevention of degradation of the quality of the cultivated waters, prediction of dinoflagellate blooms, and selection by genetic manipulation of species with resistance to disease or better growth performance.

B. TECHNIQUES AND CULTURE

Theoretical research is being performed by French scientists and specialists to test the resistance of materials to find better subsurface long-lines that can resist storms (Figure 13). Structures on the bottom at a depth of 20 m are tested for their suitability as nurseries of the young spat of *O. edulis* that settle on collectors fixed on long-lines. Each year different private companies test, with the help of research organizations, new materials for oyster culture to obtain efficient results. New types of plastic collectors, easier to transport, are used to obtain cultchless oysters. New vehicles for intertidal areas are studied to increase

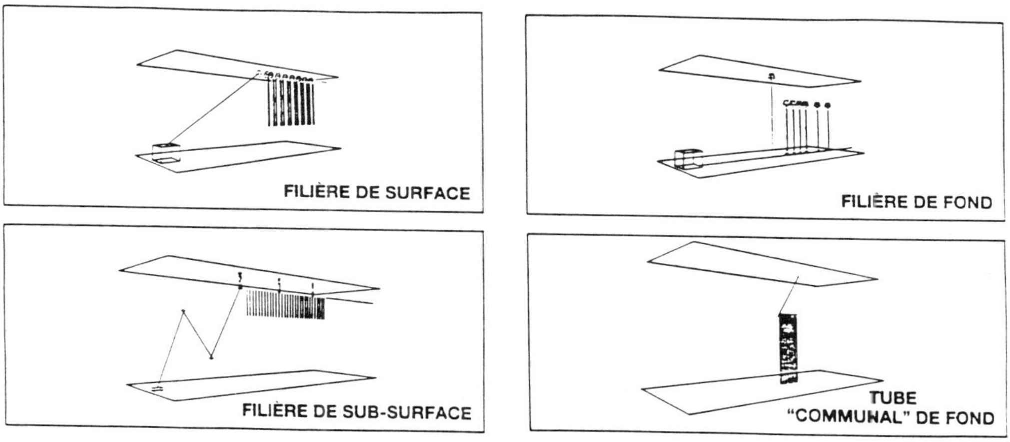

FIGURE 13. Description of different long-line techniques. (From Müller, et al., *Cultures Mar.*, 145, 4, 1987. With permission.)

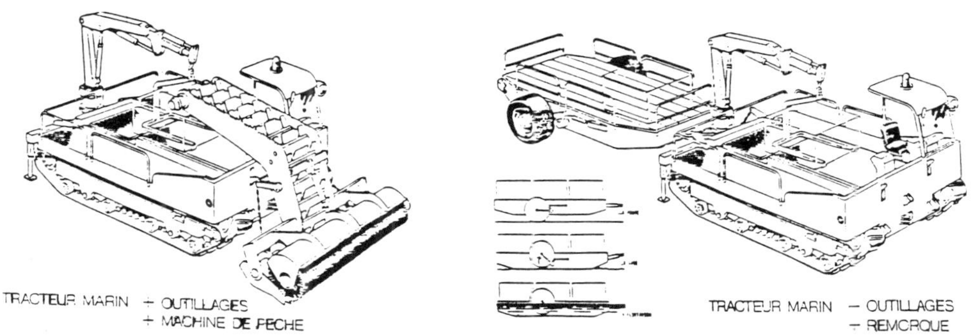

FIGURE 14. New vehicles for cultivation of oysters and mussels. (From Priour, *Equinoxe*, 9, 23, 1986. With permission.)

the mechanization of the harvest and transport of oysters and to decrease the cost of production (Figure 14).[115] Technical research has been done to optimize the final handling of shellfish (washing, grading, packaging) prior to shipment. For example, private firms and public researchers are developing machines to sort oysters by recognizing dead oysters by sound and by image analysis.

C. PATHOLOGY

After the recent diseases of the Portuguese and flat oysters, different European countries have strengthened their regulations concerning mollusk importation. The quarantine recommended by ICES is applied even to products coming from hatcheries. Only if the spat are free of disease can they be imported.

To increase the control of parasitized oysters, new techniques for diagnosis have been worked out. The techniques of monoclonal antibodies have been developed for the first time in mollusk pathology.[116] The purification protocol achieved for *Bonamia ostreae*, a parasite of *O. edulis*, is described in Figure 15. The application of these techniques for *Bonamia ostreae* was described by Boulo et al.[118] From 700 hybridomas, eight were selected for their *Bonamia*-specific antibody reaction. Two were retained to make *Bonamia* diagnosis by indirect immunofluorescence. This new technique, along with the use of the commercialized enzyme-linked immunosorbent assay (ELISA) test, is significantly less time-consuming than the traditional histological preparations. Therefore, the number of examinations will be

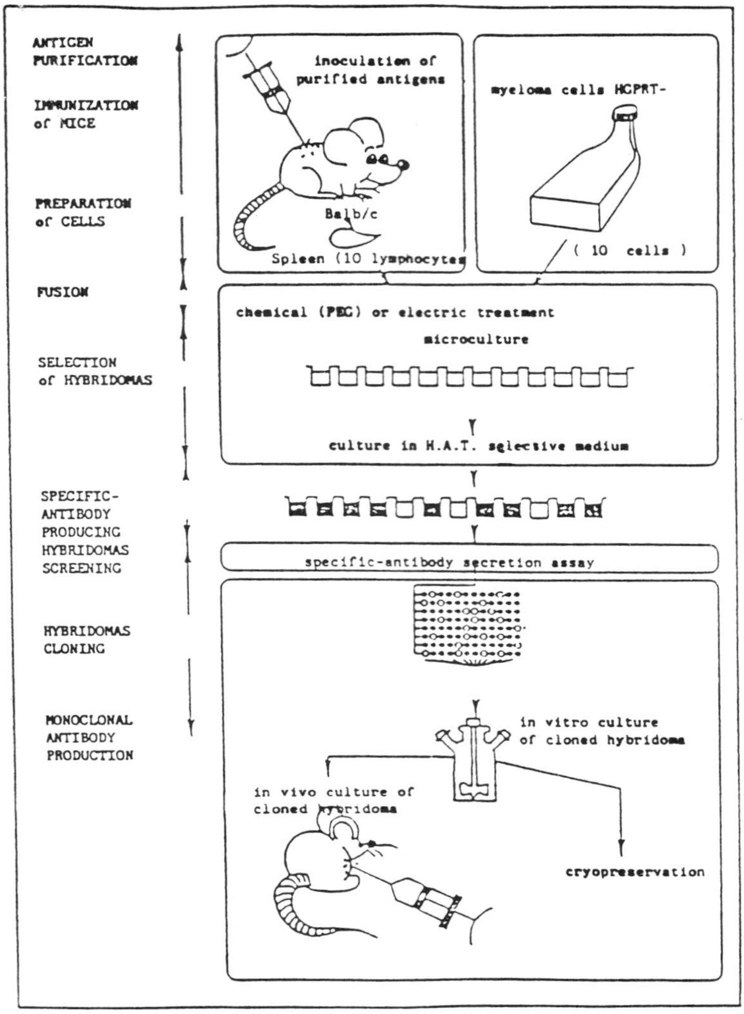

FIGURE 15. General principle of hybridoma technology. (From Mialhe, E., Boulo, E., Grizel, H., Rogier, H., and Paolucci, F., *Disease Processes in Marine Bivalve Molluscs,* American Fisheries Society, Spec. Publ. Ser., in press. With permission.)

larger, allowing for a better zoosanitary survey, and precise determination of the rate of infection to be possible.

D. CARRYING CAPACITY OF THE ENVIRONMENT

Two models have been used to estimate the carrying capacity of a body of water. The first is based on the molluscan harvest and the second on the tropic relationships. The first model uses the hypothesis that the environment is constant or its fluctuation oscillates around the same mean. The data which are necessary for the construction of the model are the growth rate of the cultivated mollusk, the survival rate, and the estimation of the total cultivated biomass.

Some sampling strategies have been developed to obtain these data. The more difficult one is to estimate the reared stocks precisely.[113] The method includes taking aerial photographs of the intertidal cultivated area and estimating the density based on subsampling in the field. Aerial photographs all over the bay measure the areas effectively. They are analyzed, either extensively for small bays or by a systematic sampling for larger areas,

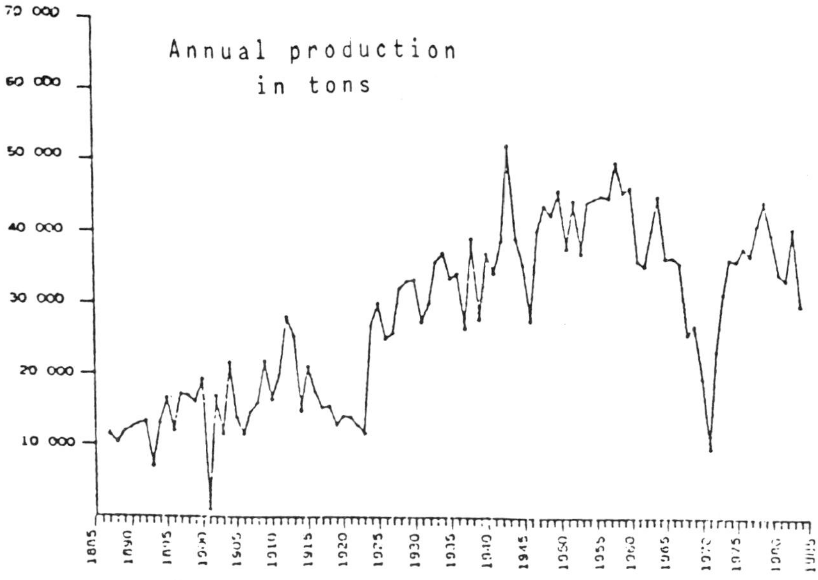

FIGURE 16. Evolution of the annual production of adult oysters grown in the bay of Marénnes-Oléron.[38]

with a precision of 3%. Research is going on with remote sensing[120] and with numerical signals obtained with a CCD video camera to optimize the cost and the performances.

Estimations of local biomasses are obtained by a random subsampling in different strata according to the modes of cultivation and the geographic areas. The stocks are computed as the product of biomass and cultivated surfaces, with a total precision of about 6%. This approach has been used since 1984 in the main large bays producing oysters along the French coast. It is only after gathering data on stocks, growth performance, and mortality rate over a long period, with a large spectrum of variation, that a precise dynamic model will be built. A model has been achieved with historical data.[6,38] The Bay of Marénnes-Oléron was chosen because it is the main European basin for oyster production and has had a large decrease in its growth performance. The evolution of the production of the cupped oyster is estimated for the last century (1885 to 1985) from three different sources of data in Figure 16.

The growth rate and the mortality rate of the population of oysters presented an increase of the duration to obtain adult oysters while the survival rate is decreasing (Figure 17).

The total biomass in culture is calculated from annual production and the yield of culture, growth, and mortality (Figure 18). This simulation of the biomass gave results that are on the same order as those achieved by sampling over the last 3 years.

The relation between the stock function of the production showed clearly a maximum plateau of 40,000 ton. This limited corresponded to the maximum production capacity of the ecosystem limited by the trophic capacities of the bay. The maximal production of the bay could be modeled by an equation of the same nature as the one used for the growth of the populations. So, the Von Bertalanffy equation $P = P_{max} (1 - e^{-KB})$ fits well with the data (P_{max} is the maximum production of the Bay of Marénnes-Oléron, with B the cultivated stock; for *C. angulata* $K = 0.026$ and $P_{max} = 41,873$ ton; for *C. gigas* $K = 0.028$ and $P_{max} = 52,450$ ton). The yield production on the stock (P/B) in relation to the stock followed a negative exponential curve as the evolution of the annual growth rate in relation to the stock (Figure 5).

The maximum production of 40,000 ton could be reached with a stock of Portuguese oysters of 130,000 ton and with a stock of Japanese oysters of 80,000 ton. This difference between the two species can be explained by the energetic demands of each oyster. For the

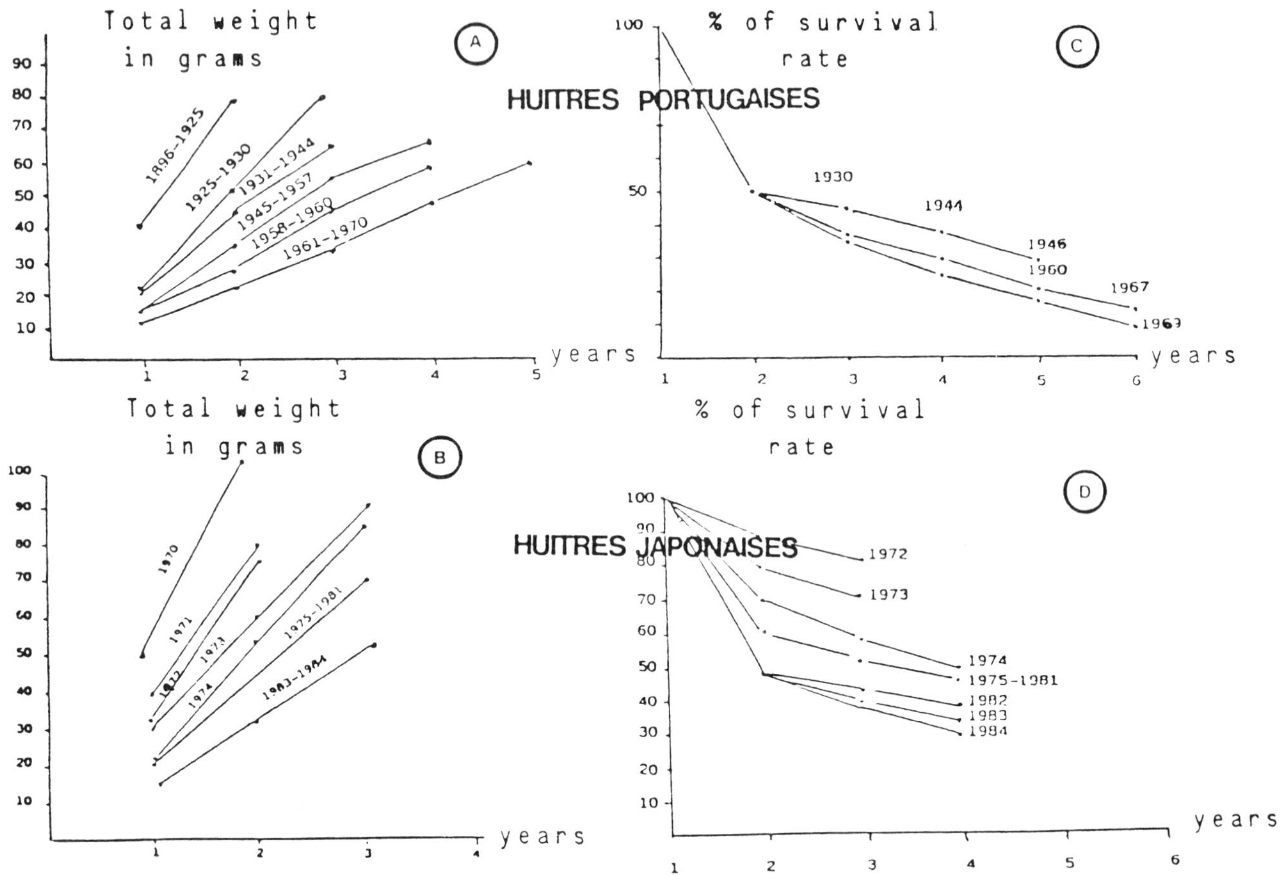

FIGURE 17. Evolution of the time necessary to obtain adult oysters: (A) Portuguese oyster, (B) Japanese oyster. Survival rate after the first breeding year: (C) Portuguese oyster, (D) Japanese oyster.[38]

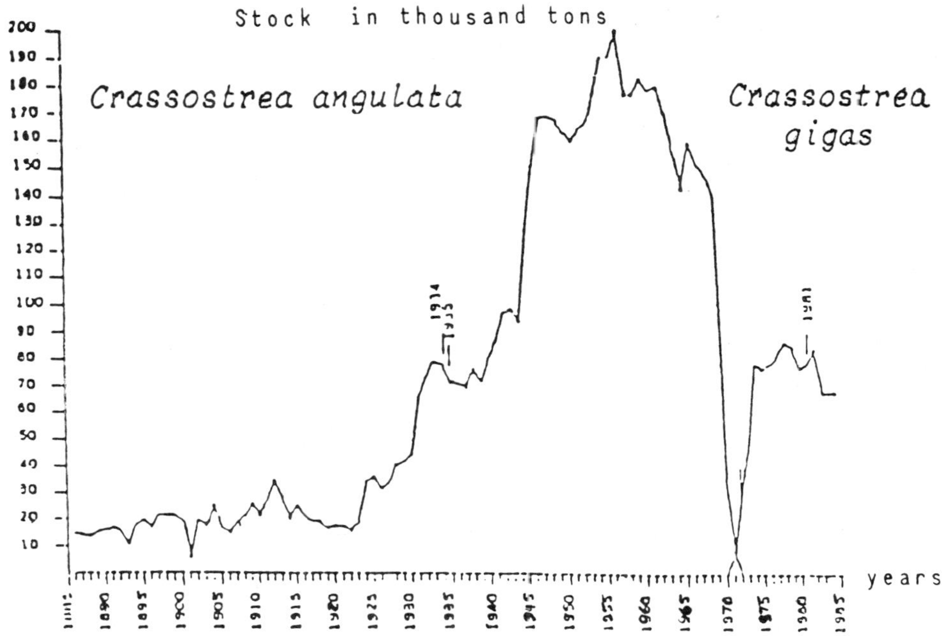

FIGURE 18. Calculated evolution of the total biomass of cultivated oysters in the bay of Marénnes-Oléron. (From Héral, M., Deslous-Paoli, J. M., and Prou, J., *ICES, Mariculture Committee,* F, 41, 22, 1986. With permission.)

same weight, the assimilation of food by the Japanese oyster is 1.7 times more than that of the Portuguese oyster.[6] If the impact of these two oysters on the ecosystem is compared, this transformation coefficient must be taken into account. This model showed that without management of cultivated oysters, the stocks tended to exceed the minimal biomass necessary to reach the maximal potential of production. If stock regulation is applied, it gives oyster farmers the advantages of decreased duration of the breeding cycle and a decrease in chronic mortalities, which brings better profits to the enterprises.

The analytical model is based, on one hand, on the energy demands of oyster populations and, on the other hand, on the quantity of available food that is transported by the currents, obtained with a numerical physical model.

A general equation of the energy budget of oyster populations has been established following the equation $A = P + R = C - (F + U)$, with A = assimilation, R = respiration, F = particular excretion (feces and pseudofeces), U = dissolved excretion, C = consumption, and P = production [$P = Pg + Pr + Ps$, with Pg = production of the flesh, Pr = production used for the reproduction, Ps = production of the secretions (shell, mucus)]. A study on oyster reefs permits evaluation of the different components of the energy budgets of 1- and 2-year populations. It appears that in the Bay of Marénnes-Oléron, the feces and pseudofeces represented 70% of the energy consumed (Figure 20), and the production of the flesh was 2.8% for the 1-year-old oyster and only 0.2% for the 2-year-old oyster.[40,121,122] During the same period, the available food was followed during tidal cycles each month for a neap tide and a spring tide. These data permitted an individual growth model, taking into account the relations between respiration assimilation and the oyster weight, the temperature, the season and the particular available food (protein, lipid, carbohydrate) (Figure 21).

To calculate the level of the available food in the whole bay, an advection-dispersion model is used. As time and spatial scales have to be consistent with the chosen biological scale (1 d), a box structure is applied to the oyster production area. The residual Lagrangian

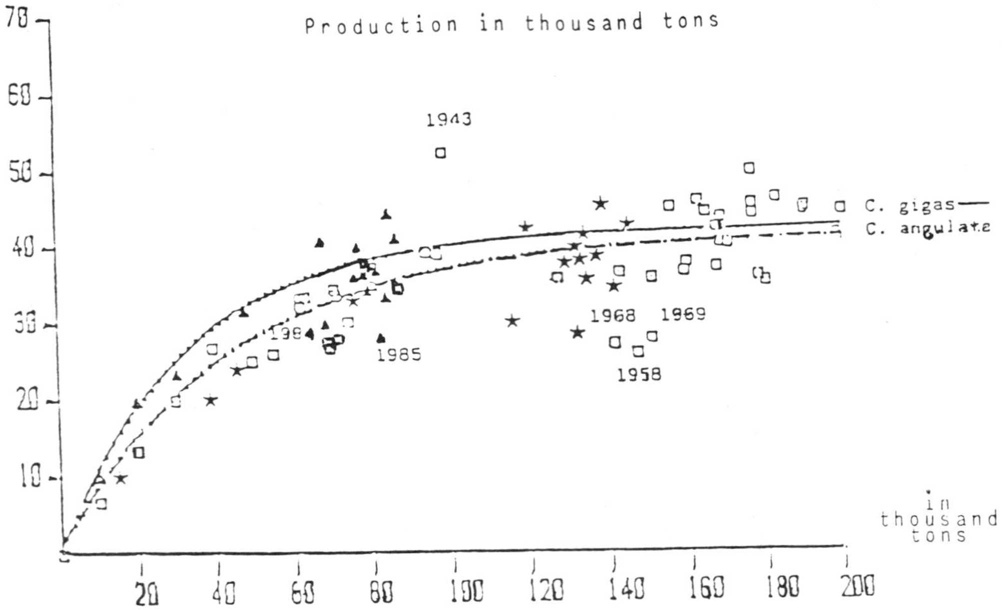

FIGURE 19. Evolution of the annual production of the stock in culture for *Crassostrea angulata* (□), *Crassostrea gigas* (▲) and for *Crassostrea gigas* converted in equivalent Crassostrea angulata (★). (From Héral, M., Deslous-Paoli, J. M., and Prou, J., *ICES, Mariculture Committee,* F, 41, 22, 1986. With permission.)

currents are calculated. The dispersion is a function of the transport and of the difference in concentration between adjacent boxes.[128] The food is actually a driving variable and is injected at the three limits of the box model as the salinity with time series on 5 years, averaged to smooth the variability. The transport model is validated by the use of predicted salinity and observed salinity data in the middle of the bay (Figure 22).

The ecosystem model is built with a stock of oysters of two ages in each box, and the growth model function of the food is transported by the physical model. The trophic molluscan shellfish competitors and their assimilation of food are introduced in each box as driving variables. This approach could permit fluctuations in the cultivated stocks of oysters and the ability to make predictions about the growth rate in the different areas. The comparison between the results of the global model (Figure 23B) and the simulation of the analytical model (Figure 23A) demonstrates the same effects of the total cultivated biomass on the annual growth rate, but with more precision for the different areas with the analytical model.

This model uses, up to now, too many simple hypotheses, but it shows the research methods we need to have a useful tool for management. It demonstrates that a multidisciplinary approach among biologists, scientists, and sedimentologists could obtain a predictive evolution of the growth rate of the cultivated species as a function of the food and of all other factors, particularly pollution, that can modify the quality and the quantity of the trophic requirements.

It is evident that to be predictive it is necessary to go farther in the study of energy demands of the oysters for particular and dissolved substances. In another way, a phytoplanktonic model, which simulates the variations of the input of nutrients from the estuary, will be helpful for the study of the consequences of the use of freshwater on oyster production.

V. FUTURE DEVELOPMENTS

A new step can be achieved in oyster culture with the development of genetic manipulation. Recently obtained triploids and tetraploids for *C. gigas* opened a new way. In

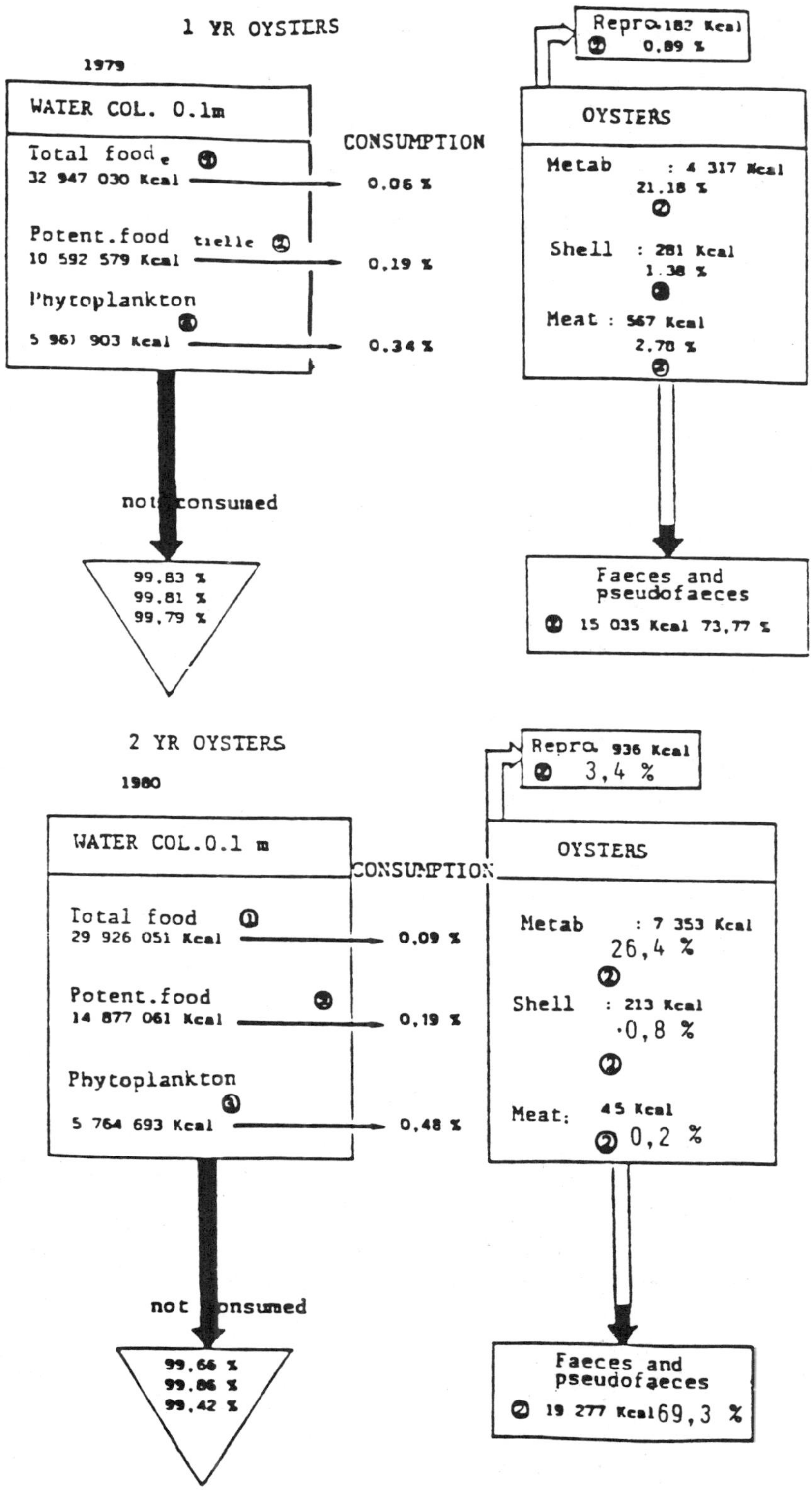

FIGURE 20. Annual energy flow between a 0.1-m water column transmitting at a current of 0.3 m/s and a population of grown oysters at a density of 200 individuals/m^2 (Héral et al.;[40] Delous-Paoli et Héral[122]) (From Héral, M., et al., *Oceanus,* 9(3), 169, 1983; Deslous-Paoli, J. M., and Héral, M., *Haliotis,* 14, 79, 1984. With permission.)

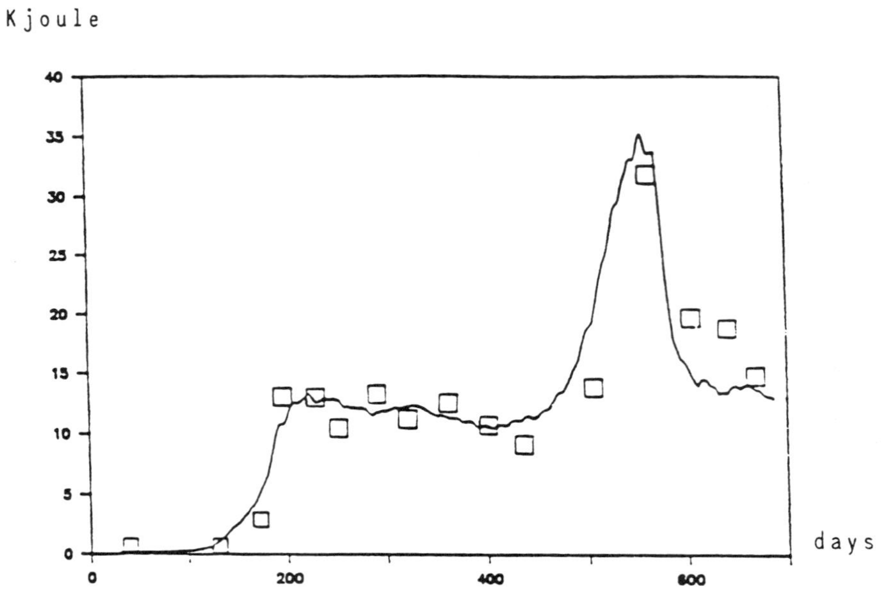

FIGURE 21. Simulation of the individual growth of oysters in the Marénnes-Oléron Bay expressed in K joule function of the assimilation and respiration (□), observed data. (From Bacher, C., Rapp. IFREMER, 1987, 12.)

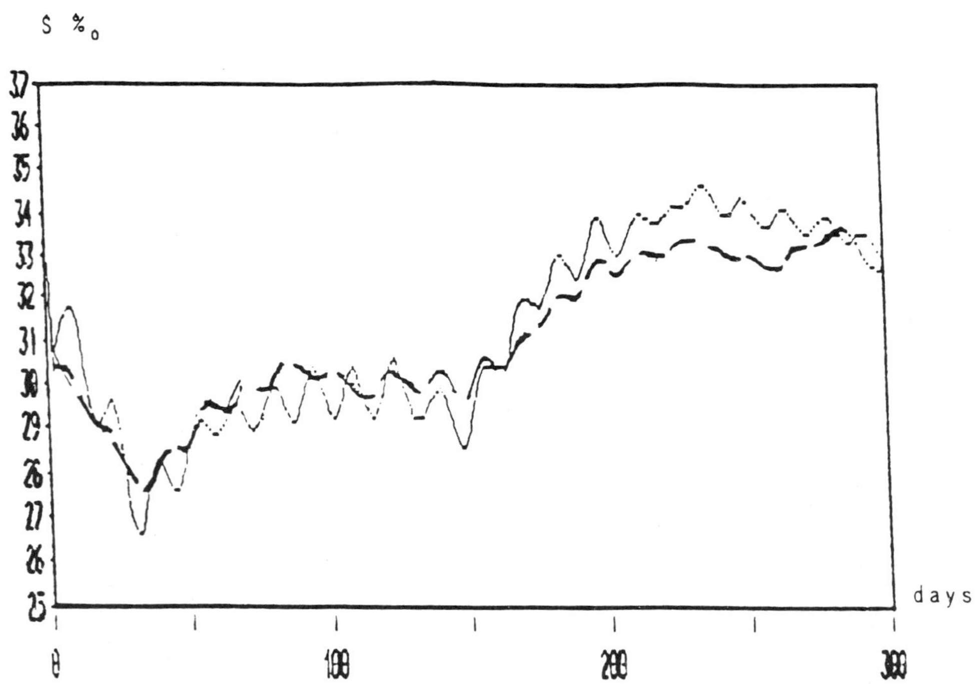

FIGURE 22. Observed salinity (—-) and simulated salinity (——) calculated by the advection-dispersion model in the middle of the bay. (From Bacher, C., Rapp. IFREMER, 1987, 12.)

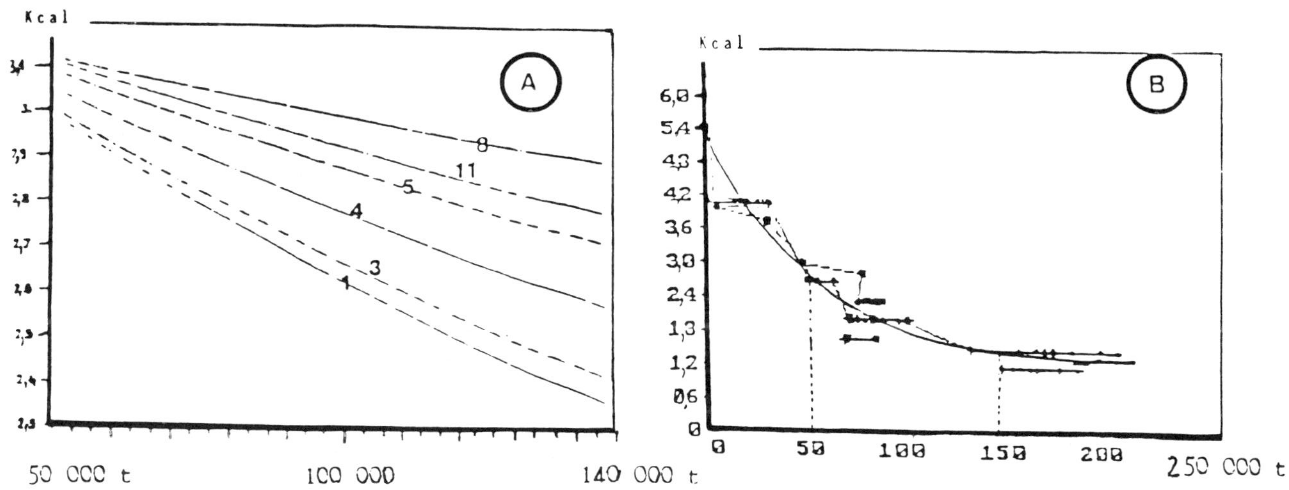

FIGURE 23. (A) Simulation of the evolution of the annual production results of the analytical model. (From Bacher, C., Rapp. IFREMER, 1987, 12.) (B) Evolution of the annual production of one oyster, results of the global model. (From Héral, M., Deslous-Paoli, J. M., and Prou, J., *ICES, Mariculture Committee,* F, 41, 22, 1986. With permission.)

summer, these oysters are losing more than half of their (dried) body weight for reproduction. It would be advantageous to produce sterile oysters that devote less effort to the reproduction process. These chromosomal manipulations can be realized by thermal shocks,[124] hydrostatic pressure, or with cytochalasin B.[125] Energy for the reproduction could be used instead for somatic growth, providing an answer for the growth decrease observed in European closed bays in the summer. Hatchery-produced seed would be a means of regulation to avoid overstocking due to natural settlement, which is too numerous in relation to the density of collectors.

Another new promising sector of research for the oyster industry in European countries is the bioeconomic analysis of molluscan shellfish. Coupling models of dynamics of production with marketing systems could demonstrate not only the different scenarios that oyster farms can choose for their individual development, but also the interest of a group strategy that can be difficult to reconcile with the individual one. Gilly and Meuriot[126] stated that "it is essential to involve disciplines such as sociology or political science in the study of the crucial elements affecting the dynamics of the shellfish sector of the economy," e.g., knowledge of the choices and strategies available to operators, either as individuals or groups; decision-making processes leading to arbitration of conflicts on use of space and the environment; and the role and conditions of using scientific information in these decision-making processes.

REFERENCES

1. **Grassé, P.,** Traité de zoologie, in *Mollusques Lamellibranches,* Vol. 5, Masson, Paris, 1960, chap. 2.
2. **Ranson, G.,** Les huîtres, biologie, culture, in *Savoir en Histoire Naturelle,* Vol. 23, Lechavalier, Paris, 1951, 260.
3. **Ranson, G.,** Les espèces d'huîtres vivant actuellement dans le monde, définie par leurs coquilles larvaires ou prodissoconques. Etude des collections des grands musées d'histoire naturelle, *Rev. Trav. Inst. Peches Marit.,* 2(3), 128 and 205, 1967.
4. **Menzel, R. N.,** Portuguese and Japanese oysters are the same species, *J. Fish. Res. Board Canada,* 31(4), 453, 1974.
5. **Buroker, N. E., Hershberger, W. K., and Chew, K. K.,** Populations genetics of the family ostreidae. I. Instraspecific studies of *Crassostrea gigas* and *Saccostrea commercialis.* II. Interspecific studies of the genera *Crassostrea* and *Saccostrea, Mar. Biol.,* 54, 157 and 171, 1979.
6. **Héral, M., Deslous-Paoli, J. M., and Prou, J.,** Dynamiques des productions et des biomasses des huîtres creuses cultivêes (*Crassostrea angulata* et *Crassostrea gigas*) dans le bassin de Marennes-Oléron depuis un siècle, *ICES, Mariculture Committee, F,* 41, 22, 1986.
7. **His, E.,** Premiers éléments de comparaison entre l'huîtres protugaise et l'huître japonaise, *Sci. Peches, Bull. Inst. Peches Marit.,* 219, 1, 1972.
8. **Bougrier, S., Gaguenes, G., Bachere, E., Tige, G., and Grizel, H.,** Essai de réimplantation de *Crassostrea angulata* en France: résistance au chambrage et comportement des hybrides *C. angulata - C. gigas, ICES, Shellfish Committee, K,* 36, 14, 1986.
9. **Marteil, L.,** La conchyliculture française. Vol. 2: Biologie de l'huître et de la moule, *Rev. Trav. Inst. Peches Marit.,* 40(2), 149, 1976.
10. **Combs, M.,** Recherches histologiques et cytologiques sur les infections intracellulaires des mollusques bivalves marins, Thèse d'Etat, Université des Sciences et Techniques du Languedoc, 1983, 128.
11. **Korringa, P.,** Experiments and observations on swarming pelagic life and setting in the European flat oyster *Ostrea edulis* L., *Arch. Neerl. Zool.,* 5, 249, 1941.
12. **Héral, M., Deslous-Paoli, J. M., Razet, D., and Prou, J.,** Essais de mises en évidence *in situ* de paramètres biotiques et abiotiques de l'eau et de l'interface eau-sédiment intervenant dans la production de l'huître, *Crassostrea gigas, Oceanus,* 10(4), 465, 1984.
13. **Deslous-Paoli, J. M., Héral, M., Goulletquer, P., Boromthanarat, W., Razet, D., Garnier, J., Prou, J., and Barillet, L.,** Evolution saisonnière de la filtration de bivalves intertidaux dans des conditions naturelles, *Oceanus,* 13(4—5), 575, 1987.
14. **Maurer, D., Comps, M., and His, E.,** Caractéristiques des mortalités estivales de l'huître *Crassostrea gigas* dans le bassin d'Arcachon, *Haliotis,* 15, 309, 1986.

15. **Mori, K.**, Effects of artificial eutrophisation on the metabolism of the Japanese oyster, *Crassostrea gigas, Mar. Biol.*, 53, 361, 1979.
16. **Perdue, J. A., Beattie, J. H., and Chew, K. K.**, Some relationships between gametogenetic cycle and summer mortality phenomenon in the Pacific oyster *(Crassostrea gigas)* in Washington State, *J. Shellfish Res.*, 1(1), 9, 1981.
17. **Neudecker, T.**, Condition index; and morphometric data of oyster *(Crassostrea gigas)* from the German Baltic coast, *ICES, Mariculture Committee F*, 9, 11, 1979.
18. **Yonge, C. M.**, *Oysters*, Collins, London, 1950, 209.
19. **Muranata, M. S. and Lannan, J. E.**, Broodstock management of *Crassostrea gigas*: environmental influences on broodstock conditioning, *Aquaculture*, 39, 217, 1984.
20. **Mann, R.**, Some biochemical and physiological aspect of growth and gametogenesis in *Crassostrea gigas* and *Ostrea edulis* grown at sustained elevated temperature, *J. Mar. Biol. Assoc. U.K.*, 59, 95, 1979.
21. **Héral, M., Deslous-Paoli, J. M., and Prou, J.**, Influence due climat sur le recrutement et sur la production d'huîtres cultivées (*Crassostrea angulata* et *Crassostrea gigas*) dans le bassin de Marénnes-Oléron (France). *ICES, Mariculture Committee F*, 40, 20, 1986.
22. **Lubet, P. E.**, Influence des facteurs externes sur la reproduction des mollusques lamellibranches, *Oceanus*, 6(5), 469, 1980.
23. **Imai, T., Hatanaka, M., Sato, R., Sakai, S., and Yuki, R.**, Artificial breeding of oysters in tanks, *Tohoku J. Agric. Res.*, 1, 69, 1950.
24. **Loosanoff, V. L., Davis, H. C., and Chanley, P. E.**, Dimensions and shapes of larvae of some marine bivalve mollusks, *Malacologia*, 4(2), 351, 1966.
25. **His, E. and Robert, R.**, Developpment des véligères de *Crassostrea gigans* dans le bassin d'Archachon. Etudes sur les mortalités larvaires, *Rev. Trav. Inst. Peches Marit.*, 47(1 and 2), 63, 1985.
26. **Walne, P. R.**, Culture of bivalve molluscs, Fishing News (Books), West Byfleet, 1974, 173.
27. **Medcof, J. C.**, Oyster farming in the maritimes, *Bull. Fish. Res. Board Can.*, 131, 154, 1961.
28. **Helm, M. M. and Millican, P. F.**, Experiments in the hatchery of Pacific oyster larvae (*Crassostrea gigas* Thunberg), *Aquaculture*, 11, 1, 1977.
29. **Askew, C. G.**, The growth of oysters *Ostrea edulis* and *Crassostrea gigas* in Emsworth harbour, *Aquaculture*, 237, 1972.
30. **Walne, P. R. and Mann, R.**, Growth and biochemical composition in *Ostrea edulis* and *Crassostrea gigas*, in *Proc. 9th Eur. Mar. Biol. Symp.*, Aberdeen University Press, 1975, 587.
31. **Hall, S.**, A multiple regression model of oyster growth, *Fish. Res.*, 2, 167, 1984.
32. **Kergariou, G.**, personal communication, 1987.
33. **Pacquotte, P. and Moriceau, J.**, Croissance et indice de condition de l'huître plate *Ostrea edulis* élevée en mer et en étang sur la côte meditérranéenne, *Haliotis*, 16, 427, 1987.
34. **Filic, Z. and Krajnovic-Ozretic, M.**, Novi nalaz Kamenice *Crassostrea gigas* (Thunberg) u Limskom Kanalu, *Ichthyologica*, 10(1), 41, 1978.
35. **Walne, P. R. and Spencer, B. E.**, The Introduction of the Pacific oyster *(Crassostrea gigas)* into the United Kingdom. Shellfish Information Leaflet, No. 21, Ministry of Agriculture, Fisheries and Food, Station Conwy, N. Wales, 1971, 14.
36. **Cooke, R. W. T. and Barry, M. D.**, Growth of the Pacific oyster *(Crassostrea gigas)* at sites on the East, South, and West coast of Ireland during 1973—1974, Bord Iascaigh Mhara Irish Sea Fisheries Board, 12.
37. **Shafee, M. S. and Sabatié, M. R.**, Croissance et mortalité des huîtres dans la lagune Oualidia (Maroc), *Aquaculture*, 53, 201, 1986.
38. **Héral, M., Deslous-Paoli, J. M., and Prou, J.**, Analyse historique de la production conchylicole du bassin de Marennes-Oléron (France), Coll. Fr. Jpn. Oceanogr., Marseille, Sept. 16 to 21, (7), 55, 1985.
39. **Hughes-Games, W. L.**, Growing the Japanese oyster *(Crassostrea gigas)* in subtropical sea water fish ponds. I. Growth rate, survival and quality indes, *Aquaculture*, 11, 217, 1986.
40. **Héral, M., Deslous-Paoli, J. M., and Sornin, J. M.**, Transferts énergétiques entre l'huître *Crassostrea gigas* et la nourriture potentielle disponible dans un bassin ostréicole: premières approches, *Oceanus*, 9(3), 169, 1983.
41. **Askew, C. G.**, A generalized growth and mortality model for assessing the economics of bivalve culture, *Aquaculture*, 14, 91, 1978.
42. **Meixner, R.**, Culture of Pacific oysters *(Crassostrea gigas)* in containers in Germans coastal waters, in *Advances in Aquaculture*, Pillary, T. V. R. and Dill, W. A., Eds., Fishing News (Books), Surrey, 1979, 338.
43. **Neudecker, T.**, Untersuchungers des Einflusses von Aufzucht vefahren und standaortbedingten Umneltfaktaren auf den Abwachs von Mollusken, Forschungsberichte Aquakulture, BMFT-FB-M, 84-0011, 1984, 120.
44. **Seamans, M. N. L.**, Ecophysiological investigations on the oyster, *Crassostrea gigas* Thunberg in Flenburg Fjord. *Veroeff. Inst. Küsten Binnenfish. Hamburg*, 89, 71, 1985.

45. **Auger, C.,** Etude de Deux Variétés de *Crassostrea gigas* Thunberg: Leur Acclimation en Rivière d'Etel (Morbihan), Thèse Eème cycle Université Paris VI, 1976, 104.
46. **Deslous-Paoli, J. M.,** Croissance et qualité de l'huître *Crassostrea gigas* Thunberg en élevage dans le bassin de Marennes-Oléron, *Tethys,* 10(4), 365, 1982.
47. **Berthomé, J. P., Prou, J., and Bodoy, A.,** Performances de croissance de l'huître creuse, *Crassostrea gigas* (Thunberg) dans le basin d'élevage de Marennes-Oléron entre 1979 et 1982, *Haliotis,* 15, 183, 1986.
48. **Landrein, S.,** personal communication, 1987.
49. **Dosdat, A.,** personal communication, 1987.
50. **Bougrier, S. and Deslous-Paoli, J. M.,** Données de base sur la génétique de l'huître plate *Ostrea edulis* et de l'huître creuse *Crassostrea gigas,* in *Génétique Quantitative,* IFREMER, Paris, 1987, 67.
51. **Grizel, H., Comps, M., Cousserous, F., Bonami, J. R., and Vago, C.,** Etude d'un parasite de la glande digestive observé au cours de l'épizootie actuelle de l'huître plate, *C. R. Acad. Sci. Paris D,* 279, 783, 1974.
52. **Pichot, Y., Comps, M., Tige, G., Grizel, H., and Rabouin, M. A.,** Recherches sur *Bonamia ostreae* gen. n., sp. n., parasite nouveau de l'huître plate *Ostrea edulis* L., *Rev. Trav. Inst. Peches Marit.,* 43(1), 131, 1980.
53. **Héral, M.,** L'ostréiculture française traditionnelle, *Aquaculture,* Barnabé, G., Ed., Lavoisier, Paris, 1986, 346.
54. **Gruet, Y., Héral, M., and Robert, J. M.,** Premiéres observations sur l'introduction de la faune associée au naissain d'huîtres japonaises *Crassostrea gigas* (Thunberg) importé sur la côte atlantique française, *Cah. Biol. Mar.,* 17, 173, 1976.
55. Anon., Yearbook of Fishery Statistics, FAO Statistics Series, Vol. 58, Food and Agriculture Organization, Rome, 1986.
56. **Ackefors, H., Mumo, A., Müller, F., and Querellou, J.,** Development of Aquaculture in Europe, FAO ERC, 4, Food and Agriculture Organization, Rome, 1986, 31.
57. **Dijkema, R.,** Assessment of size, distribution and composition of a newly developed stock of the European flat oyster *(Ostrea edulis)* in a stagnant salt water lake in the S. W. Netherlands, *ICES Shellfish Committee K,* 14, 23, 1983.
58. **Dumont, P.,** Le marché de l'huître creuse. Essai de modélisation économique, Rapport ENGREF, Rennes, 1983, 58.
59. **Loosanof, V. L. and Davis, H. C.,** Rearing of bivalve mollusks. *Adv. Mar. Biol. Acad. Press.,* 1(1), 136, 1963.
60. **Walne, R. P.,** Experimental rearing of the larvae of *Ostrea edulis* L. in the laboratory, *Fish. Invest.,* 2(2), 9 and 23, 1956.
61. **Lucas, A.,** Development Strategy for Bivalve Hatcheries in Aquaculture Shellfish Culture Development and Management, IFREMER, Brest, 1987, 249.
62. **Grizel, H., Langlade, A., and Perodous, J. B.,** Premiers essais d'une nouvelle technique de captage d'huître plate en baie de Quiberon, *ICES, Shellfish Committee K,* 24, 14, 1979.
63. **Berthomé, J. P., Prou, J., Razet, D., and Garnier, J.,** Première approche d'une méthode d'estimation prévisionnelle de la production potentielle d'huîtres creuse *Crassostrea gigas* d'élevage, *Haliotis,* 14, 39, 1984.
64. **Hamon, P. Y. and Tournier, H.,** Estimation de la biomasse en culture dans l'étang de Thau, *Sci. Peche Bull. Inst. Peches Marit.,* 318, 38, 1981.
65. **Zanette, Y. and Garnier, J.,** Etude préliminaire de l'impact des huîtres *Crassostrea gigas* (Thunberg) en élevage sur le biomasse des microorganismes des claires de Marennes-Oléron, *ICES, Environmental Quality Committee L,* 14, 17, 1981.
66. **Robert, J. M.,** Fertilité des Eaux des Claires Ostréicoles et Verdissement. Utilixation de l'Azote par les Diatomées Dominantes, Thèse d'Etat, Université de Nantes, 1983, 281.
67. **Robert, J. M., Maestrini, S. Y., Héral, M., and Zanette, Y.,** Production des micro-algues des claires ostréicoles en relation avec l'azote organique dissous excrété par les huîtres, *Oceanol. Acta* n° spécial Unesco Silco, Bordeaux, 389, 1981.
68. **Farley, C. A. Banfield, W. G., Kasnic, G., and Foster, W.,** Oyster herpes-type virus, *Science,* 178, 759, 1972.
69. **Comps, M. and Duthoit, J. L.,** Infection virale associée à la "maladie des branchies" de l'huître portugaise *Crassostrea angulata* Lmk, *C. R. Acad. Sci. Paris D,* 283, 1595, 1976.
70. **Comps, M., Bonami, J. R., Vago, C., and Campillo, A.,** Une virose de l'huître portugaise, *(Crassostrea angulata), C. R. Acad. Sci. Paris D,* 282, 1991, 1976.
71. **Comps, M. and Bonami, J. R.,** Infection virale associée à des mortalités chez l'huître *Crassostrea gigas* Th., *C. R. Acad. Sci. Paris D,* 285, 1139, 1977.
72. **His, E.,** Observations relative à l'infestation de *Crassostrea gigas* Thunberg per le copépode parasite *Mytilicola orientalis* Mori dans le bassin d'Arcachon, *ICES, Shellfish Committee K,* 33, 10, 1977.

73. **His, E., Tige, G., and Rabouin, M. A.,** *Mytilicola orientalis* Mori dans le bassin d'Arcachon, *ICES, Shellfish Committee, K14,* 12, 1978.
74. **Deslous-Paoli, J. M.,** *Mytilicola orientalis* Mori, *Crassostrea gigas* Thunberg's parasite, in the basin of Marennes-Oleron: impact on the condition and the biochemical composition of oysters during rearing, *ICES, Shellfish Committee K,* 29, 16 p, 1981.
75. **Orton, J. H.,** An account of investigations into the cause or causes of the unusual mortality among oysters in English oyster beds during 1920 and 1921, *Minist. Agric. Fish. Fish. Invest.,* London 2(6), 3 and 4, 199 and 3, 1924.
76. **Grizel, H.,** Etude des récentes épizooties de l'huître plate *Ostrea edulis* Linne et de leur impact sur l'ostréiculture bretonne, Thése d'Etat, Université des Sciences et Techniques du Languedoc, 1985, 145.
77. **Grizel, H., Tige, G., and Rabouin, M. A.,** Données sur la situation de la maladie de l'huître plate *Ostrea edulis* Lin, *ICES, Shellfish Committee K,* 18, 7, 1976.
78. **Desportes, I. and Nashed, N. N.,** Ultrastructure of sporulation in *Minchinia dentali* (Arvi) an Haplasparean parasite of *Dentalim Entale* (Scaphopoda, Mollusca), *Taxonomic Implications,* 19, 435, 1983.
79. **Comps, M.,** Observations sur les causes d'une mortalité anormale des huîtres plates dans le bassin de Marennes, *Rev. Trav. Inst. Peches Marit.,* 34(3), 317, 1970.
80. **Perkins, F. O.,** Ultrastructure of sporulation in the European flat oyster pathogen *Marteilia refringens,* Taxonomic implication, *J. Protozool.* 23(1), 64, 1976.
81. **Franc, A.,** Sur quelques aspects inédits du cycle de *Marteilia refringens* Grizel et col, 1974, parasite de l'huître plate *Ostrea edulis* L., *Cah. Biol. Mar.,* 21, 99, 1980.
82. **Cahour, A.,** *Marteilia refringens* and *Crassostrea gigas, Mar. Fish. Rev.,* 41, 19, 1979.
83. **Tige, G. and Rabouin, M. A.,** Etude d'un lot de moules transférés dans un centre touché par l'épizootie affectant l'huître plate, *ICES, Shellfish Committee K,* 18, 17 p, 1976.
84. **Comps, M., Tige, G., and Grizel, H.,** Recherches ultrastructurales sur un protiste parasite de l'huître plate *Ostrea edulis* L., *C. R. Acad. Sci. Paris D,* 290, 383, 1980.
85. **Grizel, H. and Tige, G.,** Evolution of the haemocytic disease caused by *Bonamia ostreae, Ann. Meet. Soc. Invertebrate Pathology,* September 1, 1982, Brighton, U.K., 254.
86. **Van Banning, P.,** Some aspects of the occurrence, importance and control of the oyster pathogen *Bonamia ostreae,* in the Dutch oyster culture, in *Proc. Int. Coll. Inv. Pathology,* September 1982, Brighton, U.K., 261.
87. **Polanco, E., Monte, J., Outon, M. I., and Melendez, I.,** Situation pathologique du stock d'huîtres plates en Galice (Espagne) en relation avec *Bonamia ostreae, Haliotis,* 14, 91, 1984.
88. **Ackefors, H.,** International sanitary regulations with reference to aquaculture, *ICES Mariculture Committee F,* 45, 14, 1986.
89. **Deslous-Paoli, J. M., Héral, M., et Massé, H.,** Bilan énergétique d'une population naturelle de *Crepidula fornicata* (L.) dans le bassin de Marennes-Oléron. Bases Biologiques de l'aquaculture. Montpellier 1983, *Actes Colloq.,* 1, 109, 124, 1983.
90. **Deslous-Paoli, J. M. and Héral, M.,** *Crepidula fornicata* L. (gastéropode, colyptraeidae) dans le bassin de Marennes-Oléron: composition et valeur énergétique des individus et des pontes. *Oceanol. Acta,* 9(3), 305, 1986.
91. **Fiala Medioni, A.,** Ethologie alimentaire d'invertébrés benthiques filtreurs (ascidiés). II. Variation des taux de filtration et de digestion en fonction de l'espéce, *Mar. Biol.,* 28, 199, 1974.
92. **His, E. and Robert, R.,** Action d'un sel organo métallique l'acétate de trigutyétain sur les oeufs et les larves D de *Crassostrea gigas* (Thunberg), *ICES Mariculture Committee, F,* 27, 10, 1980.
93. **Robert, R. and His, E.,** Action de l'acétate de tributylétain sur les oeufs et les larves D de deux mollusques d'intérét commercial *Crassostrea gigas* (Thunberg) et *Mytilus galloprovincialis* (Lmk) *ICES, Mariculture Committee, F,* 42, 16, 1981.
94. **Alzieu, C., Héral, M., Thibaud, Y., Dardignac, M. J., and Feuillet, M.,** Influence des peintures antisalissures á base d'organo stanniques sur la calcification de la coquille de l'huître *Crassostrea gigas, Rev. Trav. Inst. Peches Marit.,* 45(2), 101, 1982.
95. **Héral, M., Berthomé, J. P., Polanco Torres, E., Alzieu, C., Deslous-Paoli, J. M., Razet, D., and Garnier, J.,** Anomalies de croissance de la coquille de *Crassostrea gigas* dans le bassin de Marennes-Oléron. Bilan de trois années d'observations, *ICES, Shellfish Committee K,* 31, 16, 1981.
96. **Alzieu, C. and Héral, M.,** Ecotoxicological effects of organotin compounds on oyster culture, in *Ecotoxicological Testing for the Marine Environment,* Vol. 1, Persoone, G., Jaspers, E., and Claus, C., Eds., State University of Ghent and Inst. Mar. Sci. Res., Bredene, Belgium, 1983, 187.
97. **Waldock, M. J. and Thain, J. E.,** Shell thickening in *Crassostrea gigas*: organotin antifouling or sediment induced, *Mar. Pollut.,* 14, 411, 1983.
98. **Thain, J. E. and Waldock, M. J.,** The growth of bivalve spat exposed to organotin leachates from antifouling paints, *ICES, Marine Environmental Committee E,* 13, 5, 1985.
99. **Héral, M., Alzieu, C., and Deslous-Paoli, J. M.,** Effects of organotin compounds (TBT) used in antifouling paints on cultured marine molluscs: a literature study, in *Proc. Aquaculture Europe 87,* Jaspers, E., Ackefors, H., and Wilkins, N., Eds., in press.

100. **Bonnieux, F., Dauce, P., and Rainelli, P.,** Impact socio-économique de la marée noire provenant de l'Amoco-Cadiz, Rapport INRA-U.V.L.O.E., 1980, 100.
101. **Meuriot, E. and Grizel, H.,** Note sur l'impact économique des maladies de l'huître plate en Bretagne, unpublished data, 1988, 18.
102. **Gowen, R. J.,** Toxic phytoplankton in Scottish coastal waters, *Rapp. P. V. Reun. Cons. Int. Explor. Mer,* 187, 89 and 93, 1987.
103. **Tangen, K.,** The effects of phytoplankton blooms on mariculture in Norway, *Rapp. P. V. Reun. Cons. Int. Explor. Mer,* 187, poster D: 1, 1987.
104. **Fraga, S., Marino, J., Bravo, I., Miranda, A., Campos, M. J., Sanchez, F. J., Costas, E., Cabanas, J. M., and Blanco, J.,** Red tides and shellfish poisoning in Galicia (NW Spain), *Rapp. P. V. Reun. Cons. Int. Explor. Mer C,* 187, 5, 1987.
105. **Moita, M. T., Cunba, M. E., and Sampayo, M. A. M.,** Portuguese coastal red waters, *Rapp. P. V. Reun. Cons. Int. Explor. Mer B,* 187, 9, 1987.
106. **Parker, M.,** Exceptional plankton blooms, conclusions of discussions: Conener's report, *Rapp. P. V. Reun. Cons. Int. Explor. Mer,* 187, 108, 1987.
107. **Lassus, P.,** Bilan des eaux colorées observées sur le littoral française entre 1975 et 1982, *ICES, Environmental Committee L,* 29, 1987.
108. **Kat, M.,** Diarrhetic mussel poisoning, *Rapp. P. V. Reun. Cons. Int. Explor. Mer,* 187, 83, 1987.
109. **Dunne, T., Doyle, J., and Parker, M.,** Review of the incidence and effects of unusual marine blooms in Irish coastal waters 1976-1984. *Rapp. P. V. Reun. Cons. Int. Explor. Mer,* 187, poster D: 9, 1987.
110. **Caumette, P. and Baleux, B.,** Etude des equx rouges dues à la prolifération des bactéries photosynthétiques sulfoxydantes dans l'étang du Prévôt (lagune saumâtre méditerranéenne), *Mar. Biol.,* 56, 183, 1980.
111. **Coeurdacier, J. L.,** Réserves d'eau de mer et dégorgeoirs, *Equinoxe,* 8(6), 11, 1986.
112. Anon., Recommended International Code of Practice for Molluscan Shellfish, Joint FAO/WHO Standards Programme, Codex Alimentarius Commission, CAC/RCP, 18, 1978, 23.
113. Anon., Directive du Conseil du 30 octobre 1979 relative à la qualité requise des equx conchyliocoles 79/923/CEE, *J. Off. Commun. Eur.,* 10.11.79, L 281, 47, 6 p, 1979.
114. **Aure, T., Müller euga, A., Priour,** La mytiliculture en mer ouverte, *Cultures Mar.,* 145, 4, 1987.
115. **Priour,** Matériels conchylicoles: quoi de neuf? *Equinoxe,* 9, 23, 1986.
116. **Mialhe, E., Boulo, E., Grizel, H., Rogier, H., and Paolucci, F.,** Monoclonal antibodies: a new tool in molluscan pathology, in *Disease Processes in Marine Bivalve Molluscs,* Spec. Publ. Ser., American Fisheries Society, in press.
117. **Mialhe, E., Bachére, E., Chagot, D., and Grizel, H.,** Isolation and purification of the protozoan *Bonamia ostrae* (Pichot and Coll, 1979), a parasite affecting the flat oyster *Ostrea edulis* L., *Aquaculture,* in press.
118. **Boulo, V., Rogier, H., Paolucci, F., Mialhe, E., and Grizel, H.,** Immunodiagnostic of *Bonamia ostrea* (Ascetospora) parasite of *Ostrea edulis* and subcellular identification of epitopes by monoclonal antibodies, *J. Fish Dis.,* in press.
119. **Cochennec, N., Boulo, V., Mialhe, E., Grizel, H., Rogier, H., Clavies, C., Hervand, E., Pau, B., Puygrenier, M., and Paolucci, F.,** Use of monoclonal antibody in an enzyme immunoassay for the detection of *Bonamia ostrea* in the hemolymph of flat oysters, *Ostrae edulis,* 3rd Int. Conf. European Association of Fish Pathologists, September 1987, Bergen, in press.
120. **Bacher, C., Baud, J. P., Bodoy, A., Deslous-Paoli, J. M., Dréno, J. P., Héral, M., Maurer, D., and Prou, J.,** A methodology for the stocks assessments of cultivated oysters along the French Atlantic coasts, *ICES, Shellfish Committee K,* 36, 14, 1986.
121. **Deslous-Paoli, J. M., Populus, J., L'Herroux, M., and Brussieux, M.,** Estimation des stocks cochylicoles par télédétection: résultats et perspectives, *Coll. Fr. Jpn. Oceanogr.,* Marseille September 16-21, 1985, 7, 39, 1985.
122. **Deslous-Paoli, J. M. and Héral, M.,** Transferts énergétiques entre l'huître *Crassostrea gigas* de 1 an et la nourriture potentielle disponible dans l'eau d'un bassin ostréicole, *Haliotis,* 14, 79, 1984.
123. **Bacher, C.,** Modélisation de la Croissance de Huîtres dans le Bassin de Marennes-Oléron, Rapp. IFREMER, 1987, 12.
124. **Quillett, Et. and Panelay, P. J.,** Triploidy induction by thermal shocks in the Japanese oyster, *Crassostrea gigas, Aquaculture,* 57, 271, 1986.
125. **Downing, S. L. and Allen, J. K.,** Induced triploidy in the Pacific oyster, *Crassostrea gigas*: optimal treatments with Cytochalasin B depend on temperature, *Aquaculture,* 61, 1, 1987.
126. **Gilly, B. and Meuriot, E.,** Bio-Economic Analysis of Shellfish-Culture Activity in France: Prospect and Limits in Aquaculture, Shellfish Culture Development and Management, IFREMER, Brest, 1987, 367.

Chapter 14

MUSSELS IN INDIA

R. Nagabhushanam and U. H. Mane

TABLE OF CONTENTS

I. BIOLOGY

A. DESCRIPTION

Two species of the genus *Perna* (family Mytilidae) exist along the Indian coast, *Perna* (formerly *Mytilus*) *viridis* (Linnaeus) (the green mussel) (Kuriakose and Nair) and *P. indica* (the brown mussel).[1-3]

Perna is characterized by the presence of only one or two well-developed hinge teeth, the absence of the anterior adductor muscle, the wide separation of the two well-marked posterior byssal retractors, and the recurrent loop of the midgut lying at the left lateral side of the stomach and separating the crystalline style sac from the midgut.

The shells of both species are thick, equivalve, inequilateral, elongate, and triangularly oval in outline. The shell of *P. virudis* reaches 230 mm in length (commonly 180 mm) and 72 mm in height. The shell of *P. indica* reaches 121 mm in length (commonly 80 mm) and 48 mm in height. In *P. viridus* the umbo is terminal; the hinge plate is well developed and extends slightly ventrally, with two small teeth on the left valve and one large one on the right valve. In *P. indica* the umbo is terminal; the umbonal beaks are poorly developed and are terminal or slightly downward turned in adults. The hinge plate is narrow and thin, and a developed tooth on the left valve fits into a corresponding depression on the right valve. In *P. viridis* the dorsal ligament margin is curved, the middorsal margin is arcuate, the posterior margin is rounded, and the ventral margin is highly concave. In *P. indica* the dorsal ligament margin is straight, the middorsal one is highly angular with a well-developed hump where the shell is highest, the posterior one is rounded, and the ventral margin is straight. The ligament of *P. viridis* is thicker than that of *P. indica.* The external color of *P. viridis* is green, but in older mussels the anterior half may turn bluish green. *P. indica* is dark brown in color. In both species, the interior of the shell is pearlescent and shiny, and the muscle scars are deeply impressed.

In addition, the anterior adductor muscle is absent in both species. In *P. viridis* the posterior adductor is slightly elongated and located in the posterior half of the shell a little above the anteroposterior axis of the body. In *P. indica* the posterior adductor muscle is rounded and is located toward the dorsal shell margin about midway between the posterior termination of the ligament and the posterior shell margin. The anterior byssal retractor joins the shell a little behind the umbonal cavity in *P. viridis*, but in *P. indica* it inserts a little behind the umbonal cavity. The posterior byssal retractor in both the species arises as a common bundle from the base of the byssal apparatus and then splits into a "V". The anterior byssal retractor inserts on the shell below the posterior termination of the ligament, and the posterior joins the shell along with posterior adductor bundle at its anteriodorsal side.

In *P. viridis*, the mantle margin is yellowish-green, smooth, thin, and slightly extensible; and tentacles or papillae are absent. However, in *P. indica* the mantle margin is brown; the portion bordering the incurrent aperture is very thick and nonextensible; and the inner fold of the mantle margins possesses long, stout, brown branching tentacles.

P. viridis is commonly known as "kakkai", "vakundi", or "shingatale" in the Mahrashtra and Goa areas; as "kulachi" in the Malwan area; as "nilakal" or "neelkallu" in Kerala; and as "alichippalu" in Andhra Pradesh. *P. indica* is known as "chippi", "muthuva", or "muthuvachippi" in Kerala and as "kakkai", "chippi", or "kalluthodu" in Tamil Nadu.

B. RANGE

P. viridis occurs on the west coast of India in extensive beds in Quilon, Alleppey, Cochin, Calicut to Kassargade, Mangalore, Karwar, Goa, Malwan, Ratnagiri, and the Gulf of Kutch. On the east coast it occurs in small beds at Chilka Lake, Visakhapattanam,

Kakinada, Madras, Pondichery, Cuddalor, Porto Novo, and Port Blair; *P. indica* also has a very restricted distribution along the southeast coast from Cape Comorin to Tiruchendur and along the southwest from Varkalai near Quilon to Cape Comorin. Important centers are Cape Comorin, Muttom, Poovar, Vizhinham, Kovalam, Varkalai, and Quilon.

C. TYPE OF SUBSTRATE

P. viridis occurs along the rocky open coast and in the mouths of estuaries and rivers where the salinity almost equals that of seawater. It occurs from the intertidal zone to a depth of 15 m on rocks, pilings, and other hard objects. *P. indica* forms dense beds along rocky coasts from the intertidal zone to a depth of 10 m, with large mussels found in a 0.5 to 2 m depth.

D. SALINITY AND TEMPERATURE TOLERANCE

The total absence of *P. viridis* from the southeast coast and the restriction of *P. indica* to the extreme southwest are not easily explained. Population densities differ at apparently similar sites; thus, physical factors are not the sole influence. Other unknown environmental parameters must play a part.

Mass mortality of the adults of *Perna* has never been observed. *P. viridis* in the Ennore estuary at Madras has been shown to tolerate salinity as low as 16‰ under laboratory conditions,[4] with a slow decrease in byssal activity and formation below 17.5‰. In this estuary, the bar remains open throughout the year, and as a result, except during the northeast monsoon when there is a significant decrease in the salinity, salinities comparable to those of inshore waters are prevalent. In the habitat of *P. viridis* in the Ennore estuary, the salinity reaches 18‰ in monsoon season (October to December) and is otherwise near 35‰.[5] The metabolic activity of this mussel remains high in higher salinity, and this increase is due to a function of duration of exposure when the mussels are exposed to air.

The salinity tolerance range of *P. indica* is between 20.7 and 45.6‰.[6] The optimal salinity for activity and secretion of byssal threads is from 20.7 to 35.4‰. Green mussels transplanted to a raft in an estuary along the Goa coast develop an isosmotic internal medium to compensate for the considerable lowering of salinity during monsoon season (June to September).[7] The compensation is achieved by the dilution of the body fluids, resulting in higher water content in the tissues. In the postmonsoon (October to January) and premonsoon (February to May) seasons, an appropriate compensatory mechanism is developed to counteract increasing salt content in the environment. Earlier studies on this species from Ratnagiri coast have also found an increase in the water content of the whole body and different body parts during monsoon season.[8,9]

E. SEXUAL DEVELOPMENT AND SPAWNING

P. viridis from the Madras coast has been reported to spawn throughout the year.[10] In *P. viridis* the ripe, round ova measures 33 to 64 μm, with the majority around 40 μm, whereas the spermatozoan in a ripe male measures about 1 μm in head and 25 to 30 μm in length of its slender tail.[11] No change in sex is reported during the phase of redevelopment after the mussels have passed into an intermediate condition. Gonadal changes in older individuals (60 mm in shell length and 6 months old) are different from that in younger animals (below 60 mm in length). The mussels appear to breed throughout the year in the Madras region. The older ones breed from about July to December, with peak activity occurring in September to November; and younger individuals set in the monsoon season grow rapidly to start breeding from January to April with a minor peak of activity occurring in February and March.[11] Such differences have not been examined on the Ratnagiri coast where studies have been restricted to adult populations, 70 to 80 mm in length, found in Bhatia Creek at Ratnagiri.[12] Green mussels from Kakinada Bay show limited numbers of hermaphrodites (0.8%).[13]

Many different stages have been described in *P. viridis* during histological studies.[11-13] Every description states that after spawning the follicles become flaccid and the intervesicular tissue gradually enlarges. The residual gametes disintegrate and undergo cytolysis, and the gonad passes into an intermediate condition. The follicles at this stage are reduced to streaks lined by few dormant germinal epithelial cells and with much enlarged intervesicular tissue loaded with reserve materials. With the development of the gonad, reproductive elements fill the follicles, which consequently spread widely among the viscera, particularly in the mantle, leaving no intervesicular space.

In green mussels from the Ratnagiri coast, the reorganization of follicles commences in May, and a few mussels change sex, causing fluctuations in the sex ratio.[12] Hermaphrodites occur as males become female. Gametogenic activity in males commences near the end of May and in females, in early June; no such difference in timing has been found in the second peak of gametogenic activity in January. Spawning begins by the third week of July and continues until early September. Hermaphrodites occur again in September and October, when there is a sex change from female to male. Reorganization of follicles and maturation of sex products take place in succeeding months, and by February or March another partial spawning occurs in a few mussel populations.[12] The rare occurrence of individuals with gametes of both the sexes is an indication that hermaphroditism is not a regular feature and should be regarded as a purely transitional phase. In Kakinada Bay about 60% of mussels in natural beds spawn between December and July, and the peak activity occurs between January and May.[13]

At Vengurla, the salinity fluctuates from 33.2 to 37.5‰ (August to December), and the temperature varies from 26 to 31.5°C (November to March).[11] The mussels show vigorous spawning along the Goa coast soon after the monsoon season. Moderate temperature fluctuations of 26 to 29.6°C and a sudden decrease in salinity from 18 to 3‰ have been correlated with the major spawning period of mussels in Bhatia Creek at Ratnagiri. Rising salinity (from 32.5 to 33‰) and low temperature (from 24 to 27°C) combined with the effects of tides cause a further minor spawning in February and March.[12] The low-salinity period of reproductive activity, alternating with the summer period of high salinity and reproductive inactivity, lends direct support to the role of salinity change as a determining factor in controlling the gametogenic cycle of the mussel population along the Ratnagiri coast. Rising water temperature and high salinity have been correlated with spawning, and the period of decreasing temperature and low salinity coincides with the resting condition of the gonads of mussels from Kakinada.[13]

Recent studies have focused on the sexual maturation and spawning of the brown mussel, *P. indica*. Farm-grown brown mussels from Vizhinjam reach sexual maturity and spawn from the end of May until September,[14] with peak activity occurring in July and August.

Studies have been made of cultured green mussels. Mussels grown on ropes suspended at Kottalaiyer estuary, Ennore, on the east coast begin spawning by the middle of April, and spat appear on the ropes by the middle of May.[15] In green mussels (mean size 7.8 mm) that are transplanted to the Goa coast, gametogenesis commences in December, and more than 60% of the mussels reach maturation. Mature gonads in February constitute 10% of body weight at an average size of 15.5 mm and increase from 38 to 65% of body weight between March and May. Spawning in such mussels commences in February, intensifies in March and April, and attains a peak in May. The resting phase coincides with monsoon season (June and July), a period of low temperature and salinity and scarcity of food. Gonads again mature by August, and mussels spawn from September onward.[7]

Age and size at sexual maturity differ with locality. For example, in natural beds at Velsao, spatfall of green mussels occurs at least four or five times during the year. Mussels mature at about 3 months old, and the minimum size at maturity is 15.5 mm.[16]

Induction of spawning in the laboratory has been examined. Application of thermal, mechanical, and chemical stimuli has induced spawning in green mussels from Goa coast after feeding unialgal cultures of *Tetraselmis gracilis, T. chui,* and *Synecystis* spp. Raising the temperature from 26.5 to 28°C causes the mussels to spawn within 30 min. Males spawn more readily than females. Partial success in spawning has been obtained by injection of 1% aqueous solution of potassium chloride or liquid ammonia, but this causes failure in the later phases of development of larvae. The spawned eggs are spherical (45 to 50 μm in diameter) and reddish orange, filled with translucent yolk granules, and surrounded by vitelline membrane.[17]

Endogenous factors regulating sexual maturation and spawning of green mussels have been recorded.[18] Annual variation is seen in the quantity of neurosecretion from cerebral ganglia of these mussels, corresponding to the phases of gonad development and spawning. Neurosecretion is maximal in ripe mussels; the neurosecretory cells appear empty and vacuolated in spent individuals.

F. LARVAL DEVELOPMENT

Eyed larvae (those with dark pigment in the center of the body) of *Perna viridis* occur in the natural beds at Vengurla and Panjim from October to December (at the time of increasing salinity); dense setting occurs in October and November.[11] The larvae increase in number by March (at a time of suddenly rising temperature and high salinity) due to breeding by newly growing mussels from the spat that settled during monsoon and post-monsoon seasons. The eyed larvae measure 275 μm or slightly more in length. After the monsoon season, from November to April at most places, *P. viridis* sets densely from about 1 m above low-water-spring tide to varying depths depending on the substratum. The setting of the seed mussels generally coincides[3] with the colder months (November to January), at a temperature range from 26 to 31.5°C and a salinity range from 33.2 to 37.5‰.

Larval development of the artificially fertilized green mussel has been studied on the Goa coast.[17] Spawned eggs before fertilization measured 45 to 50 μm, and fertilization occurred within 30 min. Actively moving trochophore larvae measured 58 μm in length about 6 h after fertilization. The rudiments of the shell in the trochophore enlarged in 18 h to form two dorsally hinged valves. The smallest straight-hinged veliger was 75 μm long and 50.9 μm high, with a hinge 65 μm long. At the end of 48 h, the veliger had reached 88 μm in length and 65 μm in valve height, with a hinge of 64 μm. At 9 d old, larvae were still in the straight-hinged stage. A well-defined umbonal-stage veliger occurred after 14 d. On the 16th d, larvae measured 278 μm in length and 260 μm in height. The pediveliger stage occurred from day 16 onward.

Artificial development of the brown mussel from the Trivandrum coast has also been reported.[19] The ripe ovum measured 55 μm. The trochophore stage is reached in about 12 to 15 h, and the early straight-hinged larval stage in about 24 h. The early veliger stage is reached within the third day, the hinged stage on the seventh day, and the late veliger on the tenth day (325 [long] × 265 [high] μm). Metamorphosis and settlement have been observed within 1 to 3 weeks. The most suitable salinity range for early development of *P. indica* is between 27.8 and 35.5‰.[6]

G. GROWTH

Along the Indian coast, mussels grow faster than those in temperate waters. Growth is fast at higher salinity, and the species is not found in low-salinity areas. On the east coast of Madras, *P. viridis* has been shown to grow to 14.5, 19.4, 55.5, and 93 mm in shell length in 30, 84, 167, and 321 d, respectively;[10] and to reach 103 mm in 445 d. On the west coast of Ratnagiri, wild mussels grow at a rate of 7.55 mm (and 3 g)/month. Mussels that measured 8 mm in October 1973 reached 31 mm in January 1974, 61 mm in May 1974, and 90 mm by the end of a year. Those growing in hanging cages and on ropes grew faster:

8 to 10 mm and 4 to 5 g/month.[20] Offspring of the species that settle on navigational buoys in Vengurla Bay in October reach a marketable size (60 to 64 mm) in a little more than 6 months. With continued growth, they reach 96, 132, and 156 mm at the ends of the first, second, and third years, respectively.[11] Growth is faster on ropes than in natural beds. Cultivated mussels reach 85 mm in length in 11 months, an average of almost 8 mm/month, whereas wild mussels average 5 mm/month.[16] The rope-grown mussels reach marketable size in less than 5 months and also show a higher condition index than do wild ones.

The difference in growth rate between cultured and naturally occurring mussels seems to be greatest during the premonsoon season in Goa, when mussles on ropes can show an average increase in weight of 11.3 g, almost three times that reported for cultured mussels from the Ratnagiri coast.[20] During postmonsoon season, maximum growth rate is about 6.6 mm/month for both ropes and natural beds.

At Vizhinjam, mussels are cultured on ropes both in the bay and in the open sea.[24] The average annual rate of growth in the bay is 35.3 mm, but growth is faster in the open sea, where mussels can grow 25 mm in 5 months with proportionately larger gains in meat weight. A number of authors report growth rates and allometric relationships for cultured and natural beds at other localities.[21-27]

Brown mussels are cultured at Vizhinjam, and again, those cultured at sea grow faster than those in the bay, reaching marketable size in 5 as opposed to 7 months.[24] Kuriakose[6] has reported that brown mussels grown on coir ropes suspended from wooden rafts anchored in the open sea at Calicut, grow faster (reaching 80 mm in 5 months) than those cultured inshore.

The average edible portion of the meat in cultured mussels ranges from 34.8 to 40.5%, whereas the yield from natural beds is only 27.2 to 33.2% of the total weight.

II. FISHERY

A. HISTORY

The natural mussel beds along the Indian coast are being depleted rapidly. The problem is particularly noticeable along the coasts of Mahrashtra, Kerala, Karnatka, Tamil Nadu, and Andhra Pradesh[16] and is probably due to indiscriminate exploitation of the mussels. Construction of jetties, boat pens, and oil terminals and the rapid increase in industrialization along the coast also contribute to the depletion. Humam habitation along the coast and population increase have changed many of the stony and graveled beds that are ideal for spatfall into muddy and polluted areas totally unsuitable for mussel settlement. The navigation channel of Uppateru Canal near Kakinada is regularly dredged, and the dredged earth is dumped on the natural mussel beds; this practice is denuding the beds of mussels. Along the southwest coast, brown mussel landings have declined during the last few years; the probable reasons are indiscriminate fishing by local fishermen and poor settlement of these mussels in the natural beds.[23]

The scope for increasing production of mussels along the Indian coast by traditional fishing methods is not likely to exceed two or three times the present level. The only alternative is improved culture techniques. Various experiments conducted at Vizhinjam Bay since 1971 have shown that floating raft culture is ideal along the Indian coast because of the enhanced growth rate it yields and the simplicity of the technique. Under favorable conditions, raft culture or suspended culture is the most efficient method and avoids the serious problem of predation on young mussels by starfish and crabs.

B. IMPORTANCE

During monsoon season, when trawling and other conventional methods of fishing are suspended and fresh fish is not easily available in the market, mussels and other shellfish form the main seafood. The average market price of green mussels in the shell was about U.S. $0.38 to $0.61/kg from November through April 1984, and the meat was sold at a price of about U.S. $0.53 to $1.31/kg from November through March 1984 along the Maharashtra and Goa coasts. Preliminary experiments on mussel culture at Vizhinjam obtained a yield of about 200 t/ha/year.[28] In 1974 the mussel beds at Kakinada produced mussels weighing 11 t from an area of 1.47 ha with density variation from 0 to 20 mussels/m^2 and an accompanying size range of 27 to 160 mm.[13] Annual fishing production has been 5 t from these natural beds. The annual yield of green mussels from the natural beds of Goa is approximately 200 t/ha. The annual production from culture in the same area is on the order of 23 kg of mussel meat per rope or approximately 1150 kg of mussel meat per raft—equivalent to about 480 t/ha of mussels. The average annual landing of these cultured mussels at Calicut is about 900 t.[26] At Kakinada, on the east coast, these mussels form an estimated 7 t of annual catch.[29] Raft culture of brown mussels at Vizhinjam gave an yield of 150 t from 1 ha inside the bay.[30]

C. METHODS OF HARVESTING

Mussel fishing is generally a part-time occupation except during peak seasons. Low tide, clean water, and clear sunny days are preferred for fishing. Two to three fishermen go out in a wooden boat and pick the mussels by hand at low tide. A crowbar may be used to lift larger stones so that the mussels underneath can be collected, and diving is often necessary in deeper water. From the rocky intertidal zone, mussels are detached with iron implements like chisels or knives during low tide. The mussels are kept in nylon or coir bags tied around the collector's waist. The mussels are cleaned of fouling organisms, graded according to size, and sold to consumers and merchants in fresh condition.

D. SEED COLLECTION

The seed for farming operations must often be collected from areas far from the farming site. Green mussel seed suitable for transplantation (5 to 10 mm and about 4 to 6 weeks old) becomes very abundant at Velsao (Goa), about 20 km from Dona Paula, from September to November.[16] Along the Madras coast green mussel seed (10 to 20 mm) can be collected at Ennore from concrete piles and large iron buoys and transported to Kovalam.[27] Along the west coast, green mussel seed between 15 and 25 mm in length can be obtained from September to December. Along the east coast, May to June appears to be ideal season for seed collection.

Brown mussel seed ranging in length from 25 to 30 mm can be collected from July to November[31] by several methods such as (1) suspending coir and nylon ropes or spat settlers and (2) covering frilled polyester ropes or iron hapas with nylon netting during spawning periods. For transportation purposes seed can be maintained for up to 9 h out of water if the seed cages are covered with burlap soaked in seawater and periodically sprinkled with seawater, or if the mussels are transported in earthen pots or nylon net cages with seawater-soaked wood shavings.

E. USES AND PROCESSING

Mussels harvested in India are used both as food and as a source of lime. The shells are burned and processed for lime preparation, particularly along the Kerala coast. Mussels are generally sold as fresh shucked meats or in the shell. Refrigeration of shucked meats is not widely used because the mussels are not transported far inland. Studies of the biochemical and organoleptic changes in green mussel meat during preservation on ice and meat shucked from fresh and boiled mussels have shown that meats further processed into canned produce are comparatively better when canned in oil.[33] Several methods of preparing ready-to-serve, pickled products from green mussels, along with their storage characteristics, have been reported.[34] Among the various products—dried and pickled, fried and pickled, and light smoked and pickled—the latter shows the best shelf life.

F. CONSTRAINTS

Green mussels grown on floating buoys and in natural habitats at Vengurla and Goa support a variety of fouling aggregations, particularly green algae.[11] Some brown algae, hydroids, planarians, polychaetes, spat of *Saccostrea cucullata* (Born), gastropods, and barnacles also occur. Crabs cause a certain amount of damage to the mussel population. The most common species of commensal crab, *Pinnotheres* sp., is frequently found within the mantle cavity of the mussel. Green mussels cultured on rafts off Calicut show heavy settlement of barnacles, *Balanus amphitrite* (Darwin), and larvae of a bivalve, *Mytilaster arculatula* (Hanley), all over the submerged parts of the rafts and ropes. Tunicates, polychaetes, ascidians, coelenterates, and bryozoans are also common on mussels and ropes.[19] Brown mussels cultured on ropes in Vizhinjam Bay show various fouling organisms, including *Balanus amphitrite, S. cucullata,* and *Modiolus* sp. Several crabs, *Thalamita* sp. and *Porcellana* sp., are also found.[35]

Along the coast of Goa green mussels show heavy infestation of trematode parasites, particularly in the mantle,[11] which acquires a misleading orange-yellow coloration. Sexually matured digenetic trematodes in green mussels are also reported from Narakkal in Kerala.[36]

A quantitative estimation of the bacterial load of the brown mussels cultured in Vizhinjam Bay has shown that infestation can result in poor growth, thin meat, change of normal color, and failure of byssal development and can lead to paralytic shellfish poisoning problems.[37] Coliforms, fecal streptococci, and coagulate-positive staphylococci are reported. *Pseudomonas, Vibrio,* and *Micrococcus* are seen in the normal flora of the mussels. Fishes such as *Diodan hystrix* (Linnaeus), *Arius* sp., *Arothron nigropunctatus* (Bloch), *Ostracion* sp., *Chaetodon,* and *Acanthurus* sp. attack these mussels in culture.[38] The floating rafts are also attacked by shoals of silver bream, *Rabdosargus sarba* (Forskal), the stomachs of which are sometimes full of crushed mussels.

Pollution of natural mussel beds increases from day to day all along the coast. Most studies focus on estimating the quantity of any pollution substance accumulated in the body of mussels.[39,40] A study of the effect of pollution from coconut-husk netting on reproductive potential of green mussels off Elathur near Calicut showed that spawning in these mussels is fractional and incomplete. Growth is also affected.[41] As a result of the expansion of industries and the establishment of nuclear power plants, studies have been made of the concentrations of heavy metals in different parts of the mussels around Madras.[42] Similar studies have been made of mussels from the Bombay coast.[43] Titanium dioxide effluent, if present even in low concentrations in seawater, seems to affect the distribution and colonization of *P. indica.*[44] Their absence in the vicinity of the discharge site of the factory at Trivandrum is a clear indication of their reaction to this effluent stress even though a variety of suitable substrates are available for settlement and growth.

G. MANAGEMENT

The submerged rafts in the open sea at Kovalam offer the possibility of reaping a mussel harvest by seeding during June to avoid the high winds of southwest monsoons and harvesting the stock during October before the commencement of northeast monsoons, which often develop into severe cyclonic storms.[45] In Kakinada Bay the best strategy for farming of green mussels would be to seed the ropes hung from the rafts from February through April and to harvest by September before commencement of the northeast monsoon season.[13]

If the brown mussel culture is properly maintained on ropes by raft technique, the materials used can be recycled for at least a period of 3 to 5 years.[30] Heavy surf action and drift restrict open sea brown mussel culture from December to May, but within this period raft-cultured green mussels grow to the harvestable size of 60 to 65 mm length within 5 months.

III. RESEARCH AND DEVELOPMENT

Studies along both east and west coasts have yielded ways to improve mussel production, but many areas remain to be addressed before large-scale production is attempted. The most important factor is the availability of technologies for the production of seed. Until now, seed material has always been obtained from the wild. Profuse spatfall takes place on granite embankments and groins laid along the coast of Kerala,[6] and there is a need for improvement of the techniques of spat collection toward increasing production by this method. Breeding stocks of the mussels could also be kept on the mussel culture farms for production of seed. Dense spatfall could be collected on tiles.

Alternatively, artificial seed production has the advantage of assured seed supply and stock improvement through genetic manipulation. Aspects such as induced spawning, rearing of larvae, and problems of hatchery production need more attention and intensive research. In this field, considerable effort is required in India.

REFERENCES

1. **Hornell, J.,** The common mulluscs of South India—Rep. VI of 1921 (Madras Fish. Invest., 1921), *Madras Fish. Bull.*, 97, 1922.
2. **Satyamurti, S. T.,** The mollusca of Krusadai Island (in the Gulf of Mannar) II. Scaphopoda, Pelecypoda and Cephalopoda, *Bull. Madras Govt. Mus., New Ser., Nat. Hist. Sect.*, 1(2), 1, 1956.
3. **Kuriakose, P. S.,** Studies on the Mytilidae of the Indian coasts, Ph.D. thesis, Kerala University, India, 1973.
4. **Sundaram, K. S. and Shaffee, M. S.,** On the salinity tolerance of some bivalves of the Ennore estuary, *3rd All India Symp. Estuarine Biology*, University Chochin, India, Abstr. 1, 12, 9, 1975.
5. **Shaffee, M. S.,** Effect of salinity and time of exposure to air on the metabolism of green mussel, *Mytilus viridis*, L., *Indian J. Mar. Sci.*, 5, 130, 1976.
6. **Kuriakose, P. S.,** Salinity tolerance of brown mussel, *Perna indica, Proc. Symp. Coastal Aquacult., Mar. Biol. Assoc. India*, Cochin, Abstr. 249, 146, 1980.
7. **Parulekar, A. H., Dalal, S. G., Ansari, Z. A., and Harakantra, S. N.,** Environmental physiology of raft-grown mussels in Goa, *Aquaculture*, 29, 83, 1982.
8. **Mane, U. H. and Nagabhushanam, R.,** Body distribution and seasonal composition of the estuarine mussel, *Mytilus viridis* at Ratnagiri, *Riv. de Idrobiol.*, 14, 163, 1975.
9. **Nagabushanam, R. and Mane, U. H.,** Seasonal variation in the biochemical composition of *Mytilus viridis* at Ratnagiri on the west coast of India, *Hydrobiologia*, 57(1), 69, 1978.
10. **Paul, M. D.,** Studies on the growth and breeding of certain sedentary organisms in the Madras Harbour, *Proc. Indian Acad. Sci.*, Bangalore, India, 15B, 1, 1942.
11. **Rao, K. V., Krishna Kumari, L., and Dwive, S. N.,** Biology of the green mussel, *Mytilus viridis, Indian J. Mar. Sci.*, 4(2), 189, 1975.
12. **Nagabhushanam, R. and Mane, U. H.,** Reproduction in the mussel, *Mytilus viridis* at Ratnagiri, *Bull. Dep. Mar. Sci. Univ. Cochin. India*, 7(2), 377, 1975.
13. **Narasimham, K. A.,** Fishery and biology of the green mussel, *Perna viridis* (Linnaeus), *CMFRI Bull.*, Cochin, India, 29, 10, 1980.
14. **Appukuttan, K. K. and Nair, T. P.,** Fishery and biology of the brown mussel, *Perna indica* Kuriakose and Nair, *CMFRI Bull.*, Cochin, India, 29, 5, 1980.
15. **Rangarajan, K.,** Aspects of mussel culture of an estuarine environment, *Indian J. Fish.*, 24(1/2), 237, 1977.
16. **Qasim, S. Z., Parulekar, A. H., Harkantra, S. N., Ansar, Z. A., and Nair, A.,** Aquaculture of green mussel *Mytilus viridis* L.: cultivation on ropes from floating rafts, *Indian J. Mar. Sci.*, 6(1), 15, 1977.
17. **Rao, K. V., Krishna Kumari, K., and Qasim, S. Z.,** Aquaculture of green mussel *Mytilus viridis* L.: spawning, fertilization and larval development, *Indian J. Mar. Sci.*, 5(1), 113, 1976.

18. **Mane, U. H. and Nagabhushanam, R.,** Seasonal changes in the gonad and associated neurosecretion in the green mussel *Perna viridus*, in *Proc. Symp. Coastal Aquacult. Mar. Biol. Assoc. India,* Cochin, Part 2, Molluscan culture, 1983, 540.
19. **Kuriakose, P. S.,** Open sea raft culture of green mussel at Calicut, *CMFRI Bull.*, Cochin, India, 29, 33, 1980.
20. **Ranade, M. R., Raje, P. C., and Ranade, A. M.,** A note on the growth of the green mussel *Mytius viridis* (Linn.) in Ratnagiri waters, *Curr. Sci.*, 42, 584, 1973.
21. **Shaffee, M. S.,** Studies on the various allometric relationships in the intertidal green mussel, *Perna viridis* Linnaeus of Ennore estuary Madras (1976), *Indian J. Fish.*, 23(12), 1, 1978.
22. **Mohan, M. V.,** Allometric relationships in the green mussel, *Perna viridis* L., *Indian J. Mar. Sci.*, 9(3), 224, 1980.
23. **Appukuttan, K. K. and Nair, T. P.,** Mariculture of the brown mussel, *Perna indica* Kuriakose and Nair at Vizhinjam, Southwest coast of India, *Proc. Symp. Coastal Aquacult., Mar. Biol. Assoc. India,* Cochin, Abstr. No. 168, 99, 1980.
24. **Appukuta, K. K., Nair, T. P., Joseph, M., and Thompson, K. T.,** Culture of brown mussel at Vizhinjan, *CMFRI Bull.,* Cochin, India, 29, 30, 1980.
25. **Narasimham, K. A.,** Dimensional relationships and growth of green mussel *Perna viridis* in Kakinada Bay, *Indian J. Fish.*, 28(1/2), 240, 1981.
26. **Kuriakose, P. S.,** unpublished data, 1980.
27. **Rangarajan, K. K. and Narasimham, K. A.,** Mussel farming on the east coast of India, *CMFRI Bull.*, Cochin, India, 29, 39, 1980.
28. **Aqsim, S. Z. and Achari, G. P. K.,** Mussel culture—its potential and implications, *Sem. Maricult. Mech. Fish.*, Dep. Fish. Govt. Tamil India Nadu, India, (Abstr.), 13, 1972.
29. **Jones, S. and Alagaraswami, K.,** in *Proc. Symp. Living Resources of Seas Around India, CMFRI,* Cochin, India, 641, 1973.
30. **Appukuttan, K. K., Prabhakaran, T. Nair, T., Joseph, M., and Thomas, K. T.,** Culture of brown mussel at Vizhinjan, *CMFRI Bull.*, Cochin, India, 29, 30, 1980.
31. **Kuriakose, P. S. and Appukutan, K. K.,** Work details for rope culture of mussels, *CMFRI Bull.*, Cochin, India, 29, 47, 1980.
32. **Sundaram, N. and Ramadoss, K.,** Experiments on the transportation of seeds of brown mussel *Perna indica, Proc. Symp. Coastal Aquacult., Mar. Biol. Assoc. India,* Cochin, Abstr. No. 107, 62, 1980.
33. **Balachandran, K. K. and Prabhu, P. V.,** Preliminary studies on preservation and transportation of green mussel, *Perna viridis, Proc. Symp. Coastal Aquacult., Mar. Biol. Assoc. India,* Cochin, Abstr. No. 293, 172, 1980.
34. **Muraleedharan, B., Joseph, K. G., and Davadasan, K.,** Pickled products from green mussel, *Proc. Symp. Coastal Aquacult., Mar. Biol. Assoc. India,* Cochin, Abstr. No. 292, 172, 1980.
35. **Jones, S.,** Observations on the biomass and fishery of the brown mussel (*Mytilus* sp). of the Cape Region of Peninsular India, *J. Bombay Nat. Hist. Soc.*, 49(3), 519, 1951.
36. **Koya, M. S., Israil, S., and Mohandas, A.,** Rate of occurrence of a sexually matured digenetic trematode in the green mussel, *Perna viridis* Linnaeus, *Proc. Symp. Coastal Aquacult., Mar. Biol. Assoc. India,* Cochin, Abstr. No. 265, 155, 1980.
37. **Pillai, C. T.,** Microbial flora of mussels in the natural beds and farms, *CMFRI Bull.*, Cochin, India, 29, 41, 1980.
38. **Appukutan, K. K.,** Predation on mussels in culture by silver bream, *Radosargus sarba, CMFRI Bull.*, Cochin, India, 29, 44, 1980.
39. **Lakshamanan, P. T. and Nambisan, P. N. K.,** Bioaccumulation and depuration of some trace metals by the mussel, *Perna viridis* (Linnaeus), *Nat. Sem. Mussel Watch,* University of Cochin, India, 35, 1986.
40. **Nambisan, P. N. K. and Lakshmanan, P. T.,** Toxicities of selected heavy metals to the mussel *Perna viridis* (Linnaeus), *Nat. Sem. Mussel Watch,* University of Cochin, India, 132, 1986.
41. **Ajithakumar, B. S. and Alagaraswami, K.,** Effect of pollution due to coconut husk retting on reproductive potential of green mussel, *Perna viridis, Nat. Sem. Mussel Watch,* University of Cochin, India, 181, 1986.
42. **Welsey, G. and Anjeevaraj, P. J.,** Heavy metal concentration in lamellibranch, *Proc. Symp. Coastal Aquacult., Mar. Biol. Assoc. India,* Cochin, Abstr. No. 277, 163, 1980.
43. **Shah, S. M., Sastry, V. N., and Bhatt, Y. M.,** Trace elements distribution in some Mollusca from Bombay coast, *Curr. Sci.*, 42, 589, 1973.
44. **Nair, G. A., Krishnakumar, G. A., Nair, N. B., and Balasubramoniam, N. K.,** Bioassay and respiration in the mytilid *Perna indica* (Kuriakase and Nair) exposed to effluents from a titanium dioxide factory, *Nat. Sem. Mussel Watch,* University of Cochin, India, 15, 1986.
45. **Rajan, S. J.,** Experiments on submerged raft for open sea mussel culture, *CMFRI Bull.*, Cochin, India, 29, 46, 1980.
46. **Alagaraswami, K.,** Review on production of mussel seed, *CMFRI Bull.*, Cochin, India, 29, 22, 1980.

Chapter 15

OYSTERS IN INDIA

R. Nagabhushanam and U. H. Mane

TABLE OF CONTENTS

I. BIOLOGY

A. SPECIES

Many species of oysters are found in India. The shells of the Ostreida are so variable that identification is difficult. Many species are probably synonyms, and some others probably will be when the subject receives adequate attention.

Awati and Rai[1] list following *Ostrea* species: *Ostrea cucullata* Born, *O. cornucopia* Chemitz, *O. gryphoides* Newton and Smith, *O. belcheria* Sowerby, *O. discoidea* Gould, *O. quercina* Sowerby, and *O. lacerata* Hanley; *O. crenulifera* Sowerby and *O. bicolor* Hanley are also included from the Bombay coast.[2]

The nonincubatory species have long been separated from the larviparous species and placed in the genus *Crassostrea*. Two well-known backwater oysters were placed in distinct species, *Crassostrea madrasensis* (Preston) and *C. gryphoides,*[3] in addition to *C. cucullata* (see Figure 1). According to Stenzel,[4] *C. madrasensis* is a junior synonym of *C. cattuckensis* (Newton and Smith). Also Stenzel states that the species name of *C. gryphoides* is preempted by a fossil species *C. gryphoides* (von Schlotheim), and this form should also be under *C. cattuckensis*. Because of important morphological and physiological differences, Stenzel[4] elevated *Saccostrea* to generic rank and *cucullata* and similar species are now considered to be in this genus. Thus, to follow a conservative approach we recognize two main species of commercial importance, *Crassostrea cattuckensis* (Newton and Smith) and *Saccostrea cucullata* (Born), the Indian rock oyster.

B. MORPHOLOGY

Shell characteristics of the spat of *C. cattuckensis* and *S. cucullata* have been described for spat from Mulki estuary and from a typical marine habitat.[5] In 0.2 to 3.0 mm spat of *C. cattuckensis,* the lower (left) valve is flat and adheres to the substratum; the upper (right) is cupped. Both valves are translucent and brown in color. The valves are circular in spat of 20 mm in height and below, longer than high in those of 21 to 35 mm in height, and higher than long in those of greater than 35 mm in height. From 35 to 55 mm they assume an oval form in which width is less than height. Above 55 mm they are distinctly elongated. Generally, the anterior and posterior edges of the valve at the umbo region are at an angle of less than 20° to the baseline parallel to the hinge. The hinge is situated slightly interior to the umbo, forming a small "beak" underneath. In *S. cucullata* from 1 to 25 mm in height, the lower (left) valve is flat and is cemented permanently to the substratum. The upper valve has an anterior convex and an inner concave phase. Both the valves are translucent and pearly white in color up to 20 mm in height. The valves turn an opaque dull white color as growth advances. Generally, the anterior and posterior margins of the valves at the umbo region are at an angle of more than 20° to the baseline parallel to the hinge. The edges of the valves are generally sharp. The hinge is situated right at the edge of the valves at the umbo region and does not form a beak. A brownish band commonly runs from the umbo to the ventral edge on the upper valve. *S. cucullata* had dentricles around the margins of the shell, typical of the genus, whereas *C. cattuckensis* has none.

C. RANGE

C. cattuckensis occurs along the Karwar coast and the Krusadai Islands[6] and abundantly near Madras and in brackish water areas of the coast,[7] and is widely distributed along the Maharashtra and Goa coasts.[8] *S. cucullata* occurs in both inshore and estuarine environments along the Goa coast, where salinity is generally low,[9] and along various southeastern coasts.[10] Both species are found in backwaters, estuaries, and inshore seas of Tamil Nadu.[11]

D. TYPE OF SUBSTRATE

C. cattuckensis occurs in estuaries and backwaters; *S. cucullata* is found in inshore and

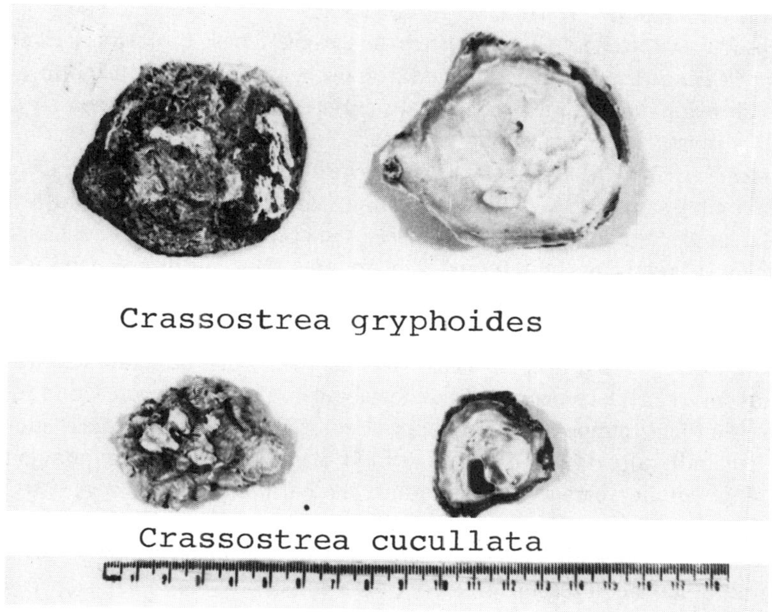

FIGURE 1. Photograph showing marketable sizes of *Crassostrea gryphoides* (Newton and Smith), and *Crassostrea cucullata* (Born) from Ratnagiri on the west coast of Maharashtra State, India.

estuarine environments, but also invades the intertidal zone. The common edible oysters usually settle on hard substrates such as boulders, coral, sandstone, or other oyster shells.

E. SALINITY AND TEMPERATURE TOLERANCE

C. cattuckensis in Bombay coastal waters tolerates large variations in salinity,[12] but oysters are eliminated from Cochin backwaters immediately after monsoons.[13] Temperature seems to play a significant role.[14] The decrease in temperature and the lowered salinity during the monsoon act synergistically and result in a general decline of oyster populations.[15] In the Ennore estuary, Madras, *C. cattuckensis* tolerated salinity as low as 7‰ in 10-d laboratory experiments.[16] Laboratory experiments on the survival and behavior of *S. cucullata* from Shirgaon Creek at Ratnagiri showed a tolerance of salinity as low as 8.6‰.[17] The low salinity during monsoons results in acclimation.[18] *S. cucullata* can tolerate salinity down to 2.5‰ during monsoon season, when the salinity at the oyster beds varies from 3 to 26‰. However, in experiments conducted in summer, when prevailing salinity is high, salinity below 14‰ caused oysters to behave abnormally.

The closed valves of *S. cucullata* proved adequate insulating mechanism against invasion by dilute seawater and evaporation during atmospheric exposure.[17] Comparative data on the survival and behavior of *S. cucullata* from Shirgaon Creek and *C. cattuckensis* from Bhatia Creek at Ratnagiri confirm the findings on acclimation.[19] In summer, *C. cattuckensis* feeds slowly down to 20‰ and *S. cucullata*, to 18.2‰ without mortality. In monsoon season, *C. cattuckensis* feeds without mortality down to 15.7‰ and *S. cucullata*, to 12.2‰. There is considerable tidal fluctuation in salinity in the monsoon season, particularly in Shirgaon Creek. Further studies have shown that both oyster species could feed only during high tide in monsoon season.[20]

F. SEXUAL DEVELOPMENT AND SPAWNING

The reproductive activity of the oysters in the east coast backwaters is periodic.[21,22] Spawning in *C. cattuckensis* is synchronized with the heavy rainfall of the northeast monsoon season, which begins in October. The rainy season differs on the two coasts, and there is corresponding divergence in spawning maxima.[2]

C. cattuckensis from Madras spawns from April to May, then again in October, at the time of maximum salinity.[23] Whether spawning, larval development, and setting in the backwaters take place in either of the two periods succeeding gametogenetic activity seems to be determined largely by salinity and temperature. These in turn depend upon whether the bar at the mouth of the Adyar estuary at Madras opens, admitting seawater, or closes, cutting off its connection with the sea.[24] *C. cattuckensis* spawns throughout the year in a marine environment, such as that of the Madras harbor.[25] Two peaks of activity occur, one in about November or December and another about May, which correspond to periods of high salinity and high temperature. In backwaters of Ennore near Madras, summer (April to June) is generally a period of high salinity (34.3 to 41.9‰) and high temperature (29.5 to 32.5°C).[26] In certain years after heavy rains, the bar remains open for a varying period. During premonsoon season (July to September) salinity falls gradually (41.9 to 13.7‰) while temperature remains high (29.5 to 32.4°C); the bar remains closed during this period. With the onset of monsoon season (October to December) the bar remains open, but salinity decreases as a result of freshets. The salinity varies from 4.6 to 28.7‰ and the temperature, from 31 to 26°C. During the postmonsoon season (January to March) the salinity increases (19.3 to 32.3‰) along with temperature (26 to 29°C). During this period the bar is open, and the salinity fluctuates widely. Spawning activity is restricted to November and December and a supplementary period in about March and April, and is induced by optimum salinity. Further descriptions of seasonal gametogenesis for *C. cattuckensis* are given by Agasthee-sapillai and Subramoniam.[27]

On the west coast, in Kelwa backwaters at Bombay, the southwest monsoon season dilutes the salinity of the backwaters from 28.6 to 13.5‰ in July and stimulates spawning in *C. cattuckensis*.[8] This spawning continues until September. In contrast, the same species in Bhatia Creek at Ratnagiri, south of Bombay, spawns not when the salinity decreases, but immediately after the close of monsoon season.[20] During August the creek salinity is about 18‰, but it increases to 23.5‰ in September, inducing spawning that continues until October (31‰). The temperature ranges from 22 to 33°C, but may not influence spawning. The salinity range required to stimulate spawning in *C. cattuckensis* from the Bombay coast is thought to be between 28.6 to 13.2‰,[8] but higher salinities, from 23.5 to 31‰ might be required for the same species to spawn at Ratnagiri.[20] *S. cucullata* from Shirgaon Creek at Ratnagiri spawns once a year, from October to January, with peak activity in November and December (salinity of 30.5 to 31.0‰ and temperature of 22 to 27.5°C).[28]

S. cucullata from the Karachi coast of Pakistan spawns with the rise in temperature (20.5°C in January to 31°C in July), but intensive spawning coincides with a steep decline in salinity.[29] On the Tuticorin coast, near the southern tip of India, *C. cattuckensis* has biannual spawning as a result of higher water temperature and higher tidal level of bay water.[30]

Hermaphrodites sometimes occur; *C. cattuckensis* from Bombay shows hermaphroditism during monsoon season, when females dominate.[8] Hermaphroditism occurs in postmonsoon and premonsoon periods; more males are found in the former and more females in the latter.[26] High water temperature and salinity favor males, whereas low water temperature and salinity favor females. Hermaphroditism in *C. cattuckensis* from Ratnagiri occurs during middle monsoon season and in late winter.[20] It has been suggested that seasonal differences in sex ratio are due to sex reversal, that higher salinity favors males; and that hermaphroditism is a transitional phase. S. cucullata is also known to show hermaphroditism.[1]

C. cattuckensis attains first sexual maturity at 12.5 mm in height[23] and an average length of 30 mm,[30] and *S. cucullata,* at 10 to 12 mm.[28] At various locations in Mulki estuary, Dakshina Kannada, *C. cattuckensis* shows sex-specific differences in size at first maturation; first maturity in males is at 12 to 14 mm and in females, at 24 to 26 mm.[31]

In addition to exogenous factors, endogenous regulation also affects maturation and release of sex products.[20,28,32] In all species, neurosecretions from the secretory cells located in the central ganglia, particularly the cerebral, show cyclic changes synchronizing with maturation and release of gametes.[32] This finding has been confirmed surgically in the cerebral ganglia of *C. cattuckensis.*[20] During the spawning period of *S. cucullata,* a very few neurosecretory cells in the cerebral ganglia possessing secretory material are found during the period of gametogenesis and maturation.[33]

Changes in water content of the whole body of *C. cattuckensis* from Bombay have been correlated with spawning; and a reciprocal relationship with fat, glycogen, and protein content of the whole body is observed.[34] These organic constituents are present in high proportions during rapid gonadal development and at low levels during spawning and recovery. Similar finds are reported on the same species from Ratnagari.[35] Water content of the whole body of *S. cucullata* from Ratnagiri has been correlated with salinity decrease in the creek and with the period prior to spawning when lipid content increases.[28]

G. LARVAL DEVELOPMENT

Very little work has been performed on the larval development of edible oysters from India. Artificial fertilization in *C. cattuckensis* has yielded trochophore and straight-hinged veliger larvae in 12 to 16 h after fertilization.[36] *C. cattuckensis* develops well in a salinity range of 30 to 40‰.

H. GROWTH

Most of the work on the growth of edible oysters pertains to *C. cattuckensis.* Spat of *C. cattuckensis* attained a length of 27 mm on spat collectors placed for two months in Ennore backwaters near Madras during the northeast monsoon season.[21] In Madras harbor this species reaches 21.5 mm in 44 d and 66 mm in 243 d.[23] Although this species occurs in fair numbers and breeds well under the almost marine conditions of the harbor, it grows to maximum size only in estuaries and backwaters.[24] In shape, the oysters conform to the substratum on which they settle.[37] Overcrowding results in a variety of irregular shapes. According to one study, the spat of *C. cattuckensis* reached a maximum height of 8.5 mm in 3 weeks, 61 mm in 6 months, 84 mm in 13 months, 96 mm in 14 months, 102 mm in 15 months, and 109 mm in 17 months. The maximum growth recorded from Adyar estuary, Madras, is 35 mm in 1 month and 3 weeks and 61 mm in 6 months.[37]

In Adyar backwaters at Madras, spat fall occurs only during periods when the bar remains open, admitting seawater into the estuary. The same is true at Ennore.[25] In *C. cattuckensis* from the Adyar estuary, growth rate varies seasonally with environmental conditions.[37] Growth rate is lower when very high or very low salinity conditions continue for prolonged periods, as happens when the sand bar cuts off communication of the estuary with the sea; growth is best under moderate estuarine conditions. Temperature fluctuations are too small (from 26 to 30°C) to be a major determining factor. Growth is most rapid from October to March, moderate from April to June, and poor from July to September. The growth rate is not affected by short, tidal fluctuations in salinity. The supply of suitable food organisms in the estuary is also higher during the periods when the bar is open and helps to account for the differences in growth rate.

A few workers have also studied the growth of *C. cattuckensis* by applying various culture techniques. In Cochin backwaters, oyster spat settle on cultch in January, and attain a length of 61.8 mm on concrete panels and 50.3 mm on the oyster shell cultches by July.[15]

Maximum growth is observed from February to March (12 to 22.2 mm), when the hydrographic conditions of the backwater are suitable. From April to June, the growth decreases gradually and reaches its lowest in July. Almost all the growing spat die by the end of July or early August, when the salinity decreases. Growth rate of *C. cattuckensis* spat from the Mulki estuary, Dakshina Kannada is 2 to 3 cm/month.[21] Spat transferred to suspended wire bags grow faster than feral spat and attain 7.0-cm shell height in about 7 months. In another study in the same estuary, spat settled on lime-coated tiles suspended from a bamboo frame after December attained a size ranging from 21 to 82 mm within 5 months.[38] The flesh weight varied from 2.5 to 10.5 g or about one sixth of the weight in the shell. In Vaigai estuary at Athankarai, *C. cattuckensis* attains an average size of 86.7 mm and a maximum size of 110 mm at the end of 1 year.[39] Average and maximum sizes of 89.6 mm and 130 mm are attained at the end of 14 months, and thereafter growth slows markedly.

II. FISHERY

A. IMPORTANCE

C. cattuckensis could be harvested at a rate of 22 t (whole weight)/ha from lime-coated tiles suspended from a bamboo frame in the Mulki estuary. A rack-culture technique along the Tuticorin coast produced 150 t/ha in 1 year.[40]

B. RECRUITMENT AND METHODS OF CULTURE

Market size for *C. cattuckensis* from Madras is 70 mm and above in height.[37] A low-budget hatchery design based on indigenous knowledge for production of oyster seeds has been proposed and includes limited facilities for seawater filtration, purification, and temperature control.[41] Induced blooms of natural phytoplankton could be used to feed oyster seed. Seed of *C. cattuckensis* is plentiful and available in the Kali estuary in Uttar Karnataka.[42]

Asbestos and wood have been used successfully as settlement substrates for several species of oysters along the Karachi coast.[43] Studies have been effected on setting of spat of *C. cattuckensis* on a variety of materials including oyster shells, concrete slabs, tiles, bamboo frames, and coconut shells in the Vaiga estuary at Athankarai.[42] Good spat fall occurs between January and April on oyster shells given a coating of cement or lime, concrete pieces with irregular surfaces, and curved tiles coated with lime. Lime-coated tiles proved to be most effective for spat collection at Tuticorin.[40] Corrugated asbestos sheets also give satisfactory results.[44] In Shimunipatnam backwaters, better settlement of spat of *C. cattuckensis* occurred on dead oyster shells kept near the bottom than on those kept suspended off the bottom.[45] Rack culture techniques seem most suitable for growing the oysters to marketable size in a short period. In Cochin backwaters floating bamboo rafts are used in culture of *C. cattuckensis*. Various cultch materials are used, but oyster shells, concrete panels, clay tiles, and aluminum sheets are the most successful. Clay tiles show the highest spat density—84 individuals on an area of 0.2 m^2. Automobile tires, rigid PVC, lime-coated tiles, and asbestos are suitable materials for settlement of spat of *C. cattuckensis* in the Mulki estuary. Cultch smeared with crude extracts of oyster tissue support more spat per unit area than untreated panels. A few authors describe details of spat collectors, breeding techniques, cages, and improved oyster farming technology[46] and discuss the economic feasibility of oyster culture.[47]

C. METHODS OF HARVESTING

Oysters are picked by hand during low-tide periods from January to May along the Kakinada coast and almost year-round, except during monsoon season, in other areas. In deeper areas, techniques somewhat similar to those for harvesting green and brown mussels are used, whereby dredges are used to dislodge the oysters from the substratum.

D. USES AND PROCESSING

C. madrasensis and *S. cucullata* from the Greater Nicobar Islands form the raw material for the manufacture of lime and cement.[48] The oyster fisheries of the Okhamandal coast and Kakinada Bay fill a considerable demand for shells used in ornamentation and lime-making and for meat to be used as a food item.[49] Badonia[50] has studied canning of *S. cucullata* in its own nectar, brine, and double-refined vegetable oil, and the method has been described for producing canned product of appealing flavor from oysters after light smoking. A plan for large-scale purification of harvested oysters has also been proposed.[51]

E. CONSTRAINTS

Oysters affected by such parasites as the pea crab, *Pinnotheres* sp., show a preponderance of males.[26] A serious impediment to culture of oysters at Karpad Creek in Tuticorin is infestations of shipworms and piddocks, which destroy the pilings and other wooden equipment used in the operations. *Teredo fucifera* (Jehoryes) is the most important borer, although others, such as *Lyrodus pedicellatus* (Quatrefages), *L. affinis* (Deshayes), and *Martesia striata* (Linnaeus), are also present.[52] Gastropods of the genus *Cymatium* cause considerable damage to the oyster stock, especially when the oysters are less than 4 months old.[53] Macroalgae, diatoms, sea anemones, serpulid polychaetes, bryozoans, spat of green mussels, *Modiolus,* and oysters as well as barnacles are also present in large numbers.

Heavy infestations of fouling organisms such as *Spirorbis, Modiolus, Balanus, Corophium, Legia,* and *Sphaeroma* are found on cultch, particularly shell cultch, in Cochin backwaters during the culture of *C. cattuckensis.*[15] High turbidity, siltation, and other changes arising from increased human activities add to the menace of fouling organisms hindering the settlement, growth, and culture of this species. Barnacles, bryozoans, and ascidians are the major fouling and boring organisms, whereas polychaetes (*Polydora ciliata* Johnson) and a sponge (*Cliona vastifica* Hancock) form minor communities in oyster farms. In the edible pearl oyster, *Placenta placenta* (Linnaeus), from Kakinada Bay, the gonad and hepatopancreas have been found to be infested by an unidentified bucephalid parasite, resulting in change of gonad color.[49] This parasite invades gonad tissue, leaving very little room for the development of gametes; thus, it affects reproductive potential and finally leads to sterility.

Municipal and industrial discharges, urban runoff, atmospheric contributions, presence of shipyards, and heavy ship traffic have led to the contamination of the estuaries and backwaters by heavy metals, which are accumulated by biota and sediments. Metals that are accumulated by oysters include zinc, copper, and iron;[54,55] concentrations in the environment and in oyster tissue vary both seasonally and spatially.[56-59]

REFERENCES

1. **Awati, P. R. and Rai, H. S.,** *Ostrea cucullata* (the Bombay oyster), *Indian Zool. Mem.,* 3, 1, 1931.
2. **Hornell, J.,** *Indian Molluscs,* The Bombay Natural History Society, 1951, 1.
3. **Durve, B. S.,** On the nomenclature of the two Indian backwater oysters, *J. Mar. Biol. Assoc. India,* 9(1), 173, 1967.
4. **Stenzel, H. B.,** Oysters, in *Treatise on Invertebrate Paleontology,* (Part N. Mollusks), Moore, R. O., Ed., 3 of 3:1953, 1971.
5. **Joseph, M. M.,** Shell characteristics of the spat of the two species of oysters *Crassostrea madrasensis* (Preston) and *C. cucullata* (Born), *J. Bombay Nat. Hist. Soc.,* 75(2), 519, 1978.
6. **Satyamurti, S. T.,** Mollusca of Krusadai Island (in the Gulf of Manner) II. Scaphopoda, Pelicipoda and Cephalopoda, *Bull. Madras Govt. Mus., New Ser., Nat. Hist. Sec.,* 1(2), 1, 1956.
7. **Venkatarman, R. and Chari, S. T.,** Studies on oysters and clams: biochemical variations, *Ind. J. Med. Res.,* 39(4), 533, 1951.

8. **Durve, V. SA.,** On the seasonal gonadal changes and spawning in the adult oyster, *Crassostrea gryphoides* (Schlotheim), *J. Mar. Biol. Assoc. India,* 7(2), 328, 1965.
9. **Rao, K. V., Kumari, C. K., and Dwive, S. N.,** Biology of the green mussel, *Mytilus viridis, Indian J. Mar. Sci.,* 4(2), 189, 1975.
10. **Pillai, C. S. G. and Appukuttam, K. K.,** Distribution of molluscs in and around the coral reefs of the southern coast in India, *J. Bombay Nat. Hist. Soc.,* 77(1), 26, 1980.
11. **Sundaram, K. S.,** Aquacultures of bivalve molluscs in Tamil Nadu coast, *Symp. Coastal Aquacult. Mar. Biol. Assoc. India,* Cochin, Abstr. No. 167, 99, 1980.
12. **Durve, V. S.,** A Study on Oysters, Ph.D. thesis, Bombay University, India, 1960.
13. **Nair, N. U.,** The settlement and growth of major fouling organisms in Cochin Harbour, *Hydrobiologia,* 30, 503, 1967.
14. **Harakantra, S. N.,** Benthon of the Kali estuary, Karwar, *Mahasagar, Bull. Natl. Inst. Oceanogr.,* Goa, India, 8, 53, 1975.
15. **Purushan, K. S., Gopalan, U. K., and Rao, T. S. S.,** Ecological studies on the seasonal production of the edible oyster *Crassostrea madrasensis* (Preston) in Cochin backwaters, Kerala, *Mahasagar, Bull. Natl. Inst. Oceanogr.,* Goa, India, 14(4), 277, 1981.
16. **Sundaram, K. S. and Shaffee, M. S.,** On the salinity tolerance of some bivalves of the Ennore estuary, *Third All India Symp. Estuarine Biology,* University of Cochin, India, Abstr. No. 1, 12, 9, 1975.
17. **Nagabhushanam, R. and Bidarkar, D. S.,** Adaptation of the common rock oyster, *Crassostrea cucullata* to low salinities at Ratnagiri, *Third All India Symp. Estuarine Biology,* University of Cochin, India, Abstr. No. 6, 34, 62, 1975.
18. **Mane, U. H.,** The adaptation of the estuarine clam, *Katelysia optima* to the salinity fluctuations, *Riv. Biol.,* 67(1/2), 75, 1975.
19. **Mane, U. H.,** Survival and behavior of oysters in water of low salinities at Ratnagiri on the west coast of India, *J. Moll. Stud.,* 44, 243, 1978.
20. **Mane, U. H. and Nagabhushanam, R.,** A study on the reproductive biology of Indian oyster, *Crassostrea gryphoides, Nat. Sci. J., Marath Univ.,* India, 15(8), 245, 1976.
21. **Hornell, J.,** Note on an attempt to ascertain the principal deteriming factor in the oyster spawning in Madras backwaters (Madras fish Investigations, 1908), *Madras Fish. Bull.,* 4, 25, 1910.
22. **Hornell, J.,** The common molluscs of south India—Rep. IV of 1921 (Madras Fish B Investigations, 1921), *Madras Fish. Bull.,* 14, 97, 1922.
23. **Paul, M. D.,** Studies on the growth and breeding of certain sedentary organisms in the Madras harbour, *Proc. Indian Acad. Sci.,* Bangalore, India, 15B, 1, 1942.
24. **Rao, K. V.,** Observations on the probable effects of salinity on spawning, development and setting of the Indian backwater oyster *Ostrea madrasensis* Preston, *Proc. Indian Acad. Sci.,* Bangalore, India, 33B, 231, 1951.
25. **Devansen, D. W. and Chacko, P. I.,** On the Madras edible oyster *(Ostrea madrasensis), Contr. Fish. Biol. Stn. Dep. Fisheries, Madras, India,* 2, 1, 1955.
26. **Rao, K. V.,** Seasonal gonadal changes in the adult backwater oyster, *Ostrea (Crassostrea) madrasensis,* Preston from Ennur near Madraw, *Proc. Indian Acad. Sci.,* Bangalore, India, 44(B), 332, 1956.
27. **Agastheesapillai, A. and Subramoniam, T.,** A comparative study on oogenesis of two marie bivalves, *Crassostrea madrasensis* and *Perna viridis, Symp. Coastal Aquacult. Mar. Biol. Assoc. India,* Cochin, Abstr. No. 59, 35, 1980.
28. **Nagabhushanam, R. and Bidarkar, D. S.,** Studies on seasonal change in the biochemical constituents of the oyster *Crassostrea cucullata, Indian J. Fish.,* 25(1/2), 156, 1978.
29. **Asif, M.,** Reproductive cycle and changes in associated parts of gonad in *Crassostrea (Saccostrea) glomerata* (Gould) from the coast of Karachi, *Indian J. Fish.,* 26(1/2), 13, 1979.
30. **Rajapandian, M. E. and Rajan, C. T.,** Studies on the gonadal maturity and spawning periodicity of *Crassostrea madrasensis* (Preston), at Tuticorin Bay, *Symp. Coastal Aquacult. Mar. Biol. Assoc. India,* Cochin, Abstr. No. 60, 36, 1980.
31. **Joseph, M. M. and Joseph, S.,** Some aspects of experimental culture of the oyster *Crassostrea madrasensis* (Preston), *Symp. Coastal Aquacult. Mar. Biol. Assoc. India,* Cochin, Abstr. No. 174, 103, 1980.
32. **Desai, K. M. and Nimavat, D.,** Neuroendocrine control of sexual maturation in some edible oysters, *Symp. Coastal Aquacult. Mar. Biol. Assoc. India,* Cochin, Abstr. No. 62, 37, 1980.
33. **Nagabhushanam, R.,** Neurosecretory control of reproduction in the oyster, *Crassostrea cucullata, Symp. Coastal Aquacult. Mar. Biol. Assoc. India,* Cochin, Abstr. No. 62, 35, 1980.
34. **Durve, V. S. and Bal, D. B.,** Studies on the chemical composition of the oyster, *Crassostrea gryphoides* Schlotheim, *J. Zool. Soc. India,* 13, 70, 1961.
35. **Nagabhushanam, R. and Mantale, B. M.,** Studies on biochemical composition of the oyster, *Crassostrea gryphoides, Marathwada Univ. J. Sci.,* India, 11, 47.
36. **Rao, K. V.,** Induced spawning of the adults and laboratory rearing of the larvae of the edible oyster, *Crassostrea madrasensis* (Preston), *Symp. Coastal Aquacult. Mar. Biol. Assoc. India,* Cochin, Abstr. No. 317, 186, 1980.

37. **Rao, K. V. and Nayar, K. N.,** Rate of growth in spat and yearlings of the Indian backwater oyster *Ostrea madrasensis* (Preston), *Indian J. Fish.*, 3, 231, 1956.
38. **Dhulkhed, M. H. and Ramamurthy, S.,** Experiments on oyster culture, at Mulky, Dakshina Kannada (Karnataka), *Symp. Coastal Aquacult. Mar. Biol. Assoc. India,* Cochin, Abstr. No. 177, 104, 1980.
39. **Rao, K. S., Sivalingam, D., and Unnithan, K. A.,** Observations on the setting of spat and growth of *Crassostrea madrasensis* in Vigai estuary at Athankarai, *Symp. Coastal Aquacult. Mar. Biol. Assoc. India,* Cochin, Abstr. No. 98, 57, 1980.
40. **Nair, K. N. and Mahadevan, S.,** Edible oyster culture at Tuticorin, *Symp. Coastal Aquacult. Mar. Biol. Assoc. India,* Cochin, Abstr. No. 172, 102, 1980.
41. **Nayar, K. N. and Easterson, D. C. V.,** On the design of a small scale hatchery system for production of oyster seed, *Symp. Coastal Aquacult. Mar. Biol. Assoc. India,* Cochin, Abstr. No. 96, 56, 1980.
42. **Nagaraj, M. and Neelakantan, B.,** Fish and shell-fish feed resources of Kali estuary along with a note on the mariculture potentialities in Uttar Kannada, *Symp. Coastal Aquacult. Mar. Biol. Assoc. India,* Cochin, Abstr. No. 108, 62, 1980.
43. **Moazzam, M. and Rizvi, S. H. N.,** Settlement of oyster larvae in Pakistan waters and its possible indication for setting up oyster culture, *Symp. Coastal Aquacult. Mar. Biol. Assoc. India,* Cochin, Abstr. No. 317, 18, 1980.
44. **Thangavelu, R. and Sundararan, N.,** Experiments on edible oyster spat collection at Tuticorin, *Symp. Coastal Aquacult. Mar. Biol. Assoc. India,* Cochin, Abstr. No. 99, 58, 1980.
45. **Reuben, S., Rao, T. A., and Manickam, P. E. S.,** Culture experiments on the edible oyster *Crassostrea madrasensis* in the Bhimunipatnam backwaters, *Symp. Coastal Aquacult. Mar. Assoc. India,* Cochin, Abstr. No. 171, 101, 1980.
46. **Achari, G. P. K.,** New design of spat collectors, breeding haps, cages and improved technology for pearl framing, *Symp. Coastal Aquacult. Mar. Biol. Assoc. India,* Cochin, Abstr. No. 182, 107, 1980.
47. **Parulekar, A. H., Nayar, A., Ansari, Z. A., and Harakantra, S. N.,** Culture of shellfish in Goa, *Symp. Coastal Aquacult. Mar. Biol. Assoc. India,* Cochin, Abstr. No. 165, 98, 1980.
48. **Daniel, A. and Rajgopal, A. S.,** Molluscs of economic value from Great Nicobar Islands, *J. Bombay Pat. Hist. Soc.*, 70(2), 394, 1973.
49. **Narasimham, K. A.,** Biology of windowpane oyster, *Placenta placenta* (Linnaeus) in Kakinada Bay, *India J. Fish.*, 31, 272, 1984.
50. **Badonia, M. R.,** Canning of edible oysters, *Symp. Coastal Aquacult. Mar. Biol. Assoc. India,* Cochin, Abstr. No. 290, 171, 1980.
51. **Nayar, K. N., Rajapandian, M. E., and Easterson, D. C. V.,** A plan for the purification system for farm grown oysters before marketing, *Symp. Coastal Aquacult. Mar. Biol. Assoc. India,* Cochin, Abstr. No. 291, 171, 1980.
52. **Nar, N. B. and Dharmaraj, K.,** Incidence of wood-boring molluscs in the oyster farms at Tuticorin, *Mahasagar. Bull. Natl. Inst. Oceanogr.*, Goa, India, 12(2), 109, 1979.
53. **Thangavelu, R. and Muthiah, P.,** Predator problem of oyster farming in Tuticorin Bay: destruction of oyster spat by *Cymatium pileare, Symp. Coastal Aquacult. Mar. Biol. Assoc. India,* Cochin, Abstr. No. 176, 104, 1980.
54. **Parameswaran, M. and Viswanathan, R.,** Ionic regulation in aquatic organisms, *Third All India Symp. Estuarine Biol.*, University of Cochin, India, Abstr. No. 3, 05, 20, 1975.
55. **Kumaraguru, A. K.,** Ph.D. thesis, Annamalai University, India, 1980.
56. **Nair, N. U.,** Studies on the backwater oyster *Crassostrea madrasensis* (Preston) of the Cochin Harbour, Ph.D. thesis, Cochin University, India, 1984.
57. **Rajendran, N. and Kurian, C. V.,** *Crassostrea madrasensis* (Preston) indicator of meta pollution in Cochin backwaters, *Nat. Sem. Mussel Watch,* University of Cochin, India, 120, 1985.
58. **Nair, N. U. and Nair, N. B.,** Seasonality of trace metals in *Crassostrea madrasensis* (Preston) inhabiting the Cochin backwaters, *Nat. Sem. Mussel Watch,* University of Cochin, India, 6, 1986.
59. **Senthilnathan, S., Mahendran, A., and Balasubramoniaum, T.,** Bioconcentration of iron in *Crassostrea madrasensis* (Preston) from Vellar estuary, *Nat. Sem. Mussel Watch,* University of Cochin, India, 160, 1986.

Chapter 16

PATINOPECTEN (MIZUHOPECTEN) YESSOENSIS (JAY) IN JAPAN

Susumu Ito

TABLE OF CONTENTS

I. INTRODUCTION

The scallop *Patinopecten yessoensis* (Jay) is a mollusk that has been of considerable commercial importance in Japan for a long time. Processed scallop meat has been exported from Japan to China for about 250 years.[1] It is a cold-water species, occurring above about 35°N latitude in the northwest Pacific Ocean on the coasts of the Kuril Islands, Sakhalin, Hokkaido, northern Honshu, and the northern Korean peninsula. Scallops are harvested from the wild, from reseeded scallop beds, and from hanging culture equipment.

II. BIOLOGY

A. MORPHOLOGY

Like most scallops, this species has a roughly circular shell with "ears" on either side of the umbo. Adult scallops reach a maximum size of 20 cm in shell length, 19.5 cm in height, and 5 cm in width, but are usually harvested when they reach a size of 9 to 15 cm in length. The left (upper) valve is purple or reddish brown, and the right, more deeply cupped lower valve is yellowish white. There are 21 to 24 radial ribs on the valve, and usually annual growth rings can be seen along with the growth striae.

The well-developed adductor muscle is slightly posterior, and the animal can swim in short skips from the bottom by quick contractions. The foot is well developed in the young spat and is used for locomotion but is vestigial in the adult. Figure 1 shows the anatomy of the soft parts.

B. SEXUAL DEVELOPMENT AND SPAWNING

This scallop is not hermaphroditic. The creamy-white male and reddish-orange female gonads are located in the visceral pouch. Some gonadal development occurs in 1-year-old scallops, but spawning does not usually occur until the scallop is 2 years old. A 2-year-old female can produce as many as 80 to 110 million eggs, and a 3-year-old, 110 to 130 million eggs.[2] Gonadal development starts in midautumn and by mid-February to mid-March the gonads are ripe and ready to spawn. There is evidence that gonadal development is promoted by exposure to low temperatures.[3,4] Spawning occurs in early March to early April and is triggered by a rapid rise in temperature of 4 to 5°C. Figure 2 shows the seasonal gonad index (gonad weight/total body weight × 100) in Mutsu Bay.

Both the eggs and sperm are discharged into the water, where fertilization takes place. After about 55 h at 8 to 9°C, the embryo develops cilia and begins its free-swimming life. In Mutsu Bay, these stages last about 35 to 40 d and are shown in Figure 3. The mature larva is 290 to 310 μm in shell length and attaches to a suitable substrate by a byssal thread. In Mutsu Bay, attachment occurs from mid-April to late May. In more northern regions, seasons for spawning, development, and attachment occur about a month later.

C. GROWTH

The attached scallop immediately begins to secrete a spat shell and to assume the adult shape. The byssal attachment continues until mid-July to mid-August in Mutsu Bay, later in more northern areas, when the size is 6 to 10 mm. Growth rate of the spat is about 100 to 150 μm/d.

After the attached byssal stage, the small spat settle to the bottom. This is often a period of high mortality.[2] Those that survive grow to 3-cm juveniles by late autumn. Standard growth curves of bottom scallops in Mutsu Bay are shown in Figure 4.[2] Shell growth and body weight are in the logarithmic phase until 3 or 4 years of age, but then slow considerably. Life span is about 10 to 12 years.[2]

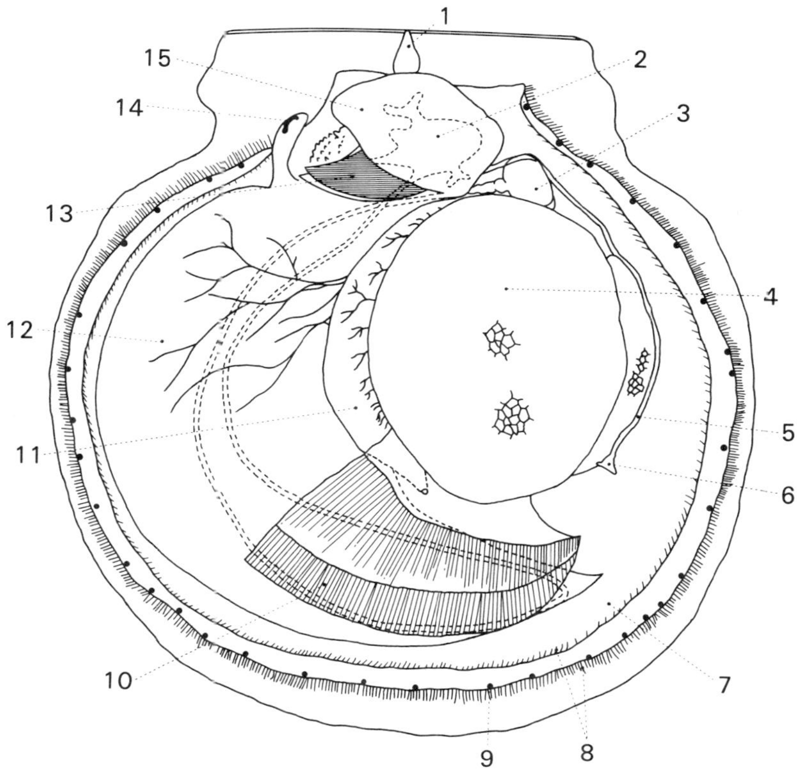

FIGURE 1. Internal anatomy of *P. yessoensis*. 1. ligament; 2. stomach; 3. heart; 4. adductor muscle; 5. intestine; 6. anus; 7. mantle; 8. tentacle; 9. eye; 10. gill; 11. kidney; 12. gonad; 13. mouth; 14. foot; 15. midgut gland. (From Yamamoto, G., Ito, S., and Fuji, A., *Through Culture in Shallow Sea*, Koseisha-Koseikaku, Tokyo, 1971, 187. With permission.)

D. SALINITY AND TEMPERATURE TOLERANCES

P. yessoensis is steno-high-haline and occurs normally between 30 and 40‰.[2] Embryos do not develop below 26.9‰, and larvae do not develop below 23.6‰.[5] Scallops have a low tolerance for high temperatures. Larval development is normal only between 5 and 20°C, and ciliary movement of excised gill tissue occurs only at these temperatures.[2] Abnormally high mortality often occurs in hanging-cultured scallops when the water temperature exceeds 23°C for a month or so during the summer.[6] The best growth occurs between 10 and 17°C.[2]

E. TYPES OF SUBSTRATE

Scallops are found on fine gravel, coarse sand, fine sand, and sandy-mud bottoms at depths from 5 to 50 m. Kinoshita et al.[7] found, from a diving survey, that scallops prefer coarse sandy bottom and fairly strong water currents. Normally, each scallop digs a shallow depression, with the left valve uppermost.[8] Yamamoto,[9] in a study in Mutsu Bay, found that scallops are found in the fourth-community area mainly located around the shallow coastal region of this bay. In addition, they survived well in such a community only when planted, and their occurrence depended not only on the sediment type and water currents, but also on the benthic community complex.

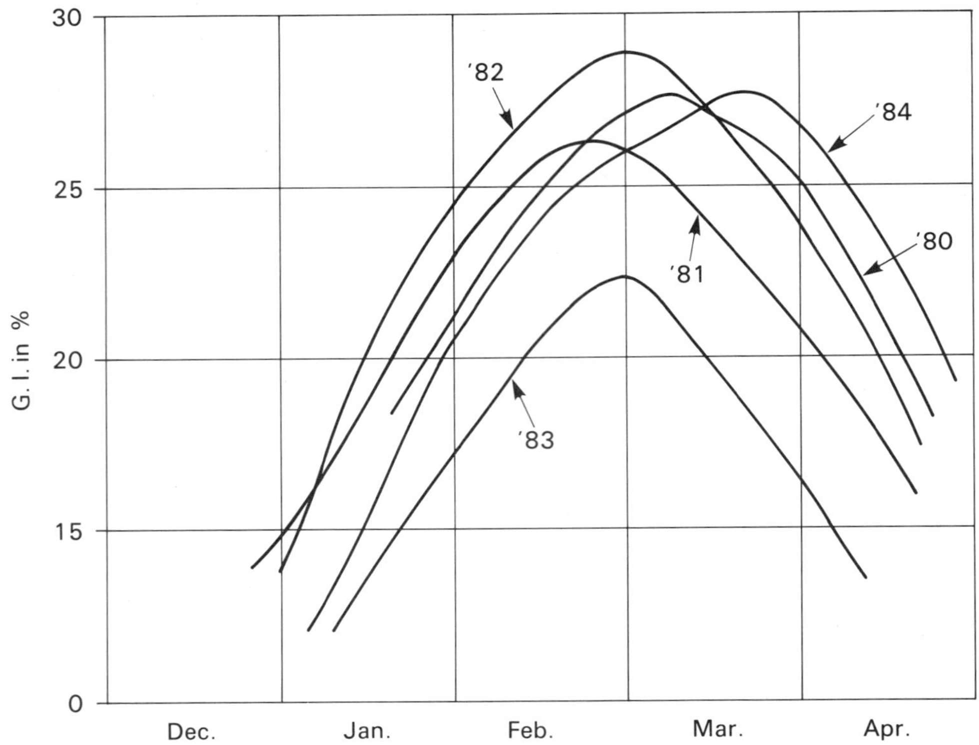

FIGURE 2. Seasonal changes of gonadal index of the scallop in Mutsu Bay. (From Hirano.)

III. FISHERY

A. FISHERY IN JAPAN

The species occurs in northern Japan. The fishery extends from Miyagi Prefecture to the Sea of Okhotsk. The main natural grounds are the Sea of Okhotsk, Lagoon Saroma, Nemuro Strait, Funka Bay, and Mutsu Bay. Hanging culture is carried out in the above-mentioned bays and lagoons and many small bays in Iwate Prefecture and Miyagi Prefecture (Figure 5).[2]

Harvesting is conducted with dredges from motorized vessels of 5 to 7 ton in weight (Figure 6). In northern Hokkaido, larger dredges and boats are used than in the southern regions. Dredging increased considerably from 1930 to 1937, and subsequently conservation regulations were introduced.[1] Regulations enacted by the prefectures apply only to bottom harvesting and to minimum shell size (10 cm). For culturing, quotas on numbers of long-lines are set by the prefectural governments and are allocated by the fishery associations.

Table 1 gives the total annual yield of scallops as well as for the main areas from 1976 to 1981. As can be seen, there has been a steady increase. For many years it has been observed that dense natural spatfall occurs in Mutsu Bay at intervals of 10 to 20 years. Collection and replanting into coastal regions resulted in fluctuating annual yields until about 1965, when hanging culture enhancement resulted in steady harvests on a yearly basis, as seen in Figure 7. Hanging culture techniques were quickly adopted in other regions, resulting in increases (Table 1).

B. PROCESSING AND VALUE

The price of scallops in the Mutsu Bay area from 1979 to 1984 can be seen in Table 2. The landed scallops are auctioned by the Aomori Fisheries Cooperative to buyers, who

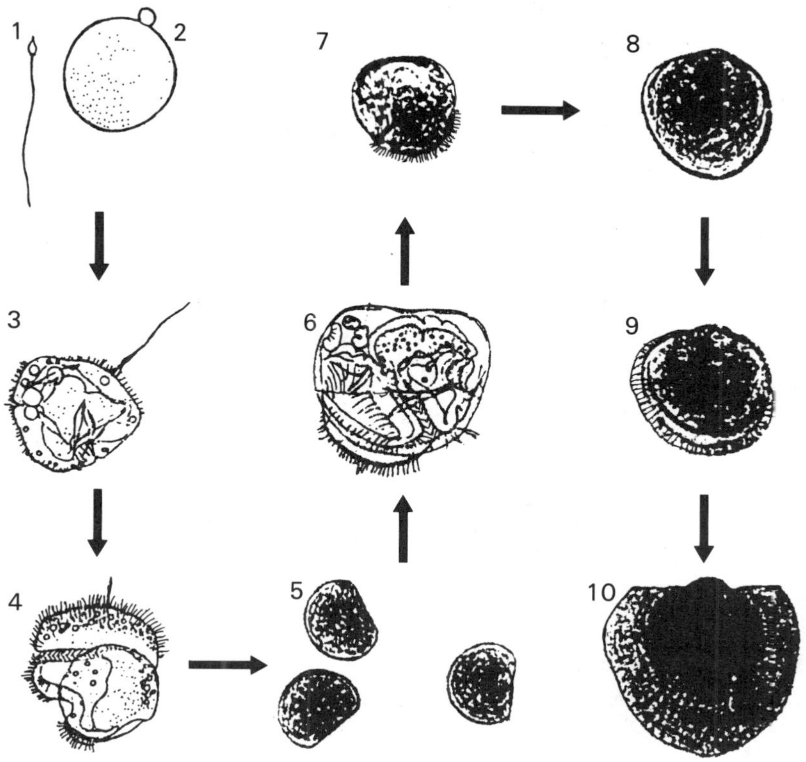

FIGURE 3. Process of larval development of *P. yessoensis*. 1. sperm; 2. fertilized egg; 3. trochophore; 4. veliger; 5. D-shaped larva (110 μm); 6. D-shaped larvae (120 μm); 7. umbo stage (180 μm); 8. full-grown larva (300 μm); 9. spat (320 μm); 10. 10-d-old spat (600 μm). (Adapted from Yamamoto, G., Ito, S., and Fuji, A., *Through Culture in Shallow Sea*, Koseisha-Koseikaku, Tokyo, 1971, 187.)

process them. Trade price of boiled meat (without shell), a typical product, is about 800 to 1000 yen/kg. Table 3 gives the various methods of processing and tonnages of shelled scallops for each process. The majority are boiled and then frozen before sale. A considerable portion are dried, some are canned, and some are sold fresh.

C. WORK FORCE

Since the development of scallop culture, farming units have increased gradually and have reached more than 7000. In Hokkaido and Aomori, these are solely for scallop culture, but in Iwate and Miyagi the farming is a side occupation to oyster and seaweed farming. Table 4 shows the extent of farming units in Japan from the 1973 census.

IV. CULTURE

A. SPAT COLLECTION

The first attempt at spat collection was done by Kinoshita[10] in 1934 in Lagoon Saroma and used the oyster cultch technique. The most successful spat collection technique was discovered in about 1965 by Kudo, a fisherman, who used an onion bag. Use of this epoch-making collector expanded quickly because of its very high efficiency.

The collectors are suspended at medium depths from a long-line (Figure 8). The spat collector is a net onion bag that is loosely filled with plastic netting or used gill netting to which the spat attaches by its byssal thread. Following detachment, the spat are retained in the net bag, which has a mesh size smaller than the scallops.

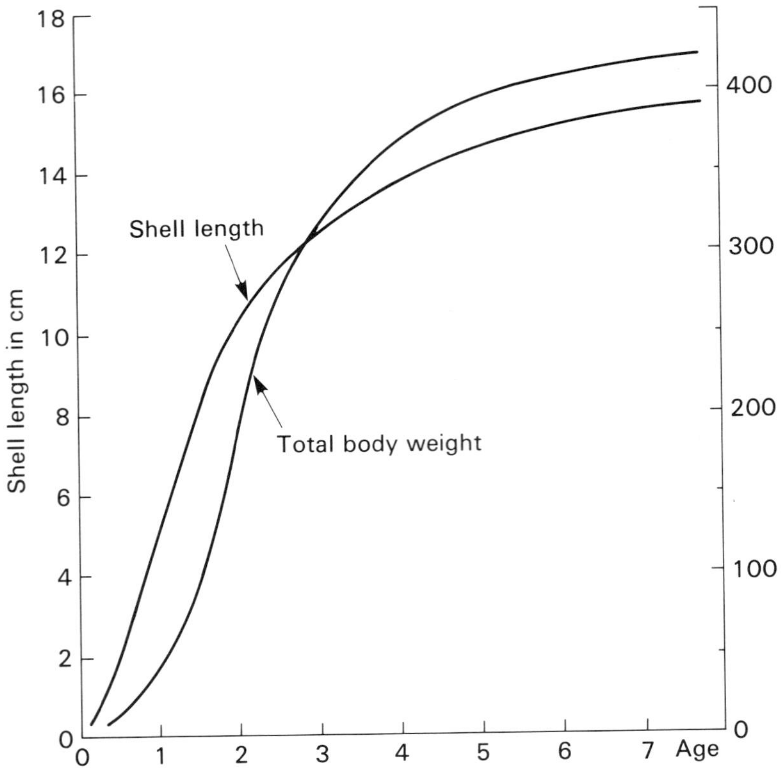

FIGURE 4. Standard growth curves of the bottom-inhabiting scallop in Mutsu Bay. (From Yamamoto, G., Ito, S., and Fuji, A., *Through Culture in Shallow Sea,* Koseisha-Koseikaku, Tokyo, 1971, 187. With permission.)

The spat collection season varies, and the Aquaculture Center of Aomori Prefecture has determined the best time to plant the collectors by several methods, including water-temperature monitoring, monitoring of the starfish larval set, and determination of the gonadal index, and the number of scallop larvae in the water. The best time for cultch planting is when the ratio of larvae over 200 μm reaches about 50%, just prior to attachment.[11] There are large annual fluctuations in density of spat attachment, which range from over 10,000 to lows of 300 per collector.

The life history and an outline of the work schedule for culture work in Mutsu Bay is shown in Figure 9. In Hokkaido, because of lower water temperatures, the schedule starts later and extends longer.

B. INTERMEDIATE REARING

The collected spat are transferred to intermediate facilities, usually between mid-July and mid-August in Mutsu Bay, a little later in Hokkaido. The collected spat are sorted into size classes with sieves, placed in pearl nets (Figure 10), and hung from long-lines.

The Aquaculture Center of Aomori Prefecture recommends 100 spat per net at first, as shown in Figure 11, later thinned to lower numbers as the spat grow. The juveniles reach a size of 4 cm by December and about 5.5 cm by March. The survival rate is more than 90%.

C. HANGING CULTURE

For final grow-out to market size, the lantern cage method (Figure 12) is the most popular in Mutsu Bay, with ten seed in each of ten levels of the cage. These are suspended

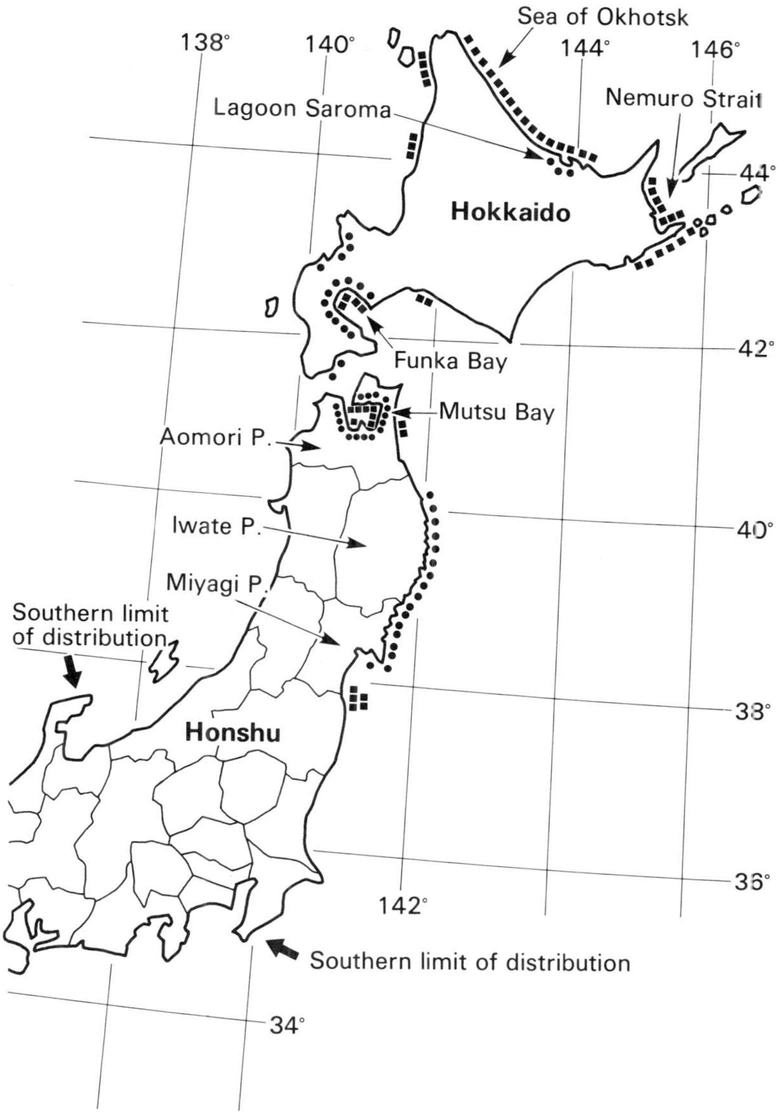

FIGURE 5. Main natural and bottom-culture grounds (■) and hanging culture grounds (●). (From Yamamoto, G., Ito, S., and Fuji, A., *Through Culture in Shallow Sea*, Koseisha-Koseikaku, Tokyo, 1971, 187. With permission.)

from the long-line for 10 to 15 months, with a survival rate of over 90%. Pearl nets are used also, and only 5 or 6 seed are placed in each net. The "ear-hanging" method (Figure 13) is becoming popular because of a greater growth rate, in spite of the greater labor involved and heavier fouling. In this method, a scallop is fastened by nylon thread from a hole drilled in one ear of its shell to a dropper rope which, in turn, is hung from a long-line.

Despite the heavy fouling that builds up in the hanging cultures, no more removal of the attached organisms is recommended than is necessary because of the effect on the growth and survival of the scallops.[12] In Hokkaido, large-scale multiple long-lines are used.[13]

D. BOTTOM CULTURE

Starfish are dredged from the bottom prior to the reseeding of scallops. Seed scallops over 3 cm in shell length are used for bottom culture, and these are released between

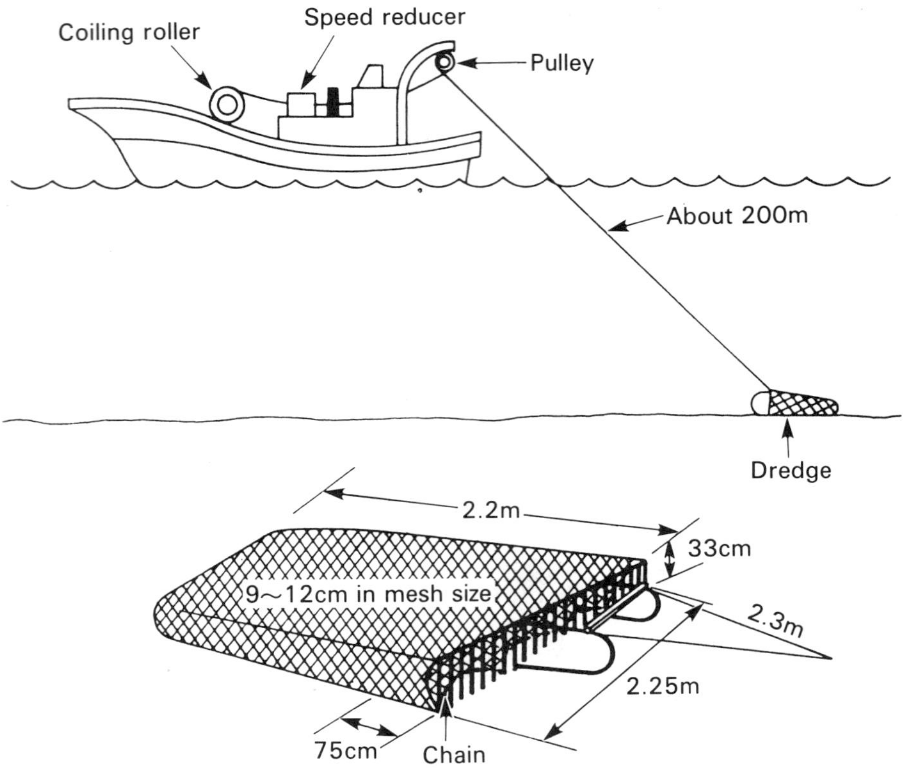

FIGURE 6. Scallop dredge and fishing boat.

TABLE 1
Annual Yield of Scallops in Japan, Wet Weight with Shell (in metric tons)

Year	All Japan	Hokkaido	Aomori	Iwate	Others
1976	95,216	61,887	26,874	5,482	973
1977	126,715	101,134	21,885	3,020	676
1978	127,414	98,938	25,569	2,592	316
1979	123,356	93,786	26,869	2,450	251
1980	123,537	88,429	31,609	3,313	186
1981	150,245	113,728	33,205	3,149	163

From Agricultural Statistics of Japanese Government.

December and April of each year at a recommended density of 5 or 6 per square meter.[2] Two to 3 years after reseeding, the survival rate is about 30 to 40%. Harvesting is conducted by dredging. The culture ground is rotated every 2 to 3 years.

V. CONSTRAINTS

Constraints on scallop production in Japan include shell-inhabiting or attached organisms, parasites, predators, occasional mass mortalities, overstocking, and shellfish poisoning caused by plankton blooms.

The shell-inhabiting polychaete *Polydora ciliata* (Johnston) often occurs in the left valve of scallops on muddy bottom, but rarely does a heavy infestation have bad effects.[14] Many

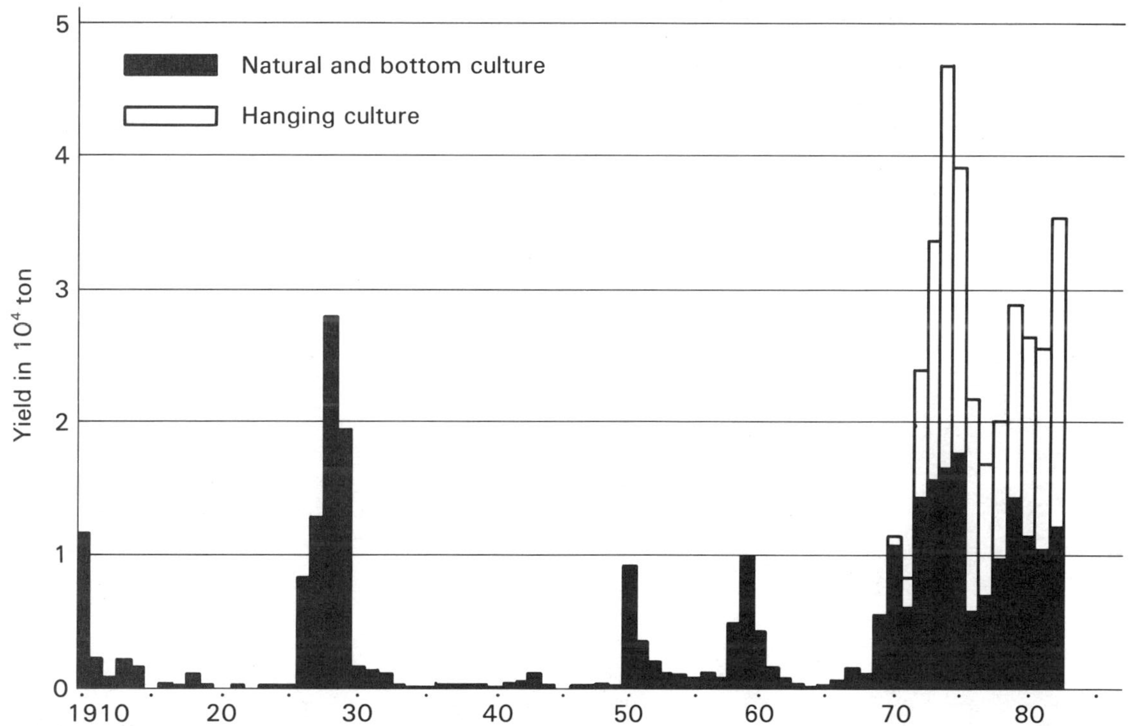

FIGURE 7. Annual yield of scallops in Mutsu Bay, wet weight with shell in 10^4 ton. (From Agricultural Statistics of the Japanese government.)

TABLE 2
Average Yearly Auction Price of Scallops with Shell in Aomori Prefecture

Year	Price (yen/kg)	Year	Price (yen/kg)
1979	233.8	1982	257.2
1980	286.8	1983	284.0
1981	295.3	1984	323.0

From Union of Aomori Fish. Coop. Assoc.

TABLE 3
Processing Products and the Amount of Raw Scallops with Shell in Tons Used in Aomori Prefecture

Year	Fresh	Frozen	Boiled	Canned	Dried
1978	64	2,009	15,190	481	1,839
1979	37	3,684	18,468	710	992
1980	90	8,856	15,880	488	2,314
1981	197	8,369	17,800	691	920
1982	380	7,031	23,007	315	2,394

From Union of Aomori Fish. Coop. Assoc.

TABLE 4
Scallop Farming Units in Japan[a]

Region	No. of villages	No. of farming units
Hokkaido	11	1663
Aomori	9	1681
Iwate	7	2894
Miyagi	5	861
Total	32	7099

[a] Villages with less than 50 units are not included.

From 1973 Fisheries Census of Japanese Government.

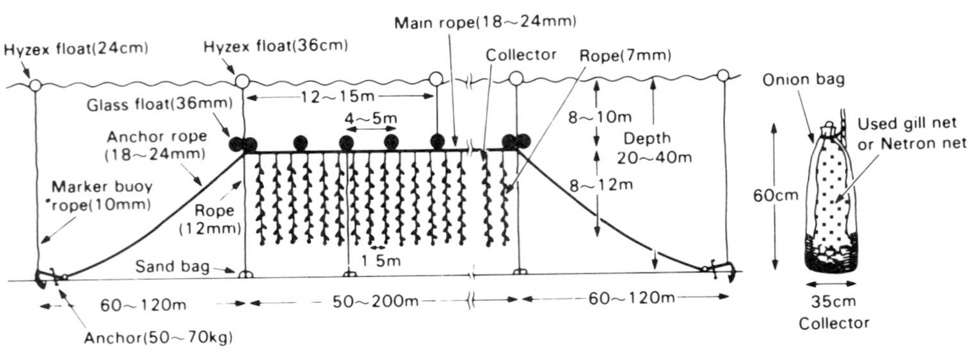

FIGURE 8. Long-line facility for spat collection in Mutsu Bay.

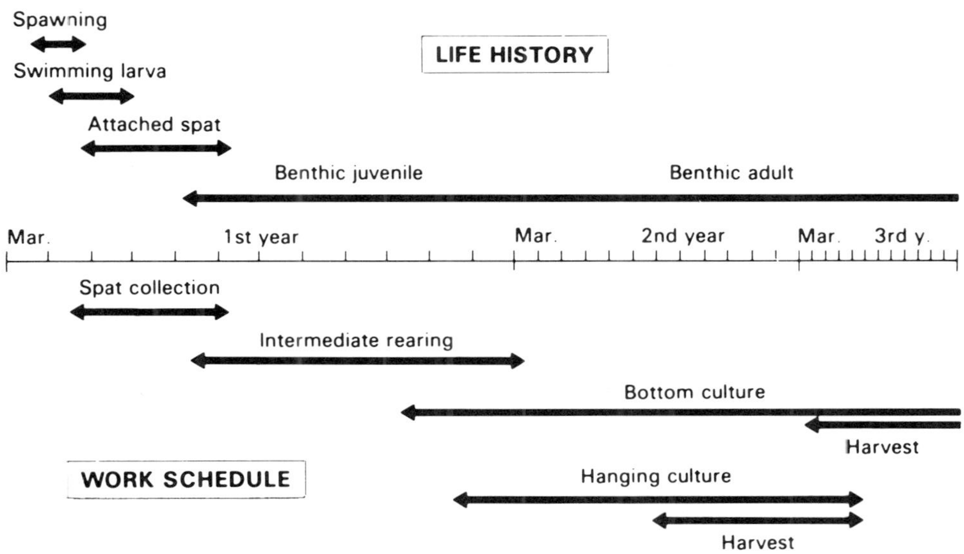

FIGURE 9. Life history of the scallop and the work schedule for scallop enhancement in Mutsu Bay.

types of organisms attach to the scallops and to the hanging culture facilities. These not only compete with the scallops for food and space, but also foul the shell and add weight equal to that of the scallops after about a year or so. Two bivalve mollusks, *Mytilus edulis* Linnaeus and *Hiatella orientalis* (Yokoyama), are often abundant, as are two or three barnacle species, solitary and compound ascidians, sponges, hydrozoans, and seaweed. Fouling is usually alleviated either by deep suspension or by planting scallops in low densities in the cages to lessen the time before harvest.

A parasitic crustacean, *Sacculina* sp., growing to 8 mm in length and living in the gill area of scallops, absorbs nutrients from the host scallop and retards its growth and fattening.[15]

Starfish, especially *Asterias amurensis* Lutekne, are the most common predator on bottom-living scallops, and sometimes the larvae attach in the collecting bags and cause severe spat mortality, starting in July. *Octopus hongkongensis* Hoyle destroys many bottom-living scallops, particularly those occurring offshore.

There have been occasional mass mortalities of scallops in hanging culture in Japan.[16] One occurred in the summer of 1975 and continued through autumn in Mutsu Bay, as well as, to a lesser extent, in subsequent years. Studies by research agencies found that the apparent reason was that scallops in intermediate culture (pearl nets) were too densely concentrated. The scallops injured their mantle edges by their swimming activities, suppressing normal shell secretion.[17-19] The recommended solution was to use lower densities in intermediate rearing.[18,19]

Scallop growth and fattening have declined in recent years in several areas, especially in closed and semiclosed bays. It has been determined that the bays have only a certain carrying capacity[20,21] and that fishermen were stocking too densely, bringing about food shortages and accelerated self-pollution by biodeposition.

In 1978 Yasumoto et al.[22] found that a dinoflagellate, *Dinophysis fortii* Pavillard, caused diarrhea in persons eating scallops with toxin concentrations in excess of 0.05 Mouse U/g. The poison accumulates in the midgut, (which is now removed before sale), and has been found in nearly all the scallop grounds from spring to late summer.[23]

Paralytic shellfish poisoning caused by *Gonyaulax tamarensis* (Lebour) Taylor and *G. catenella* (Whedon and Kofoid), which causes paralysis and sometimes death to the consumer, is found in scallop grounds, except in Mutsu Bay, from early spring to late autumn.[24] When toxin levels reach 4.0 Mouse U/g in the edible part, all scallop landings are prohibited.

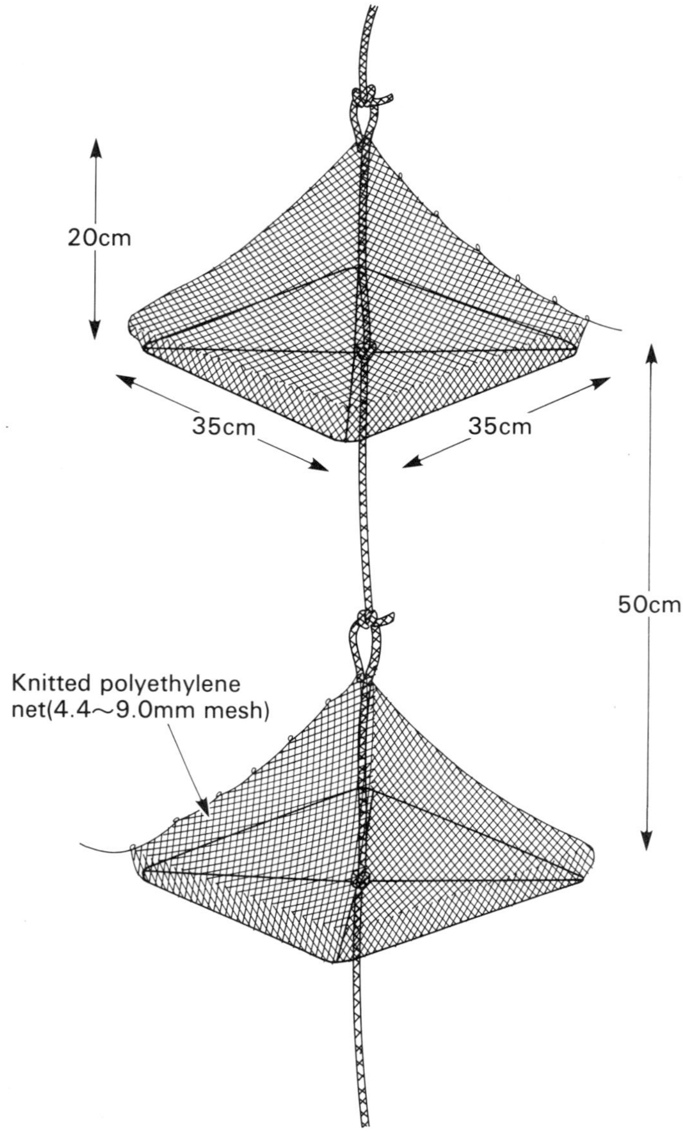

FIGURE 10. Pearl net.

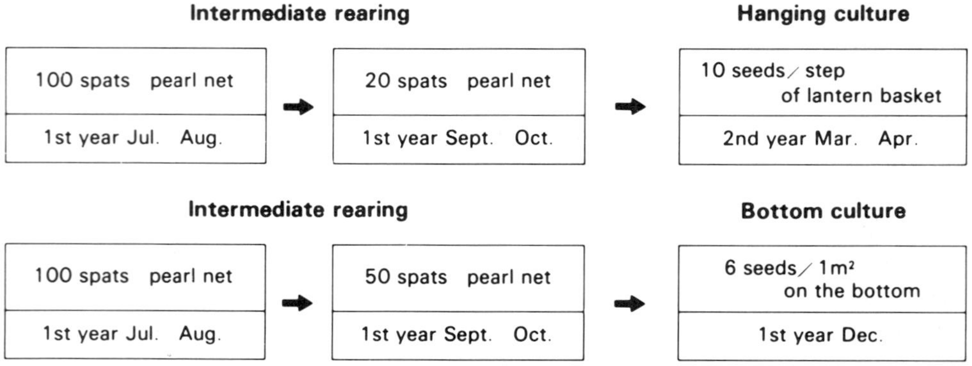

FIGURE 11. Adequate density for intermediate rearing, hanging, and bottom culture in Mutsu Bay. (From Aquaculture Center of Aomori Pref.)

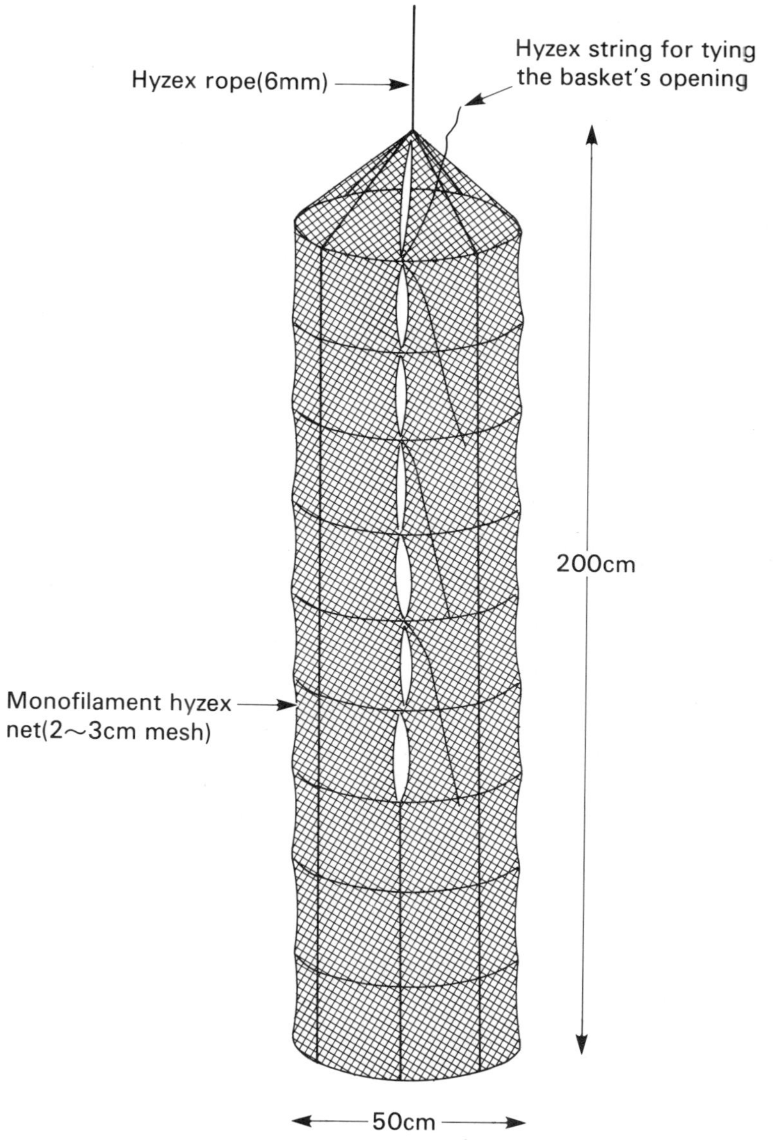

FIGURE 12. Lantern basket.

VI. RESEARCH AND DEVELOPMENT

Many studies have been done on the biology and culture of scallops, but problems remain to be solved, and new problems will arise. The immediate needs which require study are (1) to secure a healthy supply of seed, (2) to determine the capacity for maximum and continuing scallop production of culture areas, (3) to control shellfish poisoning, (4) to establish culture techniques for offshore, and (5) to establish mass production techniques for seed hatcheries.

Scallop production may continue to increase for a time but soon will reach a limit with present methods. Many bays now used for scallop culture are already overcrowded, and offshore culture is not successful. Fishermen must exercise care to eliminate overcrowding, and scientists must devise new techniques to aid this very valuable fishery.

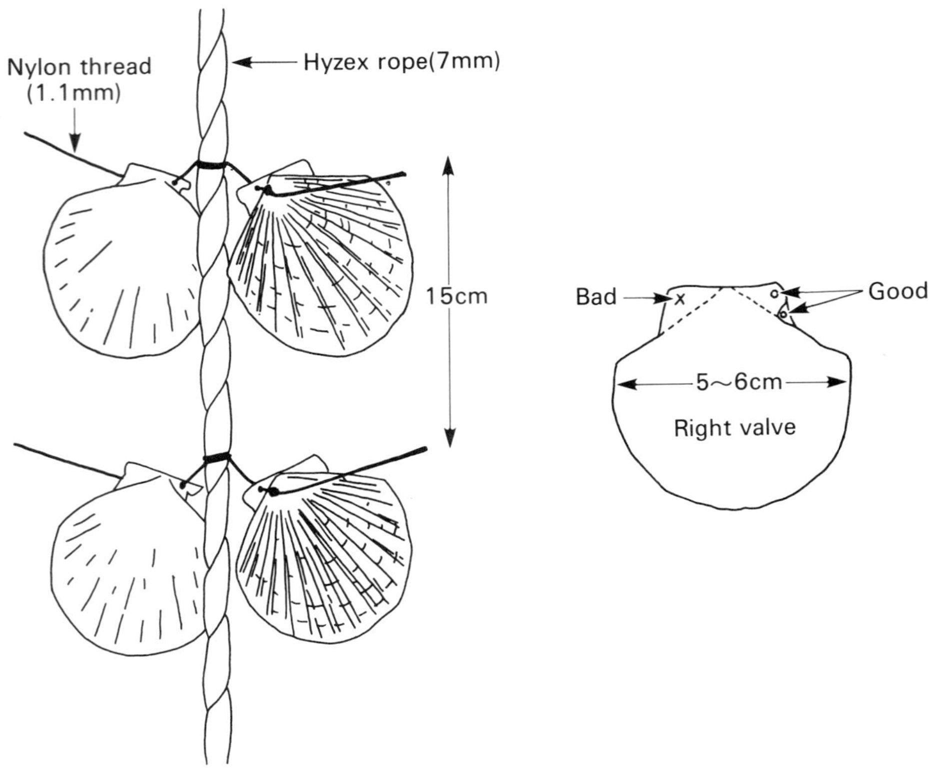

FIGURE 13. Ear-hanging method.

REFERENCES

1. **Tsubata, B.,** *Historical Review of the Scallop Fishery in Mutsu Bay,* Aomori Pref. Gov., 1982 (in Japanese).
2. **Yamamoto, G., Ito, S., and Fuji, A.,** Progression of the scallop culture, in *Through Culture in Shallow Sea,* Imai, T. Sup., Koseisha-Koseikaku, Tokyo, 1971, 187 (in Japanese).
3. **Ito, S., Kanno, H., and Takahashi, K.,** Some problems of culture of scallops in Mutsu Bay, *Sci. Rep. Tohoku Univ. Ser. IV (Bio.),* 15, 89, 1975.
4. **Ito, S.,** On the stable production of the healthy scallop seeds, in *Present Status and Prospects for the Technical Development of Fisheries,* Inoue, H., Sup., Gijutsu Joho Senta, Osaka, 1984, 109 (in Japanese).
5. **Takarada, M., Manson, C., Cooke, K., Carmichael, R., Boune, N., and Kawamura, K.,** Studies on the artificial breeding of the scallop, *J. Aquacult. Cent. Aomori Pref.,* 13, 184, 1984 (in Japanese).
6. **Ito, S. and Mitsuya, T.,** On the high water temperature observed in the summer 1975 in Mutsu Bay, *J. Aquacult. Cent. Aomori Pref.,* 6, 203, 1977 (in Japanese).
7. **Kinoshita, T., Shimizu, J., and Shibuya, S.,** Studies on the substrate of the scallop grounds, *Mon. Rep. Fish. Res. Stn. Hokkaido,* 1(1), 49, 1944 (in Japanese).
8. **Sekino, T.,** Scallop *(Patinopecten yessoensis),* in *Modern Methods of Aquaculture in Japan,* Kafuku, T. and Ikenoue, H., Eds., Kodansha, Tokyo, 1983, 143.
9. **Yamamoto, G.,** Benthic communities in Mutsu Bay, *Sci. Rep. Tohoku Univ. Ser. IV (Bio.),* 18, 482, 1950.
10. **Kinoshita, T.,** Seed collection test on the scallop, *Ten Day Rep. Fish. Res. Stn. Hokkaido,* 273, 1, 1935 (in Japanese).
11. **Kanno, H.,** Relationship between the occurrence of scallop larvae and spat attachment in Okunai district, *Aquaculture,* 17(3), 121, 1970 (in Japanese).
12. **Mikami, T.,** A viewpoint regarding mechanization techniques, in *Propagation and Utilization of the Scallop,* Jpn. Soc. Sci. Fish., Eds., Koseisha-Koseikaku, Tokyo, 1980, 66 (in Japanese).

13. **Sakai, I.**, Scallop Propagation in Japan, *Suisan Hokkaido,* Sapporo, 1976 (in Japanese).
14. **Kanno, H.**, Research on the scallop damage by *Polydora ciliata* in Karibasawa district, *J. Aomori Pref. Aquacult. Lab.*, 11, 309, 1970 (in Japanese).
15. **Takahashi, K., Tanaka, S., Ito, S., Kanno, H., and Hando, T.**, Research on the scallop parasite *Sacculina* sp., *J. Aquacult. Cent. Aomori Pref.*, 2, 231, 1973 (in Japanese).
16. **Takahashi, K., Aoyama, H., Shiogaki, M., Odagiri, A., Uemura, Y., Naoe, H., Sasaki, T., Tomabechi, S., Yokoyama, M., Kumagai, N., Asaka, N., Nishiyama, K., Kitano, E., Sugawara, I., Yamagata, M., Seikino, T., Kanno, H., Tsubota, T., and Saito, K.**, Research on the condition of the scallop and abnormal mortality in 1975, *J. Aquacult. Cent. Aomori Pref.*, 6, 1, 1977 (in Japanese).
17. **Osanai, K., Numagunai, T., Hirai, S., Tsuchiya, M., and Kakinuma, Y.**, Fundamental studies of the cause of the abnormal shape of scallop shells, *J. Aquacult. Cent. Aomori Pref.*, 8, 100, 1979 (in Japanese).
18. **Aoyama, S., Osaka, K., Odagiri, A., Sannohe, Y., Siogaki, M., Sekino, T., Takahashi, K., Takarada, M., Tanaka, S., Nagamine, F., Nakamura, T., Hirano, T., Mitsuya, T., and Yokoyama, M.**, A review of the studies of scallop culture techniques in 1976—'80, *Aquacult. Cent. Aomori Pref.*, 1987 (in Japanese).
19. **Kanno, H. and Sato, S.**, Current progress and problems with scallop propagation, in *Propagation and Utilization of the Scallop,* Jap. Soc. Sci. Fish. Eds., Koseisha-Koseikaku, Tokyo, 1980, 11 (in Japanese).
20. Aomori Prefectural Government, *Final Report on Research and Planning for the Fisheries Development in Mutsu Bay,* 1975 (in Japanese).
21. **Fuji, A.**, Environmental capacity for aquacultural productivity, in *Propagation and Utilization of the Scallop,* Jpn. Soc. Sci. Fish., Eds., Koseisha-Koseikaku, Tokyo, 1980, 26 (in Japanese).
22. **Yasumoto, T., Oshima, T., and Yamaguchi, M.**, Occurrence of a new type of shellfish poisoning in Tohoku District, *J. Jpn. Soc. Sci. Fish.*, 44, 1249, 1978.
23. **Osaka, K.**, Ecology of the shellfish poisoning plankton-Mutsu Bay, in *Shellfish Poisoning Plankton,* Fukuyo, Y., Ed., Koseisha-Koseikaku, Tokyo, 1985, 59.
24. **Osima, Y.**, Occurrence of the paralytic shellfish poison, in *Poisoning Plankton,* Jpn. Soc. of Sci. Fish., Eds., Koseisha-Koseikaku, Tokyo, 1982, 73.

Chapter 17

OYSTER CULTURE IN JAPAN AND ADJACENT COUNTRIES: *CRASSOSTREA GIGAS* (THUNBERG)

Yutaka Kusuki

TABLE OF CONTENTS

I. INTRODUCTION

Most oyster culture in the world depends on the traditional "on-bottom" methods, but in Japan "off-bottom" culture methods have been developed and perfected and have been introduced to the world by many authors.[1-4] We have some knowledge of oyster culture methods in Korea and Formosa, but little information is available on that in China. Therefore, stress will be placed on the methods used in Japan, and those in neighboring countries will be described only briefly.

II. BIOLOGY

A. MORPHOLOGY

The upper (right) valve of *Crassostrea gigas* (Thunberg) is flat and a little smaller than the lower, deeply cupped left valve. Outer shell form varies smooth to fluted, and ribs often radiate from the umbo (Figure 1).[5]

External coloration is creamy yellow with light-purple to deep-violet patches or streaks radiating from the umbo. Internal coloration is chalky white, with occasionally a purple stain over the muscle scar.

Morphological variation is great in this species. The shape of the shell varies with latitude, the southern oyster being smaller, rather oval, and more cupped.[6] Comparison of growth rates of seed oysters brought from various localities and cultured in Hiroshima Bay showed that the length-to-height ratio was about 0.7 for oysters from Hiroshima, in southern Japan, and from 0.6 to 0.65 in those from Miyagi, in northeastern Japan. The regional differences are hereditary.[7]

B. RANGE

C. gigas is widely distributed all over Japan, from Hokkaido in the north to Kyushu in the south, as well as on the coast of Korea, China, and Taiwan.

C. SALINITY AND TEMPERATURE TOLERANCES

Observations on the effects of temperature on development of the eggs of *C. gigas* to the straight-hinge stage reveal the optimum range to be between 24 and 26°C. At about 19°C, half of the eggs developed into normal straight-hinge larvae. At 15°C, abnormal development and heavy mortality occurred, although a few individuals reached the straight-hinge stage in 4 d. At the other extreme, 30°C, some of the eggs develop into early shelled larvae, but the shells in the majority of these were small and deformed.[8]

Normal development of eggs to metamorphosis is limited to a narrower temperature range than can be tolerated by early cleavage stages. At a constant temperature of 21°C, it is difficult to rear larvae to a shell height of more than 200 μm.[9] Under natural conditions in Hiroshima Bay, however, considerable numbers of oyster larvae appear as early as June, when the surface water temperature is about 20°C. At Nakaumi, a lagoon on the coast of the Sea of Japan, the water temperature in summer, when considerable numbers of oyster larvae appear and spat is collected, sometimes exceeds 30°C. Therefore, the temperature range for larval development is from 20 to 30°C, and the optimum temperature range is from 24 to 26°C.

Oyster spat on the hardening racks taken out of the water and exposed to sunlight in August, when air temperature increases from 30 to 35°C, suffer about 20% mortality in 2 h and 90% in 4 h; in the shade, only 10% are killed even in 6 h.

Surface water temperature in summer at the oyster growing areas is usually below 28°C. During severe summers, it rises above 29°C, and at Nakaumi it exceeds 30°C. Water temperatures in winter in northeastern areas of Japan and Korea fall below 10°C and in some

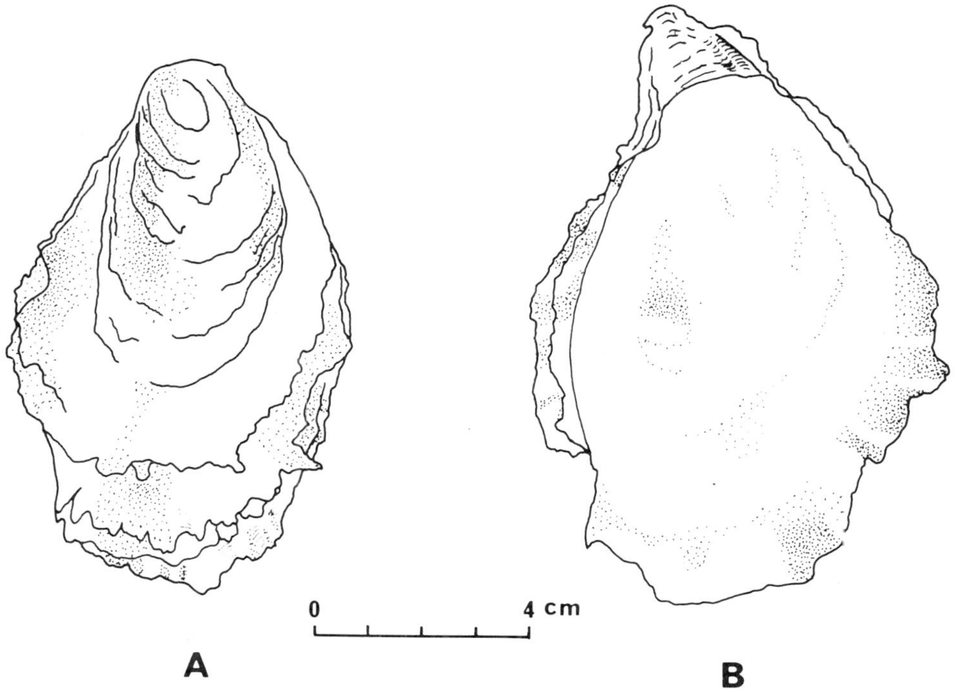

FIGURE 1. *Crassostrea gigas*. Adult oyster grown in Hiroshima Bay. (A) Exterior of upper (right) valve; (B) interior of lower (left) valve.

years as low as 5°C, but growth of oysters there continues, though the rate decreases. Lake Saroma, a lagoon in Hokkaido, is covered with ice in winter, and temperatures are as low as −2°C, but there is no mortality.

In Taiwan, spawning and larval setting occur year-round.[10] Water temperature rises to over 32°C in August. Therefore, the upper limit for larval development and adult survival of "Taiwan" oysters may be higher than that of oysters in Japan. Japanese oysters transplanted to tropical areas survive for some time, but mortality is high and transplantations generally do not succeed in the long term.

The optimum salinity range for development of eggs to the straight-hinge stage depends on the salinity where their parents were grown.[9] When adult oysters were reared at a low salinity of 10.9‰, their eggs developed normally at salinities between 15 and 22‰, whereas the salinity range was between 22 and 38‰ for eggs whose parents were reared at a salinity of 32.5‰.

The salinity range for normal development of eggs to metamorphosis has not been investigated, but, when the sea surface in a small bay near Hiroshima City was covered with freshwater for a prolonged period because of heavy rainfall, vertical distribution of oyster spat showed that the peak spatfall occurred at the depths with salinities of 15 to 20‰ (Figure 2). Above this layer, few spat attached, and below it, the number of spat decreased gradually. The lower salinity limit for larval development of the Japanese oyster is, therefore, near 15‰. Nevertheless, in Sendai Bay, where salinities are as high as 31 to 33‰, excellent spatfall occurs.

During the rainy season in Japan, a heavy rainfall causes a sudden decrease in salinity. When the large Ota River brings down an exceptionally large volume of freshwater into Hiroshima Bay, the surface salinities of the bay decrease to as low as 2 to 4‰ and can be 9‰ at a depth of 2 m on the following day. On the third day, however, salinity rises to 10‰ even at the surface, and the oysters suffer no damage. At Sendai Bay, where salinity

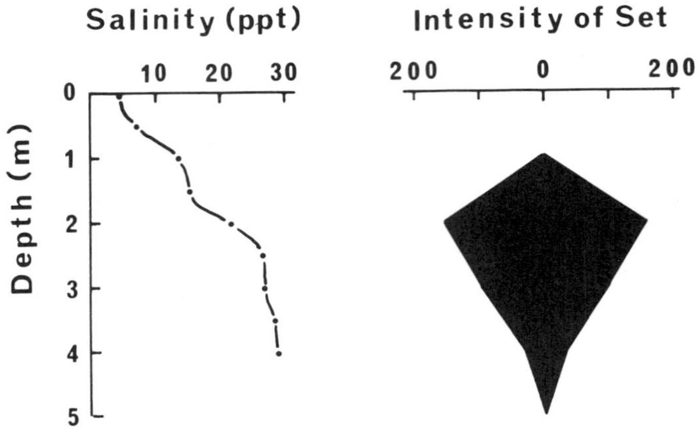

FIGURE 2. Vertical profile of oyster spatfalls in relation to salinity.

can be as high as 35‰, oysters suspended from long-lines grow well; therefore, the maximum salinity tolerated by juvenile and adult oysters must be higher than 35‰.

D. TYPE OF SUBSTRATE

Japanese oyster larvae attach themselves to many of hard surfaces, including rocks, concrete, bamboo, wood, shells of other mollusks, plastics, tin cans, and rubber. Larvae find rough surfaces more suitable than smooth surfaces, and they prefer black or brown surfaces to lighter-colored ones.

E. SEXUAL DEVELOPMENT AND SPAWNING

Two thirds of spat that settle on collectors in June and grow to more than 15 mm in shell height have gonadal development and easily determined sex when examined in mid-September.[11] A few have functional gonads. Eggs from the female oysters have been fertilized and developed normally. When spat are collected at the end of July, the sex of a few of them can be distinguished about 4 weeks after attachment, and these are all males with spermatocytes contained in their gonads.[12] In recent years, spat collections have been carried out in August. Because temperatures continue to decrease after spat are attached, such early gonadal development may not be observed.

Sexual development has been observed in detail[12] in spat and oysters transplanted to rafts. Some of the spat on the cultch are hardened and placed on rafts in October and November to start growth. The gonads of these raft oysters are in the undeveloped stage in December, but in January and February, primary ovocytes become larger and more numerous in the follicles of the females. In males, the follicles contain primary spermatocytes and, occasionally, a few secondary spermatocytes. The gonads of some females have secondary ovocytes in March. With the increase in temperature, gonad development proceeds at a rapid pace in both sexes. The gonadal layer becomes thick in May, the majority of male gonads contain primary and secondary spermatocytes and few spermatids, and female follicles contain ovocytes. In June, gonads of both sexes are full of mature eggs and spermatozoa, and the oysters spawn from June to September. After spawning, the thickness of the gonadal layer decreases, and in spent oysters, the layer becomes extremely thin. At this stage, resorption of the undischarged eggs and spermatozoa proceeds, and the gonads gradually enter into the indifferent phase.

Oyster seeds that have been kept hardening on racks since spat collection start gonadal development later in March, and in April sexes are distinct in some of the oysters. Some individuals have mature gonads in May and June, but others are still in the undeveloped, earlier spring phases. The percentage of hermaphrodites is also much higher. In July, these seed oysters are transplanted to rafts, where growth is rapid and where gonads show rapid development and reach a full ripe stage.

In Matsushima Bay in Miyagi Prefecture, water temperature is lower than in Hiroshima Bay, and gonad development is retarded.[13] In June after the setting year, gonad follicles of the male oyster contain spermatogonia and these of the females, oogonia. In May, gonad development proceeds at a relatively slow pace, but in June the gonads develop rapidly and become full of mature eggs and spermatozoa. Spawning season is from July to August, and in September the gonads are almost spent. At Ojika Peninsula, also in Miyagi Prefecture, the progress of gonad development and the spawning period are about a month later than in Matsushima Bay. A fully mature stage is reached in July and early August; spawning occurs in August and follicle shrinkage, after spawning, by the end of September.

In Taiwan, young oysters that attach in the spring season reach sexual maturity when about 5 months old, but oysters that attach in summer can have functional gonads at only 2 to 4 weeks of age.[10]

Generally, water temperatures are higher in the southern areas, so spawning periods are longer—from July to August in Sendai Bay, from the end of June to September in Hiroshima Bay, and from May to November in Ariake Sea in Kyushu. Mass spawnings and subsequent heavy spatfalls are generally observed two or three times during this period. In Hiroshima Bay in the 1960s, a combination of reduced salinity and a rapid increase in temperature induced spawning after the surface water temperature reached 23°C. Recently, in the spring and summer seasons, oysters have been suspended from rafts at depths of 5 to 10 m to avoid the red tides and fouling organisms of the surface layer. Changes in the hydrographic conditions in the subsurface layer are usually too small to induce spawning, but when a wind of over 5 m/s blows for more than several hours, the seawater is stirred up, and the temperature in the subsurface layer rises rapidly, which induces oyster spawning.[14]

In some unusually hot summers, oysters spawn vigorously, gonads are spent in September, and fattening begins earlier than usual. In cooler summers, the water temperature is lower than usual, spawning is delayed, and unspawned eggs and spermatozoa remain in the gonad layer, such that, even in November, gonads in many oysters have not been resorbed.

Generally, Miyagi oysters spawn earlier, and shrinkage of follicles occurs earlier than in Hiroshima oysters.

The sexes are generally separate, with very few hermaphroditic individuals. The highest percentage of hermaphrodites (5 to 7%) is observed among the hardening oysters in June.[12] Sex reversal of the Japanese oyster is thought to take place after spawning, when gonads are in the indifferent stage.[15]

F. LARVAL DEVELOPMENT

Eggs and sperm of *C. gigas* are discharged into the surrounding water, where fertilization takes place. The optimum temperature range for development of eggs is from 24 to 26°C, and eggs reach a straight-hinge stage in about 24 h at 25°C. After 2 or 3 d, the umbo begins to protrude, and the D-shaped larvae become almost circular. As the larvae grow, the umbo becomes prominent, and the one on the left side becomes larger than the other. Larvae at the umbo stage grow faster, and growth in shell height is about 15 μm/d.[16] Metamorphosis takes place 16 to 17 d after fertilization. The appearance and dimensions of the larva of this species have been described in great detail and compared with those of some other oyster larvae.[17]

The size of larvae at settlement was about 270 μm in shell height when spat collections were carried out on the coast of Hiroshima Bay, but the size collected from rafts anchored around islands off Hiroshima City is about 300 to 330 μm.

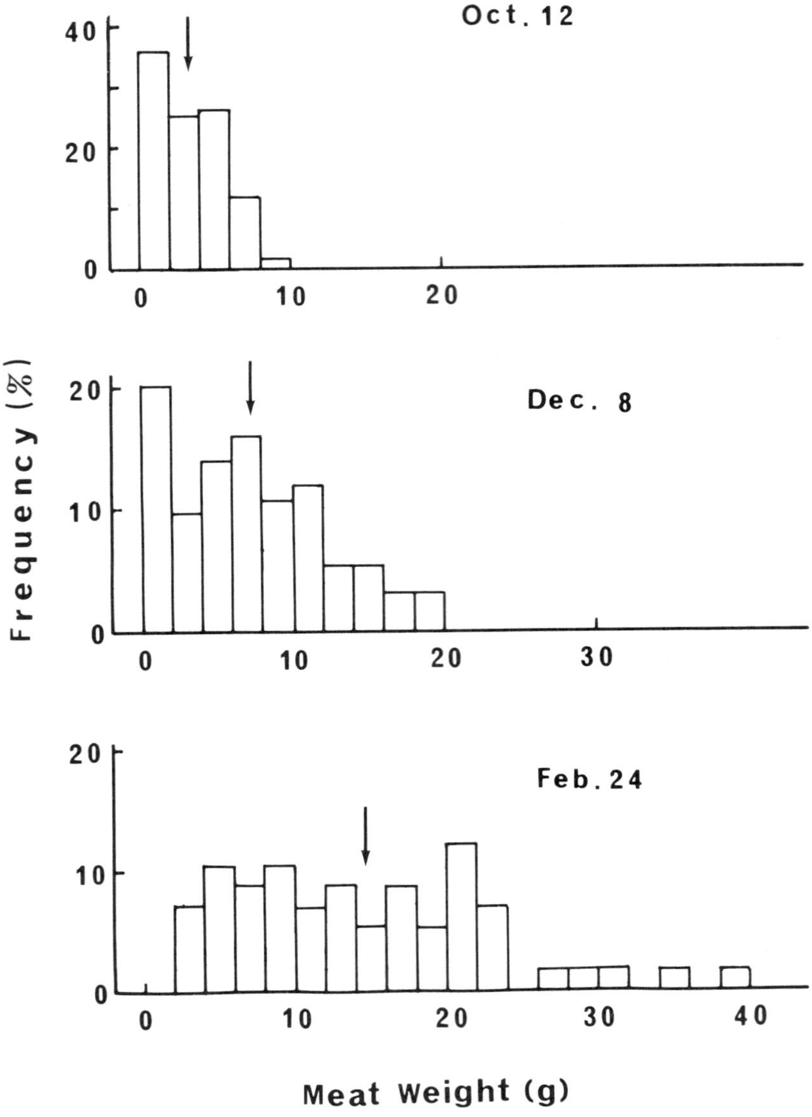

FIGURE 3. Meat weight frequency distribution showing growth of the Japanese oyster reared from a raft at Hiroshima Bay. Seed oysters were transplanted to the raft in June 1977. Arrows indicate mean weights of meat.

G. GROWTH

Growth of oysters suspended from rafts is more rapid than that of oysters cultured on the bottom. On cultch, however, growth is highly variable, largely dependent on the density of the set because of the high competition for food and space. As medium-sized oysters become surrounded by larger, faster-growing oysters, their shells become elongated, and their meats are rather poor. Smaller oysters are suppressed by others; some die, and others survive but show little or no growth. This tendency is more clearly shown by the size distribution of oyster meats than by shell size. The size distribution of oyster growth at the best growing ground in Hiroshima Bay is shown in Figure 3. Oyster seeds were transferred to rafts in early July 1977 to begin the growing phase. In October, after a period of 4 months, larger oyster meats reached 8 to 10 g, but most meats were smaller than 2 g, and average weight was only 3.2 g.[18] In February of the next year, after a period of 7 months,

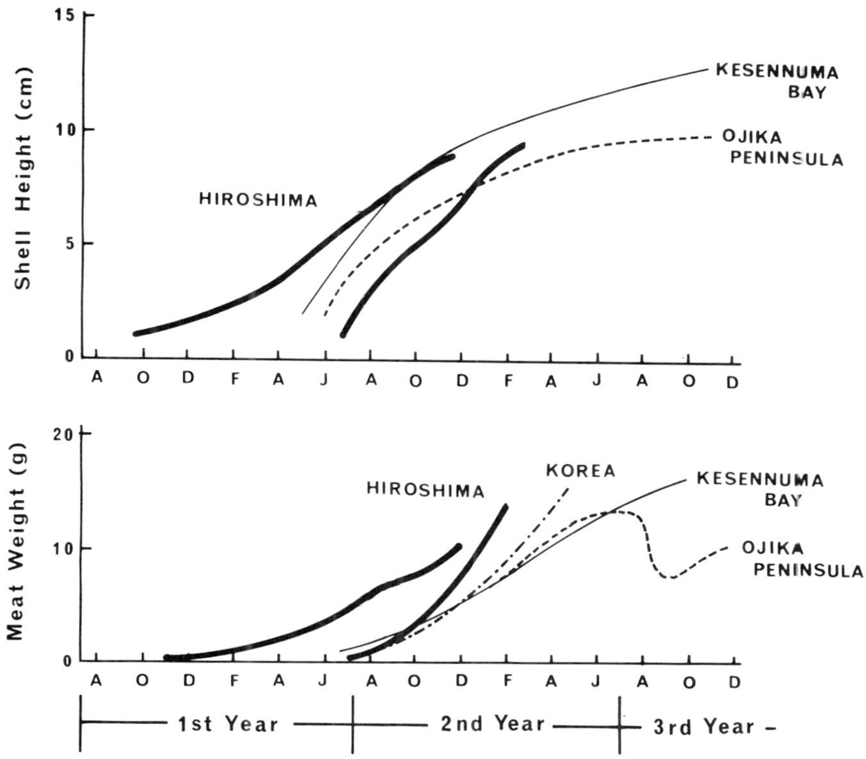

FIGURE 4. Growth curves of the Japanese oyster cultured at various areas.

a few of the larger meats had reached more than 30 g, but smaller oysters still weighed less than 5 g, and the average meat weight was 14.5 g. On cultch where oyster set is dense, as is common in Hiroshima Bay, average wet weights of meat are lower because of these small oysters.

Oyster growth curves in average shell height and wet weight of meat at various localities are shown in Figure 4. Shell growth of oysters transferred to rafts in Hiroshima Bay in October is slow during autumn and winter, faster during spring and summer, and again slow in autumn; oysters reached about 9 cm in height in December of the following year. Miyagi oysters transplanted to rafts at the Bay of Kesennuma in Miyagi Prefecture in June after the setting year grew faster in summer and more slowly in autumn and winter and reached about 13 cm in height.[19]

Monthly growth increments in wet weight of meat are nearly uniform at the best sites in Hiroshima Bay. In general, however, growth slows during the spawning period, and in some areas, such as Sendai Bay, average meat weight decreases considerably after spawning, then gradually recovers from October to December.[13] In the Bay of Kesennuma, where spawning takes place only during unusually hot summers, oysters grow continuously, and better oysters are harvested than in Sendai Bay.[19]

Growth of oysters has decreased over the years in all parts of Japan, probably because too many rafts have been anchored in the same areas. Rearing periods on rafts have become longer. For example, in Hiroshima oysters had been cultured for 1 year until the 1960s; now they are cultured for 2 years.

Growth rates of suspended oysters along the southern coast of Korea are approximately the same as those in Japan, as shown in Figure 3.[20] In China, shell height reaches about 7 to 8 cm after 1 year of age; meat conditions change seasonally much as they do in Japan.[21]

Oysters in Taiwan reach sexual maturity quickly because of the higher temperatures.

Oysters are harvested when they reach about 4 cm in length and, thus, are harvested within a year, even at 2 to 4 months of age.[10]

III. FISHERY

A. HISTORY

The shells of the Japanese oyster and *Ostrea denselamellosa* (Lischke) are found in great numbers in kitchen middens that reveal that oysters have been used as food since prehistoric times. The origin of cultivation of oysters in Japan is not known; we are obliged to rely on records written in later ages about earlier legends.

All the legends indicate that oyster culture began in Hiroshima Bay.[1] A record that describes the beginning of oyster culture was written in "The Guide to Kusatsu" (now part of Hiroshima City), issued in 1924. This report stated that "the method of oyster culture was discovered at Aki Province (now the western part of Hiroshima Prefecture) in the Era of Temmon (1532 to 1555)", but it did not describe who devised it or any details of the method.

There are many records of legends that report in some detail how the stone and stick culture methods came to be discovered and practiced in the 17th Century. The stone culture method, still used in Korea and China, was abandoned in Japan partly because the beaches are rather steeply sloping, with narrow tidal flats, and partly because branches of trees and bamboo, which flourish everywhere, proved preferable to stones, which are heavy and hard to transport. Thus, the stick culture method was developed and widely used in Japan.

In 1923 to 1924, Seno and Hori were the first to experiment with immersing oysters in the sea in cages suspended from a raft, and they obtained good results for both growth and fattening.[22] From 1925 to 1930, experiments on raft-hanging culture were carried out by a number of experimental fishery stations at various localities. Clusters of oysters were fastened to rope with fine wire in these experiments, and spat collection using shells of other mollusks as cultch was tried and in practical use by 1935.

Japan is situated in the path of typhoons, which severely damaged early, wooden oyster rafts; therefore, the raft culture method was adopted only in small, well-sheltered bays and inlets. The rack-hanging method was developed in Hiroshima Bay, because the River Ota had wide tidal flats at its estuary. After World War II, bamboo replaced wooden poles for constructing rafts, as it is flexible and is less likely to be damaged in strong waves. Raft culture spread, and oyster production increased, especially in Hiroshima Bay, where the seabottom is flat, with an average depth of 15 to 20 m. In northestern Japan, the long-line method was devised and developed.

China is generally believed to be the area where oyster culture was first practiced, but little is known of the development of the Chinese method of cultivation. Several species of oysters are cultivated in China, of which *C. gigas* and *C. rivularis* Gould are widely distributed along the coast of China.[21]

The beginning of oyster culture in Taiwan and Korea is also uncertain. The western coast of Taiwan is shallow to a great distance from the shore in the areas where oysters are cultured. Inhabitants of Taiwan say that a man came over with transplanted seeds from China and began culture using the on-bottom method and the stone culture method.[23] The stone, sowing, and simple stick culture methods were used at the end of the 19th Century. After World War II, the raft and long-line methods were introduced, using *C. gigas*.

The tidal range of the western coast of Korea, where the stone culture method is used, is very large. On the southwestern coast of Korea, where there are wide tidal flats, the sowing method was adopted at the end of the 19th Century. Experiments with the rack method began in 1924. In 1957, the Korean government initiated a plan to increase the yields of marine fisheries, including that of oysters. Thus, shores were exploited for sowing

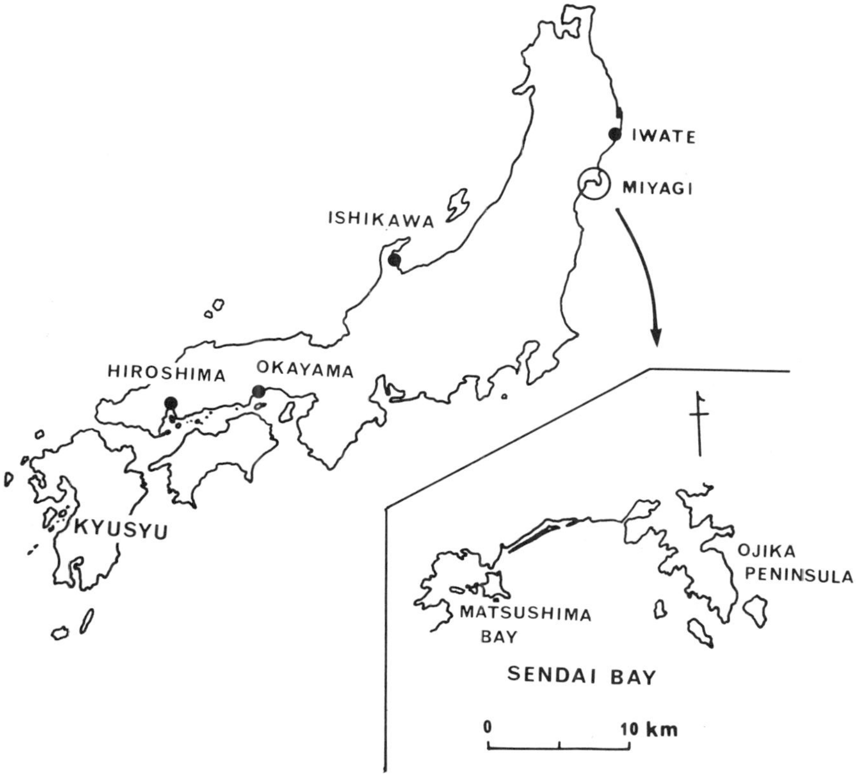

FIGURE 5. Major oyster production areas in Japan.

and stone methods, methods of raft and long-line culture were also introduced from Japan into the southern areas of Korea, and the yield has increased at a rapid rate. At present, the long-line method is the main method used to culture *C. gigas*.

B. IMPORTANCE OF THE INDUSTRY

Japan is second in the world in oyster production, after the U.S., producing annually about 250,000 t with shell, which accounts for approximately 25% of annual world production. Korea is third, with a rapidly increasing production of some 200,000 t, comprising about 20%. Production in Taiwan has gradually increased in recent years and now reaches 26,000 t. (No data from China are available.) Together these countries produce approximately half of the world oyster production.

Japan was once an important exporter of seed oysters of *C. gigas*. Miyagi Prefecture used to export seed oysters to the U.S., Korea, France, and other countries, but *C. gigas* flourished in these countries, and except in the U.S., spat are successfully collected. On the northwest coast of the U.S., commercial hatcheries have been erected to produce seed oysters. Thus, demand for seed oysters of this species had decreased, and seed oyster exports here decreased year after year and are now almost zero.

Cultured marine products such as shellfish, sea vegetables, and fish are popular in Japan. Among them, oysters are especially important, because they are cultivated all along the coast of Japan and account for about 25% of the total production of cultured marine products and about 75% of the total shellfish culture. The main oyster producers are the prefectures of Hiroshima, Miyagi, and Okayama (Figure 5). Of these, Hiroshima Prefecture is most important and averages 160,000 t annually, accounting for about 60% of the total production of Japan (Table 1). Miyagi Prefecture is second to Hiroshima in oyster production; its average

TABLE 1
Oyster Production with Shell by Prefecture, Japan

Prefecture	1965	1970	1975	1980	1984
Hiroshima	125,617	112,802	129,037	166,006	153,823
Miyagi	33,850	41,347	33,401	42,281	44,540
Okayama	5,917	6,585	17,615	18,976	25,712
Iwate	9,811	11,822	8,871	11,058	9,289
Ishikawa	9,811	3,393	3,787	5,971	5,005
Others	32,142	14,850	8,462	17,031	18,757
Total	217,148	190,799	201,173	261,323	257,126

Note: Weight given in metric tons.

From Annual Report of Production on Fishery and Aquaculture, Statistics and Survey Division, Ministry of Agriculture, Forestry, and Fishery, Japan.

TABLE 2
Oyster Production with Shell by Cultural Method and Number of Farmers in Japan in 1984

Prefecture	Facilities: Raft[a]	Facilities: Long-line[b]	Facilities: Rack (× 1000 m²)	Production with shell (ton)	No. of farmers	Production per farmer (ton)
Hiroshima	9,946	0	9	153,823	623	246.9
Miyagi	785	18,225	241	44,540	2,053	21.7
Okayama	3,579	0	0	25,712	303	84.9
Iwate	1,150	3,710	0	9,289	1,254	7.4
Ishikawa	22	350	85	5,005	128	39.1
Others	1,646	4,377	2,485	18,757	1,420	13.2
Total	17,128	26,662	2,820	257,126	5,781	44.5

[a] Standard size of a raft is 18.2 × 10.9 m (= 198.4 m²).
[b] Standard length of a long-line is 54.5 m.

From Annual Report of Production on Fishery and Aquaculture, Statistics and Survey Division, Ministry of Agriculture, Forestry, and Fishery, Japan.

annual production is approximately 44,000 t, about 17% of the total. Okayama Prefecture has steadily increased its production and now contributes about 10% of the total.

There were 5781 oyster farmers operating oyster culture in 1984. In Miyagi Prefecture, there were 2053 oyster farmers, about 35.5% of the farmers in Japan (Table 2), whereas Hiroshima Prefecture, the main producer, had only 623 farmers. Therefore, the annual production per oyster farmer is highest in Hiroshima Prefecture, reaching 246.9 t, and the economy of the farmers is based solely on oysters. As the oyster production per farmer in the northeastern part of Japan, including Miyagi and Iwate Prefectures, is low, oyster farmers usually culture other products, such as scallops, sea vegetables, and sea squirts. Formerly, Nagasaki Prefecture in Kyushu was one of the major producers, but the oyster-farming beds are also suitable for the culture of nori, a sea vegetable, and oyster production has decreased considerably.

Normal oyster culture operations are generally carried out by members of the farmer's family, and during the harvesting period (from October to March) a few extra women are employed to shuck oysters.

The most important species of shellfish in culture in Korea is the oyster, which accounts

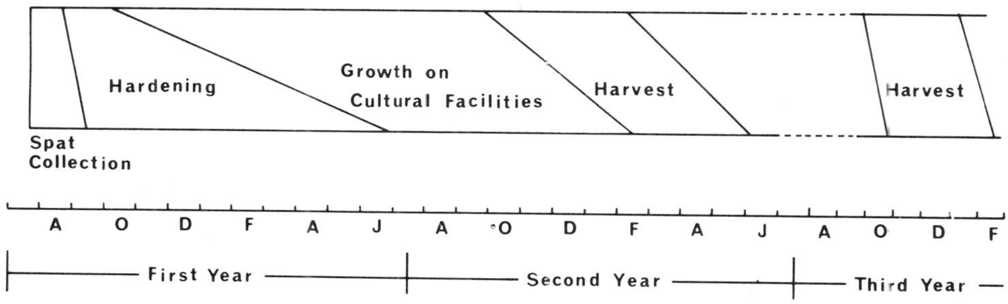

FIGURE 6. Seasonal activity for oyster culture in Japan.

for approximately 70% of total shellfish production.[24] Oyster production in Korea was 203,312 t in 1984, of which 96% was produced along the southern coast. Kyongnam Province, in particular, on the southeastern coast, produced 168,980 t, comprising 83.1% of the total; the long-line method was the major procedure. There were 1062 licensed areas in 1984, of which 460 were operated by private farmers and the others by groups of farmers. Only about ten were operated raft culture, but each of those had more than 100 rafts.

Culture of the Japanese oyster is also the most important shellfish culture in Taiwan.[10] About 50% of the shellfish production is of oysters. Oyster production in 1984 reached 28,503 t, produced along the western coast.

C. METHODS OF CULTURE

Methods of oyster culture in Japan have been described elsewhere in considerable detail.[1,3,4] Therefore, they will be described here only briefly.

The methods of oyster culture used at present in Japan are the hanging, sowing, and stick methods. The sowing method is used only in Hokkaido, and the stick method only in Kyushu, and oyster production by these methods accounts for only 0.2% of the total production. As a result, the oyster industry in Japan can be regarded as solely dependent on the hanging culture method. There are three kinds of hanging method, the raft, long-line, and rack methods. Each locality uses one or two of these methods, according to the depth and strength of wave action. The raft method is used everywhere in Japan, but the long-line method is confined to northern Japan, and the rack method is only practiced on a small scale, as shown in Table 2.

1. Seasonal Activity

Seasonal procedures for oyster culture vary according to the geographic and hydrographic conditions of different areas, but one part of these different procedures is standard: after spat are collected in August, the strings of collectors are transferred to intertidal racks for hardening. As oyster harvesting begins in October, seed oysters that have been hardening are gradually transplanted to the vacant rafts or long-lines from November onward, but a great part of the seedlings are transplanted in March and April of the following year when the harvest period is over. Two-year-old oysters can be harvested from October onward in Hiroshima, Okayama, and Ishikawa Prefectures, but in Miyagi and Iwate Prefectures oysters are harvested after 3 years, although in the northern part of Japan they are called 2-year oysters. Figure 6 shows the oyster culture schemes.

In Hiroshima Bay, heavy sets of mussels, *Mytilus edulis* L., are observed from January to June. Therefore, the period of transferring seed oysters to rafts is divided into two, the first from October to December and the second in June.

2. Spat Collection

Spat collection takes place at every growing area from Miyagi Prefecture southward.

However, except in Hiroshima and Miyagi Prefectures, the numbers of seed oysters collected do not fulfill demand, and farmers must, therefore, purchase seed oysters from these two prefectures.

Spat collectors consist of about 80 scallop shells strung on 1.6 m of galvanized wire with 2-cm plastic spacers between them. Two or three collector strings, each folded double, are bunched together and hung from the rafts. Spat must attach within 2 or 3 d after the collectors are set out, before the surfaces of the collectors become covered with mud or detritus making attachment impossible. Therefore, it is very important to know the best time to set the collectors out. Time, duration, and intensity of spatfalls can at present be calculated from the sizes and numbers of oyster larvae present. Oyster farmers are informed of these settling trends and can plant cultch accordingly.

3. Hardening

If spat attach to the collectors in sufficient numbers within a week, the strings are transferred to intertidal racks to harden the spat. In Hiroshima Bay, where the tidal range is around 3 m, the height of the racks is adjusted to the highest lower level of neap tides. In Miyagi Prefecture, where the tidal range is as small as 30 cm, the strings are laid horizontally on the racks. In other areas, where seed oysters are bought from Miyagi and Hiroshima Prefectures, the purchased seedlings are hardened on racks or hung densely from racks until they are transferred to cultural facilities.

4. Growth on Rafts

In Hiroshima Bay, the largest rafts, made of bamboo, are 20 m long and 20 m wide. A raft has 43 transverse bamboos, each 10 m long, and each bamboo has 16 oyster strings, making 700 oyster strings per raft. An oyster string is 9 m long, with 40 collectors spaced by 20-cm plastic tubes.

Rafts are kept afloat by Styrofoam buoys. Five rafts are strung together with a 150-m wire and anchored at each end of the line by 200-m wires to 5-ton concrete weights.

In other areas, the rafts are smaller, usually 10 m long and 6 m wide, and are anchored singly or in pairs.

5. Long-line

A long-line is 55 m or 100 m long. A typical long-line unit consists of about ten Styrofoam buoys arranged in a row, 6 or 7 m apart, with two parallel polyethylene ropes fastened to them, one along each side of the row. The row of buoys is anchored at both ends. Oyster strings 8 to 9 m long are hung from the ropes, 30 cm apart.

The standard length of a long-line unit in Korea is 100 m. Glass spheres 30 cm in diameter are used as buoys, placed 5 m apart. The oyster strings are relatively short, 4 m long, because the growing areas are as shallow as 7 m.

The long-line method has also been introduced in Taiwan, where Styrofoam buoys are used, placed 6 to 8 m apart. Oyster strings hold 15 to 20 seedlings each and are suspended 40 to 60 cm apart.

6. Rack Method

The rack method is used in areas shallower than 3 or 4 m. Vertical wooden or bamboo poles are driven into the bottom about 2 m apart in a long row. The tops of the poles are connected by horizontal ones, and strings of seed oysters are suspended from the horizontal poles. This method is also used in Taiwan, Korea, and Fujian Province in China.

7. Stone Culture Method

Stone culture is a traditional method that is still widely used in Korea and China. It is the major method used in Shandong Province, China, where several variations of it have developed: (1) large stones, about 2 m long and 20 × 20 cm, are erected on the bottom like tombstones; (2) five or six elongated concrete blocks or stones are stood against each other; and (3) stones are piled in heaps.[21]

8. Stick and Sowing Methods

The stick method is used in China, Taiwan, and Korea. Usually, tree branches or bamboo are erected on tidal flats. In a variation used in Taiwan, deft bamboo sticks, the length of which depends on the depth of the water, are inserted into the bottom of shallow areas with one or two empty shells wedged into each one.[10,23]

D. METHODS OF HARVESTING

Oysters are harvested for the fresh market from October to March and for canning from February to May.

As oyster strings are as long as 9 m in the Hiroshima area, strings of oysters grown to market size are very laborious to collect by hand, and it is difficult to collect all the oysters without losing them. Therefore, about a 20-m working boat is used, equipped with a 150-hp engine, a power winch, and a 10-m mast in the bow. The winch can hoist more than ten oyster strings at a time to the top of the mast. As the oyster strings consist of galvanized wires, the oysters slip off into metal baskets under their own weight when the bottom ends are cut.

The shucking houses on shore are equipped with special quays that enable working boats to moor alongside even at low tide. A traveling crane that connects the shucking house and the quay lifts the metal baskets of oysters and brings them to the hopper of the washing machine. The washer has a slowly rotating cylindrical steel cage set at an angle to the hopper. Water flows through it at high pressure. The clustered oysters introduced into the washing machine from the hopper emerge gradually, detached and cleaned, and are dumped onto a conveyor belt that transports them into the shucking house.

Although oyster farmers in areas other than Hiroshima harvest oysters by hand, hoisting up oyster strings by winch is becoming popular elsewhere in Japan. Because ropes are used for strings in Miyagi Prefecture, the strings are reeled onto a rotating drum on board the working boat there.

In Korea, raft-grown oysters are harvested by boats larger than those used in Hiroshima and shucked on board. Oysters raised by long-line or rack methods are harvested by hand. Oysters grown on stones, cement blocks, and sticks are chipped off with a scraper.

E. USES AND PROCESSING

In Japan and neighboring countries, oysters are mostly sold in shucked form. In Hiroshima district, shucking is done with a special tool that consists of a wooden handle with a narrow steel blade and in other districts with knives. Shucked oysters are carefully washed with chlorinated seawater, packed into zinc cans, and stored or transported below 10°C. Some of the shucked oyster meat is sent directly to the markets. It is packed in cans holding 2, 5, or 20 kg of meat. These cans are collected from farmers in the evening and carried by refrigerated railcars or trucks to the markets of big cities, arriving there the next morning.

The great majority of oyster meat is packed in 18-l zinc cans, weighing 20 kg each, and transported to broker's factories, where the meat is washed and packed in cans of the various sizes for markets or in small vinyl packages for supermarkets.

From February through May, shucked oysters are processed. In Japan, most of the processed oysters are from Hiroshima, where approximately half of the oysters produced are processed. Processed oysters are exported to Hong Kong, Australia, and the U.S.

Some of the shucked oyster meats in Korea are also canned and frozen,[24] but export of these processed oysters has decreased in recent years. Some of the fresh, raw oyster meat is exported to Japan, but the amount is not known.

The empty shells of the shucked oysters are shipped to shell factories, where they are roughly crushed for chicken feed or finely ground for neutralizing soil, as Japanese soil has a tendency to be acidic.

F. CONSTRAINTS

Oyster production in Japan is, in my opinion, approaching its highest level. The following discussion will concern principally the problems that plague oyster farmers in Japan, and many of these problems will, in the future, prevent or limit expansion of oyster production in Korea and other countries where oyster production could be increased by application of the hanging methods.

First of all, the hanging culture methods, particularly the raft method, are liable to become overcrowded. Cultural facilities such as rafts or long-lines become more numerous annually, while growth of oysters has gradually been decreasing in all the growing areas in Japan. The problem is that the oysters are overcrowded, and natural food supplies cannot meet the requirement of the oysters for rapid growth and adequate fattening. The growing period is becoming longer, so the cost of production is increasing.

The great number of oysters suspended from the rafts and racks produce vast amounts of biodeposits, which accumulate on the bottom as sediments rich in organic matter.[25] Oxygen consumption by these sediments rapidly increases at high temperatures in summer, resulting in reduction of dissolved oxygen in bottom-layer water. In Hiroshima Bay, the rafts are towed to offshore growing grounds around the island, where available oxygen is plentiful but food for the oysters is sparse. If the offshore grounds were not available, the oyster industry would have to be abandoned.

Oyster spat are collected mainly in Hiroshima and Miyagi Prefectures, where the spatfall period has been delayed from July to August. In recent years, year-to-year variation in the natural setting of oysters has been high; the daily set of oyster spat on a cultch is as high as 1000 in some years but only 20 or 30 in others. In recent years, farmers failed to collect sufficient numbers of spat on cultches and had to place collectors many times. The farmers fear that spat collection might fail in the future. Artificial rearing of oyster larvae in Japan is still in the experimental stage, but in Fujian Province, China, oyster hatcheries have developed as a response to failure to natural reproduction.

The most important fouling organisms are mussels and barnacles. Mussel set lasts from winter to June in Hiroshima and to July or August in northern parts of Japan. In the Hiroshima district, seed oysters are not transferred to rafts until June, solely to avoid the heavy set of mussels. In northern waters, seed oysters are transferred in the spring, and the strings are dipped into hot water in August and September to kill young mussels attached to oyster cultches.

The period of the barnacle set is longer in the warm southern waters. Generally, it begins in June and lasts till September. Therefore, in years when a heavy set of barnacles occurs, cultches for collecting oyster spat are occupied by barnacles. In some areas of heavy barnacle set, vast numbers of them attach to the shells of live, adult oysters, severely limiting growth.

In Hiroshima Bay, the serpulid worm *Hydroides norvegica* Gunnerus settled profusely on raft-grown oysters in September 1969, causing great damage.[26] In 1970, the time of set of the worm was determined, and rafts were towed back to these growing grounds only after the set of the worm had ended.

Predators such as drills and starfishes are not a serious threat to off-bottom culture oysters. Flatworms, *Stylochus ijimai* Yeri et Kaburaki, attack adult oysters,[27] but cause only occasional and local damage.

Mortality ranges from 20 to 90% depending on time and place,[28] but is highest from August to October, that is, during and after the spawning season. In northeastern Japanese waters, mass mortality was not observed before 1960. However, it has occurred in Matsushima Bay annually since 1961. It is ascribed to a low water temperature in summer.[29,30] In Hiroshima Bay, where eutrophication has also occurred, mass mortality occurs only occasionally and ranges only from 20 to 30%. The cause of mass mortality is not known. Use of hardened seed oysters seems to decrease the mortality rate, but this method is ineffective when the growing period on rafts is extended.

Mass mortalities in Korea occurred in 1973, and oyster production in the shell suddenly dropped to 65,000 ton, from 93,000 ton in the preceding year. In Taiwan, mass mortality occurred from 1974 to 1976.[10]

At the coasts of Japan and China are in the path of typhoons, only well-sheltered bays and inlets have been exploited for oyster culture. However, because population in these areas has increased and industries have developed, the growing grounds have shrunk, and the number of private oyster farmers in Hiroshima Prefecture has decreased from 800 in the 1970s to about 600 in more recent years.

In Hiroshima Bay, Matsushima Bay, and Kesennuma Bay, the municipal and industrial pollution resulting from coastal development has increased the frequency of red tides. When a red tide occurs, the organisms die and rapidly disintegrate, causing a reduction of oxygen content of water at the bottom, which does serious damage to bottom oysters but little damage to raft-grown oysters. Nevertheless, oysters did not grow after the large-scale red tide of *Chattonella antiqua* (Hada) Ono occurred in Hiroshima Bay in 1969.

Water quality is measured as the amount of organic matter, by the biochemical-oxygen-demand method. Pollution levels have been successfully lowered, and red tides have decreased.

Shellfish poisons, both paralytic and diarrhetic, have been found in mussels, scallops, and oysters, particularly in northern Japanese waters. Their effects on scallop culture are serious, but on oysters they are negligible because the periods of their appearance are usually in the spring. However, in some districts, sale of oysters has been prohibited for some time. As shellfish poisons occur all over Japan, they may do great damage to the oyster industry in the future.

Shells not used for chickenfeed, lime, or reclamation materials are left piled in heaps on the ground. These shells offer a breeding place for flies and are, therefore, public health hazards.

The shucking procedure is labor intensive, and about half of the sale price is used as wages. Labor costs are increasing more rapidly than the price per kilogram of shucked oysters. This situation has reduced the profit, especially when overcrowding and longer growing periods reduce the growth of oysters.

G. MANAGEMENT

Shallow sea areas around Japan belong to the Japanese government. All the fishermen are organized in the fisheries' cooperative associations. Those who intend to engage in oyster cultivation must submit applications to the prefectural authorities through the cooperatives, with culture methods to be used and locations of grounds. Licenses are then given to the cooperatives, which pay a nominal fee for registrations. Permissions are renewed every 5 years. The cooperatives must take charge of the leased areas and collect from each farmer fees to cover management costs, employing a watchman to prevent vandalism, and maintaining small rafts on which lights are set so that ships sailing at night will be able to avoid the culture facilities. Numbers of culture facilities (racks, rafts, or long-lines) in the leased area, allotments, and arrangement of facilities in the area are decided by conferences of farmers.

The prefectural public health authorities control the hygenic conditions of oyster farming and processing. The laws and regulations that control sanitary conditions were strengthened in 1967. There regulations set forth requirements for the shucking, handling, and packing of shucked oyster meat. Shucked oyster meat must be carefully washed in clean seawater or sodium chloride solution of more than 2% concentration. When shucked meat is stored or shipped, it must be kept at a temperature of 10°C or less.

Toxic levels of shellfish poisons are tentatively set at 2 mU/g for paralytic shellfish poisons and 0.05 mU/g for diarrhetic shellfish poison. Analyses are carried out by each fishery experimental station and public health service. If poison concentrations higher than these are detected, sale of oyster meats is not permitted until lower levels are reached.

IV. RESEARCH AND DEVELOPMENT

A. BIOLOGICAL RESEARCH

Growth rates of cultured oysters have decreased over the years. One of the causes is that oysters are overcrowded as a result of the increase in production. Good-quality oysters will result only if the production is kept within the range of productivity of the area in which they are grown. Many ecological data must be obtained before the carrying capacity of a bay can be determined. When annual oyster production in a certain bay stops increasing and, despite farmers' efforts, reaches a constant level, it is thought that the maximum productivity of the bay has been reached. Methods for determining the number of cultural facilities a bay can support within this level have been studied.

Another important problem of oyster culture is the mortality and poor meat quality that result from gametogenesis and spawning. Nonreproductive oysters would presumably have lower mortality rates and higher quality. Thus, triploid oysters, if sterile, have potential for benefit to oyster farmers. Treatment of newly fertilized eggs of both American and Japanese oysters with cytochalasin B or hydrostatic pressure has successfully produced triploids,[31,32] and the search has begun in Japan for a genetic means of producing triploid oysters.

B. FUTURE DEVELOPMENT

The hanging culture method, developed in Japan and now used extensively in Korea, greatly increases production, but its proliferation has resulted in overcrowding and vast amounts of biodeposit. Investigations must continue into determining the density that will yield maximum production and still produce quality oysters.

Because natural spatfall is successful in Japan and neighboring countries, artificial rearing of oyster larvae and spat collection have been only experimental. Culture of sterile triploid oysters, however, will require artificial rearing of oyster larvae.

Shucking is laborious. In Hiroshima, about half of farmers' income is taken up by wages to shuckers, and wages are increasing more rapidly than the price of oysters. In some areas in the U.S., the "shock method" has been adopted to facilitate shucking.[33] Many oyster-shucking concepts have been experimented with, but none has yet found commercial acceptance. Research must continue.

REFERENCES

1. **Cahn, A. R.,** Oyster culture in Japan, Rep. Nat. Resour. Sect. G.H.Q., No. 134, 1950.
2. **Fujiya, M.,** Oyster farming in Japan, *Helgol. Wiss. Meeresunters.,* 20, 464, 1970.
3. **Korringa, P.,** *Farming the Cupped Oysters of the Genus Crassostrea,* Elsevier, Amsterdam, 1976, chap. 2.
4. **Ventilla, R. F.,** Recent developments in the Japanese oyster culture industry, in *Advances in Marine Biology,* Vol. 21, Blaxter, J. H. S., Russell, F. S., and Yonge, M., Eds., Academic Press, New York, 1984, 1.
5. **Torigoe, K.,** Oysters in Japan, *J. Sci. Hiroshima Univ., Ser. B, Div. 1 (Zool.),* 29, 291, 1981.

6. **Seki, H.,** On the difference between *Ostrea gigas* from Hiroshima and Sendai Bay, *Fish. Invest. Imp. Fish. Exp. Stn. Suppl. Rep.*, 4, 45, 1937 (in Japanese).
7. **Imai, T. and Sakai, S.,** Study of breeding of Japanese oyster, *Crassostrea gigas, Tohoku J Agric. Res.*, 12, 125, 1961.
8. **Seno, H., Hori, J., and Kusakabe, D.,** Effects of temperature and salinity on the development of eggs of the Japanese oyster, *J. Imp. Fish. Inst.*, 22, 169, 1926 (in Japanese).
9. **Numachi, K.,** Progress in oyster culture, in *Aquaculture in Shallow Seas,* Iami, T., Ed., Kouseisha-Kouseikaku, Tokyo, 1971, 95 (in Japanese).
10. **Ting, Y. Y.,** Shellfish culture in Taiwan, *Tungkang Mar. Lab. Conf. Proc.*, 1, 129, 1984.
11. **Amemiya, I. and Nakamura, N.,** Is the Japanese oyster protandrous?, *Nihon Gakujutsu Kyokai Houkoku,* 15, 338, 1940 (in Japanese).
12. **Ogasawara, Y., Kobayashi, U., Okamoto, R., Furukawa, A., Hisaoka, M., and Nogami, K.,** The use of the hardened seed oyster in the culture of the food oyster and its significance to the oyster culture industry, *Bull. Naikai Reg. Fish. Res. Lab.*, 19, 1, 1966 (in Japanese, summary in English).
13. **Koganezawa, A.,** Ecological study of the production of seeds of the Pacific oyster, *Crassostrea gigas, Bull. Jap. Sea Reg. Fish Res. Lab.*, 29, 1, 1978 (in Japanese, summary in English).
14. **Kusuki, Y. and Aratani, Y.,** Massive spawning of oysters suspended in the sub-surface layer owing to a rapid increase in water temperature, *Bull. Hiroshima Fish. Exp. Stn.*, 16, 19, 1986 (in Japanese, summary in English).
15. **Amemiya, I.,** On the sex-change of the Japanese common oyster, *Ostrea gigas* Thunberg, *Proc. Imp. Acad.*, 5, 284, 1929.
16. **Kusuki, Y.,** Rearing larvae of Japanese oyster, *Crassostrea gigas* (Thunberg), *Bull. Hiroshima Fish. Exp. Stn.*, 6—7, 27, 1976 (in Japanese).
17. **Chanley, P. and Dinamani, P.,** Comparative descriptions of some oyster larvae from New Zealand and Chile, and a description of a new genus of oyster, *Tiostrea, N. Z. J. Mar. Freshwater Res.*, 14, 103, 1980.
18. **Kusuki, Y., Kimura, T., Mekuchi, T., and Hashimoto, T.,** Growth of oysters in relation to quantities of food, *Bull. Hiroshima Fish. Exp. Stn.*, 13, 7, 1983 (in Japanese).
19. **Obata, K., Suda, Z., and Gouko, T.,** Studies on the decline of productivity of the growing grounds in Kesennuma Bay, *Bull. Kesennuma Miyagi Pref. Fish. Exp. Stn.*, 3, 1, 1977 (in Japanese).
20. **Bae, P. A., Kim, J. E., Kang, P. A., and Kim, Y.,** Studies on suspended culture of oyster, *Crassostrea gigas* in the Korean coastal waters. III, *Bull. Fish. Res. Dev. Agency,* 15, 73, 1976 (in Korean, summary in English).
21. Shandong Province Revolutionary Committee, Oyster Culture, in *Marine Shellfish Culture,* Shandong People's Publishing Firm, 1974, 11 (in Chinese).
22. **Seno, H. and Hori, J.,** A new method of fattening oysters, *J. Imp. Fish. Inst.*, 22, 211, 1927 (in Japanese).
23. The Public Welfare Bureau, the Government-General of Taiwan, General Situations of Fisheries in Taiwan, 1896, 12 (in Japanese).
24. Fisheries Bureau, Korean Government, A Fishery Industry White Paper in 1985, 1986 (translated into Japanese by Fisheries Bureau, the Japanese Government).
25. **Kusuki, Y.,** Fundamental studies on the deterioration of oyster growing grounds, *Bull. Hiroshima Fish. Exp. Stn.*, 11, 1, 1981 (in Japanese, summary in English).
26. **Arakawa, K. Y.,** Notes on a serious damage to cultured oyster crops in Hiroshima caused by a unique and unprecedented outbreak of a serpulid worm, *Hydroides norvegica* (Gunnerus) in 1969, *Jap. J. Malacol.*, 30, 75, 1971 (in Japanese, summary in English).
27. **Arakawa, K. Y.,** Notes on damage to cultured oysters in Hiroshima caused by predation of a species of turbellarian, *Stylochus ijimai* Yeri et Kaburaki, *Jap. J. Malacol.*, 29, 65, 1970 (in Japanese, summary in English).
28. **Koganezawa, A.,** Present status of studies on the mass mortality of cultured oysters in Japan and its prevention, Proc. 3rd U.S.-Japanese Meet. on Aquacult., Tokyo, 1975, 29.
29. **Kan-no, H., Sasaki, M., Sakurai, Y., Watanabe, T., and Suzuki, K.,** Studies on the mass mortality of the oyster in Matsushima Bay. I. General aspects of the mass mortality of the oyster in Matsushima Bay and its environmental conditions, *Tohoku Reg. Fish. Res. Lab.*, 25, 1, 1965 (in Japanese, summary in English).
30. **Imai, T., Numachi, K., Oizumi, J., and Sato, S.,** Studies on the mass mortality of the oyster in Matsushima Bay. II. Search for the cause of mass mortality and the possibility to prevent it by transplantation experiment, *Tohoku Reg. Fish. Res. Lab.*, 25, 27, 1965 (in Japanese, summary in English).
31. **Stanley, J. G., Allen, S. K., Jr., and Hidu, H.,** Polyploidy induced in the American oyster, *Crassostrea virgincia*, with cytochalasin B, *Aquaculture*, 23, 1, 1981.
32. **Chaiton, J. A. and Allen, S. K. Jr.,** Early detection of triploidy in the larvae of Pacific oysters, *Crassostrea gigas,* by flow cytometry, *Aquaculture,* 48, 35, 1985.
33. **Pringle, S. B.,** The South Carolina report on the "hot dip" or "shock" method of opening oysters, in Proc. Shellfish Sanitation Workshop, Jensen, E. T., Ed., U.S. Department of Health, Education, and Welfare, Washington, D.C., 1961, 62.

Chapter 18

THE PEARL OYSTER, *PINCTADA FUCATA* (GOULD) (FAMILY PTERIIDAE)

Katsuhiko T. Wada

TABLE OF CONTENTS

I. INTRODUCTION

Pearls have long been of economic importance as jewelry. Many shelled mollusks produce pearls, which are the animals' means of excluding or walling off foreign bodies by secreting shell around them. Even our edible oysters produce pearls, but they lack the luster of those produced by species in the family Pteriidae and some other mollusks, e.g., some freshwater mussels. Until methods were developed to induce pearl formation, the entire supply of pearls was from wild animals. These, often called natural pearls, command a higher price than so-called cultured pearls, probably because of their scarcity.

II. BIOLOGY

A. MORPHOLOGY

The body and shell of the pearl oyster, *Pinctada fucata* (Gould), are asymmetrical. The shell is shaped like a horseshoe. The left valve is deeper than the right one, and there is a byssal notch on the anterior side. Anterior and posterior ears are located at the ends of the hinge line, and the former is larger. The hinge line is longer than the shell height at juvenile stages, as in other species of Pteriidae, but they are almost equal in adults (Figure 1). The growth processes form scale-like growth lines on the surface of the shell, which have a tapering structure and finely rounded tips. They are arranged in radial rows and are well developed in rapidly growing specimens of younger stages. The ligament connects the two valves at the center of the hinge line. The hinge teeth are well developed in young individuals, and the posterior tooth of the left valve is above that of the right. The most common external color patterns are well-defined continuous rays and small, more or less regularly dispersed flecks of color. The external color is brown, green, red, yellow, or white. The nacre is of a hard white metallic luster and sometimes yellow, silver, gold, or pink. The shell is composed of three layers: the periostracum, the prismatic layer, and the nacreous layer. The thin outer layer, the periostracum, is sometimes eroded in old or dead specimens. The prismatic layer forms the major portion of the shell, and the nacreous layer is the inner pearly lining of the shell cavity.

B. RANGE

Pinctada fucata is mainly distributed in the equatorial zone between the Tropic of Cancer and the Tropic of Capricorn of the Pacific Ocean. An exception is Japan, where the distribution is to 35°N, probably because of the warm Kuroshio current. Japanese taxonomists describe a subspecies, *P. fucata martensii,* for Japan. In ancient times, the species was reported from the Red Sea, Persian Gulf, and Arabian Sea; it was then called *P. vulgaris,* now synonymous with *P. fucata*. The present distribution of fisheries is limited to India, East Africa, Sri Lanka, southeast Asia, Japan, Korea, China, Papua New Guinea, Palau, and Australia.[1-3]

C. SALINITY AND TEMPERATURE TOLERANCES

The values given here are from experiments and observations in Japan and may differ somewhat in other geographic areas.

The majority of swimming larvae die within 24 h in seawater diluted to a salinity of 14.5‰ at temperatures of 26 to 28°C.[4] The lower limit of tolerance was estimated to be 17.2‰. Larvae exposed to a salinity of 17.4‰ for 24 h, then observed 72 h after replacement in normal seawater, showed an average of 12.2% mortality. Mortality was 91% at a salinity of 9.0%.[4]

The effect of low salinity on mortality, percent attachment, and heartbeat of spat was evaluated. The lower limit of tolerance was deduced to be 13 to 14‰ for 24 h of exposure, and the lower limit of optimum salinity to be above 22.7‰, from exposures of 25 d.[5]

FIGURE 1. Typical color pattern of shells of *Pinctada fucata*. (Magnification × circa 0.8.)

In the adult, effects of salinity on ciliary movement on the gills, heartbeat, mass of water filtered, disappearance of the crystalline style, and mortality were studied. Reported lower tolerance limits ranged, among reports, from 8.0 to 14.5‰, probably because of the different methods used. Ciliary movement in the gills was abnormal in seawater diluted to a salinity of 13.2‰ and stopped at 9.5‰.[6,7] The heartbeat was reported to be affected by 80% diluted seawater.[8] With respect to mass of water filtered, optimal salinity was estimated to be between 27.5 and 34.0‰, and the critical lower limit between 12.0 and 14.5‰.[9] The crystalline style disappeared in salinities of 10.5 to 14.5‰ after 57 to 114 h.[10] Finally, 50% mortality was observed when adults were exposed for 48 h to salinities of 12.0 to 14.0‰.[11]

Temperature tolerances of the larval stages have not been determined experimentally. The temperature range has been deduced to be from 17.2 to 29°C, although wild spat have been collected in water temperatures higher than 30°C. In hatcheries, larvae are usually reared at temperatures of 25 to 28°C. From observations on mortality, growth rate, attachment rate, and heartbeat rate of (3 mm) spat raised at different temperatures, the critical levels are thought to be a low of 15°C, a high of 32°C, and an optimum between 17.2 and 29°C.[12]

The critical temperature reported for the adult ranges from 7 to 29°C, and the estimated optimum temperature from 13 to 25°C. Mortality increases when animals are kept at temperatures below 7 to 8°C.[9] Experiments on ciliary movement of gills showed the same result; movement stopped at 7 to 8°C in spring.[8] Upper critical temperatures for ciliary movement and hearbeat are above 35°C. It seems that the effect of high water temperature on the adult varies, depending on factors like oxygen content, salinity, and the condition of animals. Data are scarce.

D. TYPES OF SUBSTRATE

Pinctada fucata is epifaunal, attaching to rock, floating substrate, or algae by the byssus. The animals sometimes detach and reattach elsewhere. In Japan, natural spat are collected on the suspended leaves of the Japanese cedar (*Cryptomeria*). Plastic films are used as collectors in hatchery production of spat.

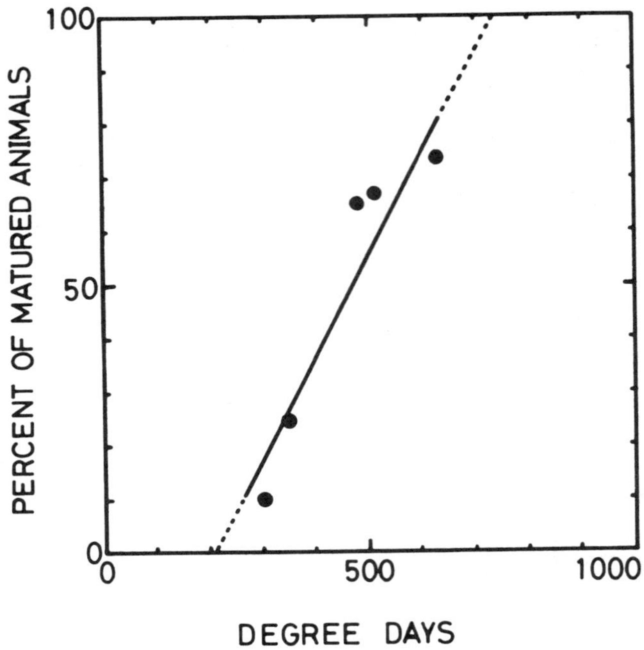

FIGURE 2. The relation between the percent of histologically matured animals reared in the laboratory and degree days exposure above 13°C.[14]

E. SEXUAL DEVELOPMENT AND SPAWNING

Pinctada fucata is dioecious. The gonad surrounds the digestive diverticula, intestines, and retractor muscles. Sometimes, sexual change is observed between different spawning seasons, and protandry is observed in many cases.

Gametogenesis is almost the same as in other bivalve mollusks.[13] Minimum shell height at sexual maturity is reported to be 17.5 mm for males and 26.0 mm for females, and the age at maturity is usually 2 years.[6] Fast-growing individuals sometimes mature in 1 year and slow growers in 3 years.

The spawning season is usually from June to July in the middle region of Japan and from winter to spring in the Red Sea (Figure 2). Temperature in degree days [$\Sigma(T_i - 13)$] required for maturation was estimated to be between 700 and 800 from the results of rearing experiments in tanks.[14]

F. LARVAL DEVELOPMENT

The average size of spawned and unfertilized eggs is about 50.7 μm. The germinal vesicle is broken down in the spawned eggs. The length of sperm is about 60 μm. The eggs complete meiosis when fertilized, and the first and second polar bodies are extruded before the polar lobe extrudes. The first cleavage is completed in 40 min.[15] At 3 h after fertilization, the embryo has developed to the morula stage and to the trochophore after about 4.5 h. The embryo grows to a D-shaped veliger by 24 h and to the full-grown stage by 13 to 25 d after fertilization. The mean shell length is 72 μm at the D-shaped stage and 214 μm at the full-grown stage of planktonic larvae, when it is ready to attach to the substrate.

G. GROWTH

The spat of *P. fucata* usually grow to 1 to 3 mm in shell length by 1 month after fertilization and from 2 to 6 mm by the second month, depending on environmental factors. There are many reports on the growth rate of Japanese cultured specimens. One study reports average size (shell height) as follows: 1 year, 4.64 (3.03 to 5.45) cm; 2 years, 5.82 (3.78 to 6.67) cm; 3 years, 7.27 (6.67 to 7.58) cm; 4 years, 7.82 (7.42 to 8.12) cm; 5 years, 8.06 (7.58 to 9.08) cm (mean and range of measurements).[6] Average weight at successive ages is as follows: 1 year, 9.4 g; 2 years, 18.8 g; 3 years, 43.1 g; 4 years, 51.8 g; 5 years, 69.4 g; 6 years, 76.1 g.[6] The growth rate of the adult varies more than that of the juvenile. Data show that the most growth is in the first year and that the growth during the first 2 to 3 years is much greater than that of the next 4 to 5 years. The maximum size recorded is about 10 cm in shell height, and the longevity is about 10 years. Recent hatchery studies have resulted in improved growth rates through selective breeding[16] as well as in the production of seed out of season.

The growing season is from spring to autumn and depends on temperature.[6] Shell growth is greatest in summer and autumn, but either is slow or ceases entirely in winter and early spring in the middle region of Japan. Temperature may be the most important factor that affects the growth rate. It is reported that the species grows continuously in winter in Japan in temperature-controlled tanks when fed rice powder.[17] Meat weight increases mostly from autumn to winter, perhaps as a result, as in most other bivalve species, of the physiological seasonal rhythm of the accumulation of glycogen after the spawning season.

III. FISHERY

A. HISTORY

The natural pearl was an important jewel or gem in ancient times. The earliest record of a pearl fishery appeared in India in 4000 BC. Fishermen have collected wild pearl oysters from the sea and harvested the natural pearls since antiquity over almost the entire distribution of the species in the Mediterranean Sea, the Red Sea, the Persian Gulf, India, Sri Lanka, the Middle East, Japan, and China. The resource decreased partly because of overfishing in many places and partly because of the low yield of pearls from wild oysters. For example, the results of experimental harvesting reported that natural pearls from 200 Japanese pearl oysters weighed 9 ''rin'' (about 3.75 mg) and consisted of 186 pearls from muscle and 19 from visceral parts.[18] Pearls were also harvested from the tropical regions of both the Atlantic and Pacific regions of the New World in earlier times.

In Japan, the national and prefectural governments have issued regulations many times to control overfishing, and some prefectural experiment stations conducted transplantation and culture experiments from the end of the 19th to the early 20th centuries.

Alagarsuwami and Quasim[19] described the history of the pearl fisheries in two regions of India: the Gulf of Kutch and Gulf of Mannar. At the beginning of this century, the pearl fishery of the Gulf of Kutch was under the control of the government. Harvesting was allowed every year or in alternate years from 1913 to 1939. After 1940, it was allowed every 3 to 4 years, and since 1967, there has been no fishery at all, as the population of the oyster was found to be seriously depleted. The Gulf of Mannar, on the other hand, was better known as a fishery area. The fishery was intricately connected with the power and might of the Nagah of Madura, the Nawab of Carnatic, and the Portuguese during the 16th century. After many battles, the pearl fishery came under Dutch control in 1658. In 1796, the British took control of the pearl fishery of the Gulf of Mannar. In the post-Independence period, the fishery is being fully exercised by the government of Tamil Nadu. Since 1962, the pearl oyster has been very sparse.

Culture of pearls started in Japan around the end of the 19th century. In 1893, Kokichi Mikimoto did the first experiments in which he succeeded in producing semispherical pearls. The names of two other Japanese, Tatsuhei Mise and Toukichi Nishikawa, are also often cited as figuring in the history of invention of methods for producing spherical cultured pearls. The main techniques of the insertion operation methods for producing spherical pearls are said to have been invented by these three Japanese pioneers.[20] The hanging method for rearing the animals in the sea was introduced with the beginning of pearl culture.

Improvements in the pearl culture industry have been made in the operation for insertion of nucleus and graft, rearing of the animals, wintering, and cleaning of the shell. Wooden or bamboo rafts have been devised for collecting the natural spat and rearing the juveniles or adults with inserted nuclei. Tools have been devised for insertion of the nucleus.

Artificial production of seed has been developed in governmental and private hatcheries in Japan and India. Other Asian countries such as China and Korea seem recently to have developed their own pearl culture industry, although the details have not been published.

B. IMPORTANCE

Other species of pearl oysters are *P. margaritifera* (Linnaeus), *P. maxima* (Jameson), *P. cheminitz* (Philippi), and *Pteria penguin* (Roding), but *Pinctada fucata* is most often used for producing cultured pearls.

Production of pearls cultured by *P. fucata* in Japan reached a record level of 4 ton in 1928. After World War II, the pearl industry in Japan flourished, recovering to the prewar level by 1951 and increasing more than 12% per year until 1966, when the record level of 130 ton was achieved. There were about 4700 pearl cultivators at that time. The production has decreased since 1966, probably because of the insufficient supply of labor, wage increases, and past overproduction. The latest statistics indicate that the production of cultured pearls, including the pearl industry of *P. margaritifera* and *P. maxima,* was 6.4 ton and that the number of cultivators was about 2000 in 1984 in Japan.

Alagarsuwami and Quasim[19] described the present status and future of the pearl industry of *P. fucata* in India. The Central marine Fisheries Research Institute established a pearl oyster farm in 1972 and collected native pearl oyster species for experiments on pearl culture. The pearl trade of India was 10,396 × 1000 rupees for export in 1972 to 1973.

C. RECRUITMENT AND METHODS OF CULTURE

1. Collecting Spat

The rapid growth of the pearl culture industry led to a need to devise new methods for collecting the animals (Figure 3). The naturally grown adult pearl oyster had been collected from the seabottom until the newer method was devised to collect spat. Cedar leaves are usually used for collecting the natural spat in Japan. Twigs of the Japanese cedar are bunched together with rope and suspended in the sea from floating wooden or bamboo rafts in the spawning season at a depth of 1.5 to 3.0 m. Edible oyster and abalone shells are also used in some cases in Japan. Another collector is old cages or nets. Recently, the long-line style of raft has been used, particularly in bays facing the ocean.

The spawning season is from mid-June to late August in Japan, depending on environmental conditions. The collectors are set a little after spawning is observed or when larvae are detected in plankton surveys (Figures 4 and 5).

Hatchery production of seed is reported in Japan and India. In Japan, methods of artificial spawning and rearing of the larvae and spat in tanks are similar to those for other species of bivalves. Induced spawning or dissection of the gonad are the usual ways to get sperm and eggs. The latter method involves treatment of the eggs and sperm with ammonia seawater to promote fertilization and development. The larvae are fed cultured phytoplankton, such as *Pavlovia lutheri* (Droop) and *Chaetoceros calcitolans* Takano.[21] Tanks of 500 to 10,000 l are usually used to rear larvae, and filtered seawater is supplied through plankton nets (Figure 6). Knitted films are used for collectors, and they are moved to the sea from the tank one month after fertilization, when the spat have reached 2—3 mm in shell length. The collectors with spat are placed in mesh nets of about 1 mm, suspended from rafts.

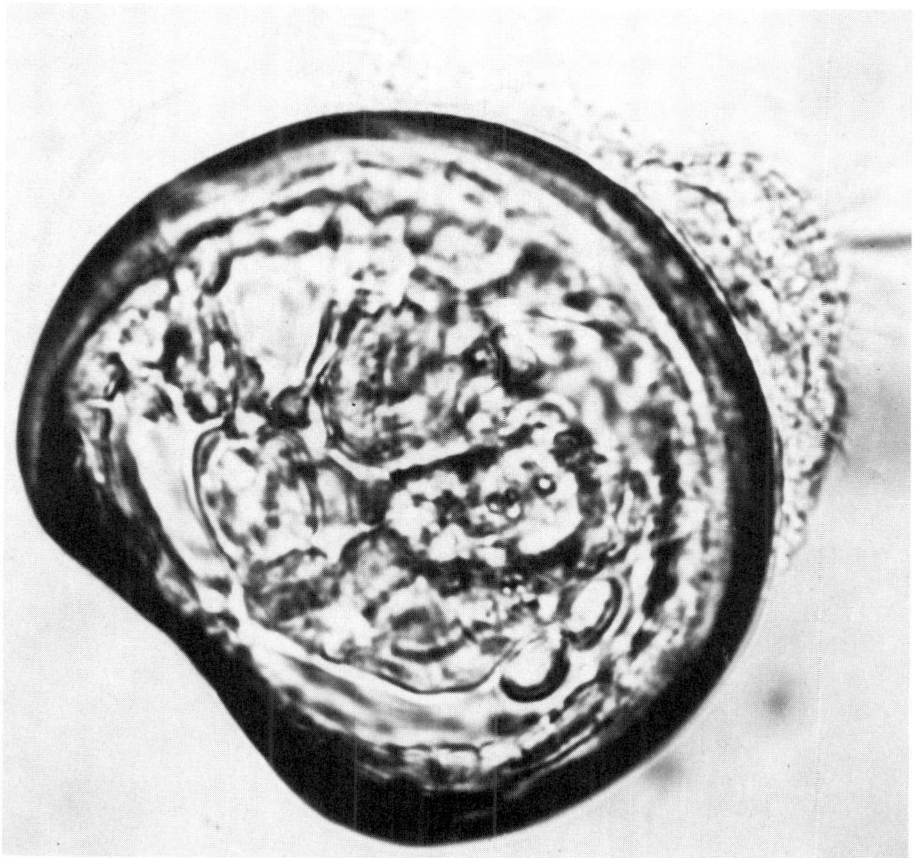

FIGURE 3. D-shaped veliger larva of *P. fucata* (shell length 70 μm).

2. Rearing the Juvenile and Adult Pearl Oyster

The young oysters (1 to 2 cm) are removed from the collector and are transferred to rearing nets of nylon or cremona (Figure 7). The nets are called lantern jets because they are shaped like lanterns. The young oysters grow so fast that the farmers must change the nets for ones with larger mesh and must periodically clean the shells, reduce the number per net, and treat the oysters with freshwater to kill fouling organisms (Figures 9 to 11).

3. Preparation of the Pearl Oyster for the Insertion Operation

At between 6 months and 1 year of age the pearl oysters are moved from the nursery area to the farms. Before the insertion operation and the graft of mantle tissue, the host pearl oyster should be conditioned to tolerate the shock of the operation. The nucleus is usually so large that the oysters might suffer great shock. Physiological conditioning is the most important procedure for the farmers after they choose the best animals on the basis of morphological characteristics. The most important part of this procedure is forced spawning or prevention of maturation. The oysters are tightly packed in baskets in appropriate condition in order to adjust the physiological level.[22]

4. Insertion of the Nucleus and Graft

Grafting is the technique that produces round pearls. This is a delicate surgical operation that is conducted by skilled workers. Wedge-shaped wooden plugs are inserted between the valves before the operation, and the oysters are inspected for physiological condition. The

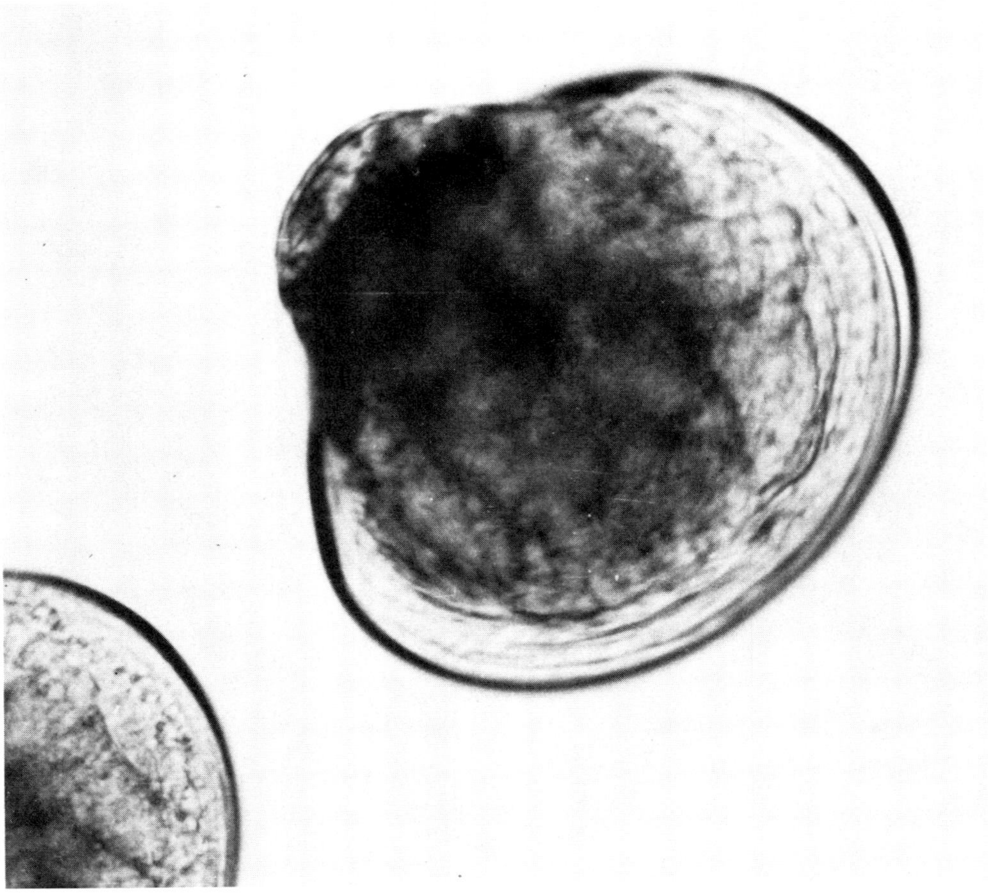

FIGURE 4. Umbo-stage larva of *P. fucata* (shell length 180 μm).

graft tissue, a small piece of the mantle from another individual, is inserted into the body together with the nucleus.

The graft tissue is prepared from opened healthy oysters. The adductor muscle is cut, and a strip about 7 cm long is removed from the frilled mantle edge. The strip must be stored carefully on a wet graft-trimming block, and the adhering slime or mucus wiped off. Graft tissue pieces about 3-mm square are cut from this strip.

The nuclei are made of shell from the freshwater pig-toe mussel (*Pleurobema* sp.), which is supplied from the Mississippi River. The shells are cut into small cubes of the appropriate size and then polished to make them perfectly spherical. The size and number of nuclei depend on the size of the oyster to be used and on the size of pearl to be produced. A highly skilled operator is said to be able to operate on as many as about 300 oysters a day.

The oyster epithelial cells of the mantle grow up to be a pearl sac completely surrounding the nucleus after the operation.[23]

5. Cultivation of the Oysters after Insertion

The oysters, with nuclei inserted, are put in cages, which are suspended 1 to 3 m from rafts anchored in calm water near the operating facility. The cages are usually left undisturbed for a few weeks after nucleus insertion. During this convalescence period, the pearl oysters recover from the operational shock and repair any injuries they have sustained. The oysters are then inspected for physiological condition, and the dead ones are removed.

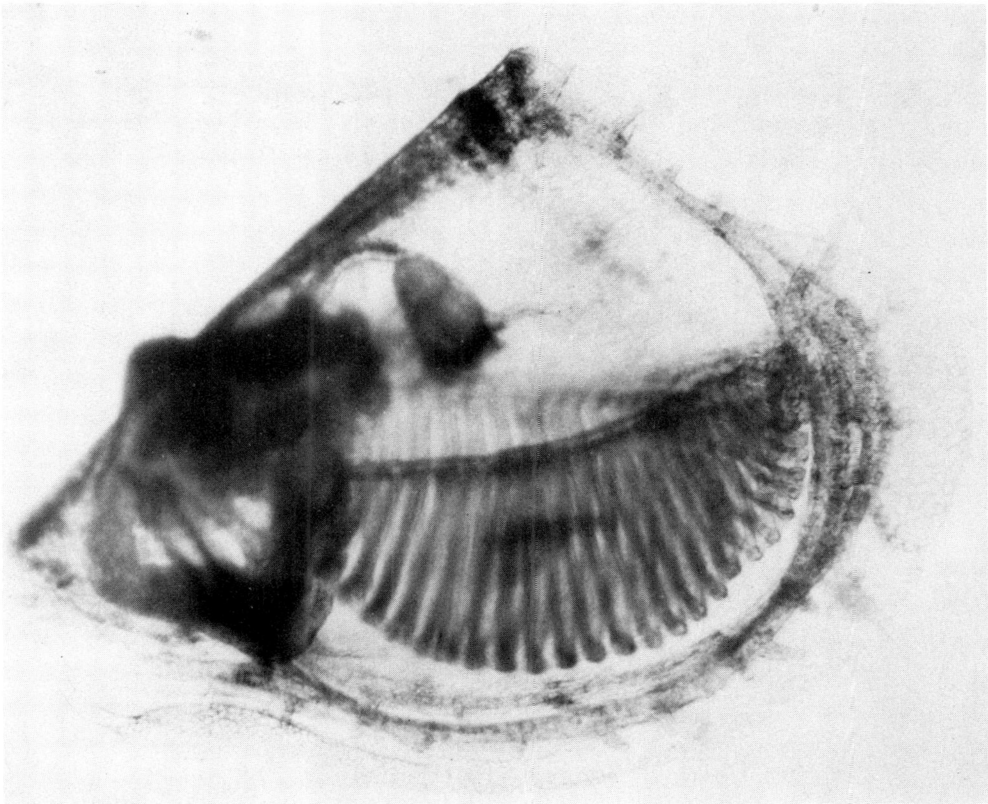

FIGURE 5. The spat; the semitransparent shell shows the viscera, gill, and mantle (shell length about 1 mm).

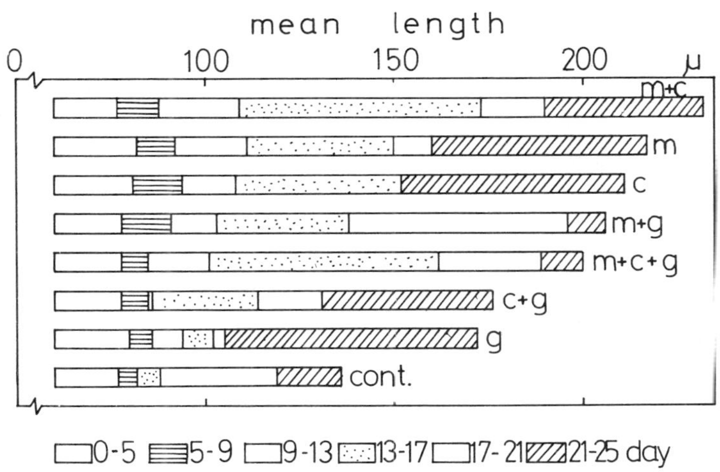

FIGURE 6. Growth in culture of *P. fucata* larvae fed with different species of microalgae. m, *Pavlovia lutheri;* c, *Chaetoceros calsitolans;* g, *Chlorella* sp.; cont., unfed culture.[21]

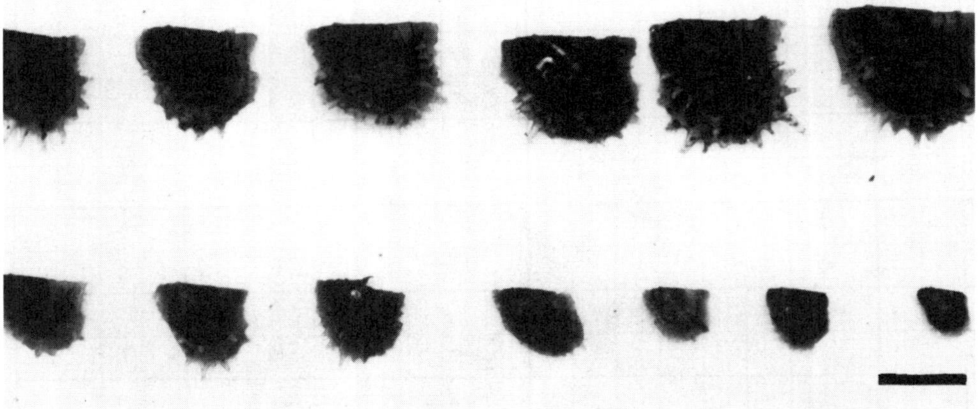

FIGURE 7. Sequential changes of the shell shape in the growth process of juvenile *P. fucata*. Scale bar = 1 mm.

FIGURE 8. General view of the farm for collecting and culturing natural spat of the Japanese pearl oyster, showing the wooden (B) and the long-line (A and C) rafts.

After recovery, the oysters are put in a net basket or pearl net and suspended from rafts like those the juveniles and adults are cultured in (Figure 12). The depth varies from 1 to 7 m depending on the physiological condition of the animals and environmental factors such as temperature, salinity, food plankton, and others. Rafts are set at appropriate locations in a calm bay. About 60 nets are suspended from each raft, and each net contains from 50 to 60 oysters. Shells are cleaned and the nets changed periodically to avoid the mortality caused by fouling organisms. Many methods are used for cleaning the shells, but the oysters are

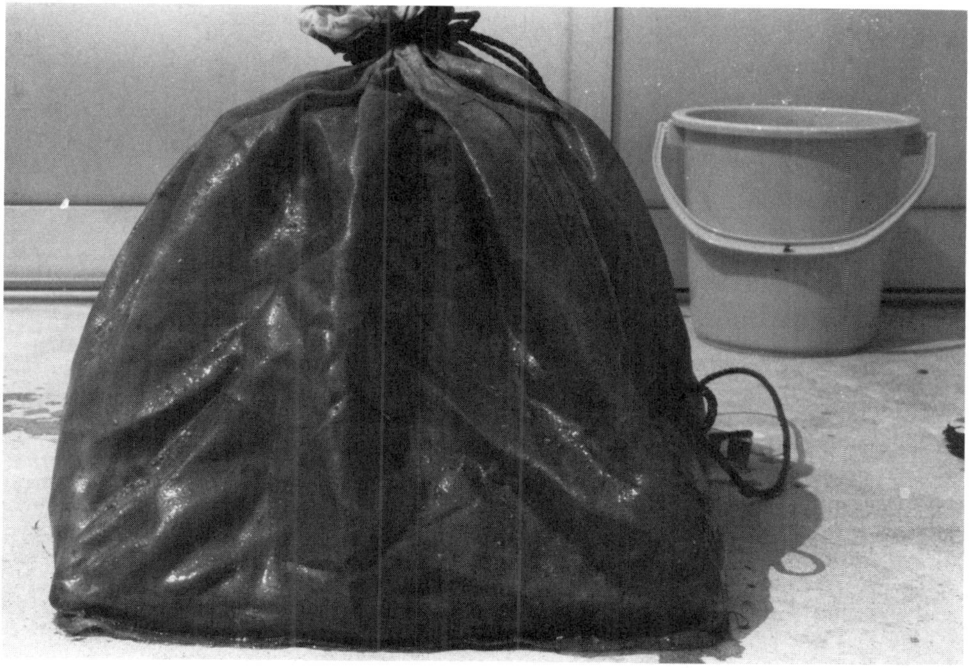

FIGURE 9. The net to hold the collectors with spat of *P. fucata*.

FIGURE 10. The collectors with *P. fucata* spat, which grow to 1 to 2 cm in the net and are ready to move to the lantern net.

FIGURE 11. The net to hold the *P. fucata* spat moved from the collector.

usually cleaned in the net by a cleaning machine devised for this purpose. Sometimes the shells are cleaned by hand with the help of knives. The nets are cleaned for reuse. Dead oysters are removed at that time, and some oysters are opened to determine the growth of the pearls, particularly just before harvesting season.

D. CONSTRAINTS

1. Parasites

Parasites of *P. fucata martensii* have been known since the early stages of development in the Japanese pearl industry. Damage to the pearl industry by parasites has been of particular significance since 1952. The rapid increase in damage over the whole oyster-producing pearl area seems to have been caused by the transplantation of the pearl oyster and to be correlated with lack of biological data on the parasites. Mudworms, sponges, and a trematode are the main parasites affecting production of pearls.

Mudworms of the genus *Polydora* produce a mud tube and blister in bivalve shells. Of the several species of *Polydora* found in shells of *P. fucata martensii, Polydora ciliata* (Johnson) is recognized as the most important.[24] Damage to the industry by mudworms has occurred since 1960 in the whole pearl-culturing area. Oysters are fatigued, resulting in heavy mortality. Spawning of mudworms occurs in May, June, and October in Ago Bay of Japan. Larvae settle on the shell, and mud tubes and blisters are formed in summer and winter. To control mudworms, the pearl oysters are dipped into freshwater for about 15 minutes.[25] There seem to be no reports on the control of sponges.

FIGURE 12. Modern rafts to culture pearl oysters with nuclei in place.

Sakagushi[26] reported the life history of the trematode *Bucephalus varicus* Manter and the extent of the damage caused by this parasite in the pearl industry. Pearl oysters infected by larvae of the trematode cannot be used for pearl culture. In Japan, the infection has spread over the whole pearl-culturing area, and a rate of infection as high as 40% is reported in some farms. The pearl oyster is the first intermediate host of the cercaria larvae. A proposed method for control of this parasite is not to culture pearl oysters during the season when trematodes invade oysters.

2. Fouling Organisms

Many species of invertebrates inhibit growth and injure pearl oysters by attaching to the shell, particularly at the hinge, often interfering with the opening of the shell, and may cause death. Fouling organisms are barnacles, sponges, worms, edible oysters, and seaweed. The dominant species of barnacles are *Balanaus variegatus tesselatus* Darwin and *B. amphitrite* Broch. *Hydroides norvegica* Gunnerus, *Dexiospira forminosus* Bush, *Bugula neritina* Linn., and *Dakaria subovoidea* D'Orbigny are also fouling organisms in Japan. Periodic cleaning of the shell is the only practical defense against these organisms. Shell cleaning is carried out in the period from April to November.

3. Predators

Some fish, the eel *Anguilla japonica* Temminck and Schlegel, the black porgy *Sparus* sp., and the globe fish *Sphoeroides* sp., as well as the octopus, *Octopus* sp., are known to feed on pearl oysters. They attack the oysters especially just after the nucleus operation.

4. Red Tide

Red tides have been known to be important constraints on marine culture organisms. Many records describe the damage by the red tides (a dinoflagellate) in the pearl culture industry. The species reported are *Gymnodinium* sp., *Gonyaulax* sp., *Cocurodium* sp., and

Prorocentrum sp. The actual mechanism by which these microorganisms cause death is not entirely clear. Japanese pearl oyster culturists usually tow the rafts out of the area of red tides to avoid their damage.

E. METHODS OF HARVESTING, USES, AND PROCESSING

The oysters are suspended in the sea from rafts for 1 to 4 years, depending on the size of the introduced nucleus and the quality of the pearl desired. After the prescribed time, the nets are lifted in late autumn to winter, when the pearls acquire their best luster.[27] The matured pearls are removed. The meat, except for the adductor muscle ($2.0 \times 1.0 \times 0.5$ cm), which is edible and used for Japanese dishes, is churned to the consistency of a thin soup. The smaller pearls and pearly fragments settle to the bottom and are used for costume jewelry.

The pearls are washed to remove the slime and dirt and then dried. They are classified with respect to size, color, and other traits. Pearls are drilled to make necklaces, rings, pendants, brooches, earrings, bracelets, and other jewelry.

IV. RESEARCH AND DEVELOPMENT

Japanese scientists have conducted extensive research projects on stable production of good-quality pearls. The production of Japanese cultured pearls had a peak in 1966 and then decreased; production in 1970 had decreased by 27% from that in 1966. The quality of pearls also tended to deteriorate. In addition, high mortality has been reported, particularly in oysters recently operated on. Reduced production can be attributed to serious problems of pollution, overcrowded pearl farms, and increased numbers of parasites. Studies on the physiology of the pearl oyster and on the oceanographic aspects of pearl farms are aimed at providing suggestions for the management of pearl oyster farms.[28,29] The results of pathological studies have also been beneficial for the prevention and control of the parasites.[24-26] Another factor is the overproduction of low-quality pearls. Consequently, it is clear that control of the quality of pearls and not simply their mass production should be of importance in the Japanese pearl industry. Only a high grade of pearls will be required in the future in the pearl market. Therefore, research should be conducted to control qualities of cultured pearls such as size, color, and luster from the standpoint of many biological aspects of cytology, biochemistry, biominerology, genetics, physiology, and ecology. These factors are related to pearl calcification,[27] inheritance of the traits that may control pearl production,[30] selective or crossbreeding,[16,31] tissue culture of epithelial cells of the mantle,[32] and physiology of hemolymph.[32]

Thus, the most important goal of research activity on pearl culture must be the high productivity of the various types of desirable pearls. Future development should be concentrated on improving the quality of pearls, which is related to the thickness of the pearl layer and the kinds and arrangement of crystals of nacre. Pearls of many sizes, shapes, and colorations can be developed by new techniques.

REFERENCES

1. **Cahn, A. R.,** Pearl Culture in Japan, Natural Resources Section Report No. 122, General Headquarters Supreme Commander for the Allied Powers, Tokyo, 1949, 91.
2. **George, C. P.,** Cultivation of pearl shell and pearls in Indo-Pacific region, *Lapidary J.*, 4, 72, 1979.
3. **Hynd, J. S.,** A revision of the Australian pearl shells, genus *Pinctada* (Lamellibranchia), *Aust. J. Mar. Freshwater Res.*, 6, 98, 1955.

4. **Ohta, S.,** Physico-ecological studies on the free swimming larva of the pearl oyster *Pinctada martensii* (Dunker). I. On the resistability of the larva for the low salinity sea water, *Bull. Natl. Pearl Res. Lab. (Kokuritsu Shinju Kenkyusho Hokoku),* 3, 207, 1957 (in Japanese).
5. **Numaguchi, K. and Tanaka, T.,** Effect of temperature on mortality and growth of the spat of the pearl oyster, *Pinctada fucata martensii, Bull. Natl. Res. Inst. Aquaculture,* 9, 41, 1986 (in Japanese).
6. **Kobayashi, S. and Watabe, T.,** *Studies on Pearls,* Gihohdoh, Tokyo, 1955, 280 (in Japanese).
7. **Kobayashi, H. and Matsui, J.,** Studies on the resisting power of the pearl-oyster (*Pinctada martensii*) against the change of environment. I. On ciliary movement of the gill, *J. Shimonoseki Coll. Fish.,* 3, 123, 1953 (in Japanese).
8. **Kobayashi, H.,** Studies on resisting power of the power of the pearl oyster, *Pteria (Pinctada) martensii* (Dunker), against changes of environment. II. On the heart beats, *J. Shimonoseki Coll. Fish.,* 4, 95, 1955 (in Japanese).
9. **Miyauchi, T.,** Studies on the water filtration by the pearl oyster, *Pteria (Pinctada) Martensii* (Dunker). II. The influence of temperature and density of sea water on the water filtration, *Aquaculture (Suisan Zoushouku),* 10, 7, 1969 (in Japanese).
10. **Wada, K. T.,** Studies on the crystalline style of Japanese pearl oyster. I. Effect of diluted sea water on the crystalline style, *Bull. Jpn. Soc. Sci. Fish.,* 35, 133, 1969.
11. **Katada, S.,** The influence of low salinity sea water on death and growth of pearl oysters (*Pinctada martensii*), and quality of the cultured pearls, *Bull. Natl. Pearl Res. Lab. (Kokoritsu Shinju Kenkyusho Hokoku),* 5, 489, 1959 (in Japanese).
12. **Numaguchi, K. and Tanaka, T.,** Effect of temperature on mortality and growth of the spat of the pearl oyster, *Pinctada fucata martensii, Bull. Natl. Res. Inst. Aquaculture,* 9, 35, 1986 (in Japanese).
13. **Tateishi, S. and Adachi, I.,** Histological observation of the gonad of the pearl-oyster *Pinctada martensii* (Dunker), *Bull. Fac. Fish. Nagasaki Univ.,* 5, 75, 1957 (in Japanese).
14. **Wada, K. T.,** Temperature requirement for maturation of gonads in Japanese pearl oyster, *Bull. Natl. Pearl Res. Lab. (Kokoritsu Shinju Kenkyusho Hokoku),* 20, 2244, 1976 (in Japanese).
15. **Kobayashi, S.,** On the study of pearl culture. I. On development of *Pinctada martensii* in tanks, *Bull. Jap. Soc. Sci. Fish.,* 17, 65, 1948 (in Japanese).
16. **Wada, K. T.,** Genetic selection for shell traits in the Japanese pearl oyster, *Pinctada fucata martensii, Aquaculture,* 57, 171, 1986.
17. **Kuwatani, K., Nishii, T., and Wada, K. T.,** Growth and maturation of Japanese pearl oyster reared in the tank in winter, *Bull. Natl. Pearl Res. Lab. (Kokoritsu Shinju Kenkyusho Hokoku),* 18, 2118, 1974 (in Japanese).
18. **Ogushi, J.,** *Studies on Pearls,* Ito Bunshindo, Tokyo, 1938, 218 (in Japanese).
19. **Alagarsuwami, K. and Quasim, S. Z.,** What are pearls and how are these produced, *Seafood Export J.,* 6, 1, 1974.
20. **Matsui, Y.,** On the origin of pearl-culture and the history of it in Japan, *Bull. Fac. Agric. Kinki Univ.,* 8, 1, 1975.
21. **Wada, K. T.,** Growth of Japanese pearl oyster larvae fed with three species of micro-algae, *Bull. Natl. Pearl Res. Lab. (Kokoritsu Shinju Kenkyusho Hokoku),* 17, 2075, 1973 (in Japanese).
22. **Uemoto, H.,** Effect of shitate-sagyo and the operation of nuclear insertion on some serum components and on the weight of crystalline style in the pearl oyster, *Bull. Jap. Soc. Sci. Fish.,* 33, 705, 1967 (in Japanese).
23. **Machii, A.,** Histological studies on the pearl-sac formation, *Bull. Natl. Pearl Res. Lab. (Kokoritsu Shinju Kenkyusho Hokoku),* 13, 1489, 1968 (in Japanese).
24. **Mizumoto, S.,** Parasites affecting the pearl industry in Japan, in Proc. 3rd U.S.-Japan Meet. on Aquaculture, Tokyo, 1975, 39.
25. **Funakoshi, S.,** Studies on the methods using saturated salt solution to kill Polychaeta. I. On killing Polychaeta inhabiting mother oyster shells, *Bull. Natl. Pearl Res. Lab. (Kokoritsu Shinju Kenkyusho Hokoku),* 9, 1156, 1964 (in Japanese).
26. **Sakaguchi, S.,** Studies on the life-history of the trematode parasitic in pearl oyster, *Pinctada fucata,* and on the hindrance for pearl culture, *Bull. Natl. Pearl Res. Lab. (Kokoritsu Shinju Kenkyusho Hokoku),* 13, 1635, 1968 (in Japanese).
27. **Wada, K.,** Relation between calcium metabolism of pearl sac and pearl quality, *Bull. Natl. Pearl Res. Lab. (Kokoritsu Shinju Kenkyusho Hokoku),* 16, 1949, 1972 (in Japanese).
28. **Sawada, Y. and Taniguchi, M.,** The oceanographical studies on the pearl culture ground. VI. On the relation between the rafts density in pearl culture ground and the contaminated degree of bottom mud, *Bull. Natl. Pearl Res. Lab. (Kokoritsu Shinju Kenkyusho Hokoku),* 14, 1719, 1969 (in Japanese).
29. **Seki, M.,** Studies on environmental factors for the growth of pearl oyster, *Pinctada fucata,* and the quality of its pearl under the culture condition, *Bull. Mie Pref. Fish. Exp. Stn.,* 1, 32, 1972 (in Japanese).
30. **Wada, K. T.,** Breeding study of the pearl oyster, *Pinctada fucata, Bull. Natl. Res. Inst. Aquaculture,* 6, 79, 1984 (in Japanese).

31. **Wada, K. T.,** Genetic variability at four polymorphic loci in Japanese pearl oysters, *Pinctada fucata martensii,* selected for six generations, *Aquaculture,* 59, 1986.
32. **Machii, A.,** Organ culture of mantle tissue of the pearl oyster *Pinctada fucata* (Gould), *Bull. Natl. Pearl Res. Lab. (Kokoritsu Shinju Kenkyusho Hokoku),* 18, 2111, 1974 (in Japanese).
33. **Funakoshi, S.,** Effect of dissolved oxygen contents on the hemocyte constituents and heart rate of the pearl oyster, *Pinctada fucata, Bull. Natl. Pearl Res. Lab. (Kokoritsu Shinju Kenkyusho Hokoku),* 21, 2309, 1977 (in Japanese).

Chapter 19

THE CULTURE OF MARINE BIVALVE MOLLUSKS IN CHINA

Zong-Qing Nie

TABLE OF CONTENTS

I. INTRODUCTION

China is famous for the culture of aquatic organisms, especially freshwater fishes and marine algae, but it has only been in recent years, since the regime of the People's Republic of China, that substantial progress has been made in marine mollusk culture, although harvesting has occurred for some species since ancient times. Now there are many stations and universities at which Chinese scientists study and develop the aquaculture of bivalves.

About 18,000 km of coastline with about 2×10^6 ha of intertidal and shallow water are available for mollusk culture, and about 148×10^3 ha were under culture by 1978. The total yield of cultivated bivalves was about 200×10^3 t in 1978. Of these, mussels (90×10^3 t), razor clams (40×10^3 t), and oysters (30×10^3 t) were the most important (weights are of mollusks in the shell, except for oysters, which are meats only). Scallops, small-neck clams, hard clams, cockles, and pearl oysters are also major culture crops.

The Chinese coast ranges from temperate to subtropical to tropical (Figure 1). Bivalve culture is most extensive in the Provinces of Liaoning, Shandong, Zhejiang, Fujian, and Guangdong.[1-3] Each taxon will be covered separately. A separate chapter by Wang and Shien discusses Farrer's scallop.

II. OYSTERS: FAMILY OSTREIDAE

The most common species of oyster cultivated is *Crassostrea plicatula* (Gmelin) (= *Saccostrea cucullata* (Born) (Figure 2)), which accounts for over 80% of the yield. This species is cultivated mainly in Fujian and Zhejiang and occurs in salinities of 13 to 33‰. Two types of cultch are used to collect seed oysters, the "stone bridge" and bamboo stakes. The former is used mainly on semihard bottom and the latter on soft bottom.[4-10]

In the stone bridge method (Figure 3), stone bars ($80 \times 20 \times 8$ cm) are placed in spaced rows in the form of bridges on the midtide flats in May and June. After the spat settle, the bars are dispersed, at a density of about 15×10^3 stones per hectare to insure better growth rates. The growth period to market size is 7 to 12 months. The average yield is 3 ton/ha but can be up to 8 ton. Bamboo stakes (1.2 m long, 1.5 cm diameter) (Figure 4) in bundles of 4 or 5 are planted in rows at midtide level, usually on mud flats. There can be 18 to 21×10^4 stakes/ha. These are thinned once or twice during the growing periods, to about 12 to 15×10^4/ha by time of harvest. Spat attached in May require about 8 to 11 months when cultivated in the lower intertidal zone to reach harvesting size and those attached in September about 16 to 18 months. If they are raised in the midintertidal zone, 17 to 20 months is required. The yield averages about 6 ton/ha and can be up to 11 ton. Harvest usually takes place in winter and spring.

Crassostrea rivularis (Gould) (Figure 2) occurs in estuaries of low salinity (range 1 to 32, optimum 10 to 28%) and is cultivated to a lesser extent. Spatfall occurs from June to August, the period of highest temperature and lowest salinity. Cultch material to collect spat has traditionally been oyster shell and gravel. Since the 1960s cement plates (17 to 24 $\times$ 14 to 19 cm) or bars (40 to 80 $\times$ 4 to 6 cm), reinforced with embedded bamboo stakes, have been used increasingly. These are easier to handle, provide more surface for spat attachment, and are not so readily covered by silt. Oyster larvae in the water are monitored and the farmers informed of the best time to plant the cultch. Spatfall occurs from the low-tide line to a depth of 10 m, with the maximum at $\pm$ 0.4 m. Spat collectors are placed in rows in rectangular blocks, usually 30 to 37.5×10^3 bars or 100 to 135×10^3 plates per hectare. The cultch with attached oysters is moved a few months before harvest to more fertile waters for fattening (Figure 5). Harvest size is reached in 3 to 4 years and the yield is 4.5 to 11 ton/ha.[1,11]

In recent years, experiments have begun on raft culture in the southern regions, similar

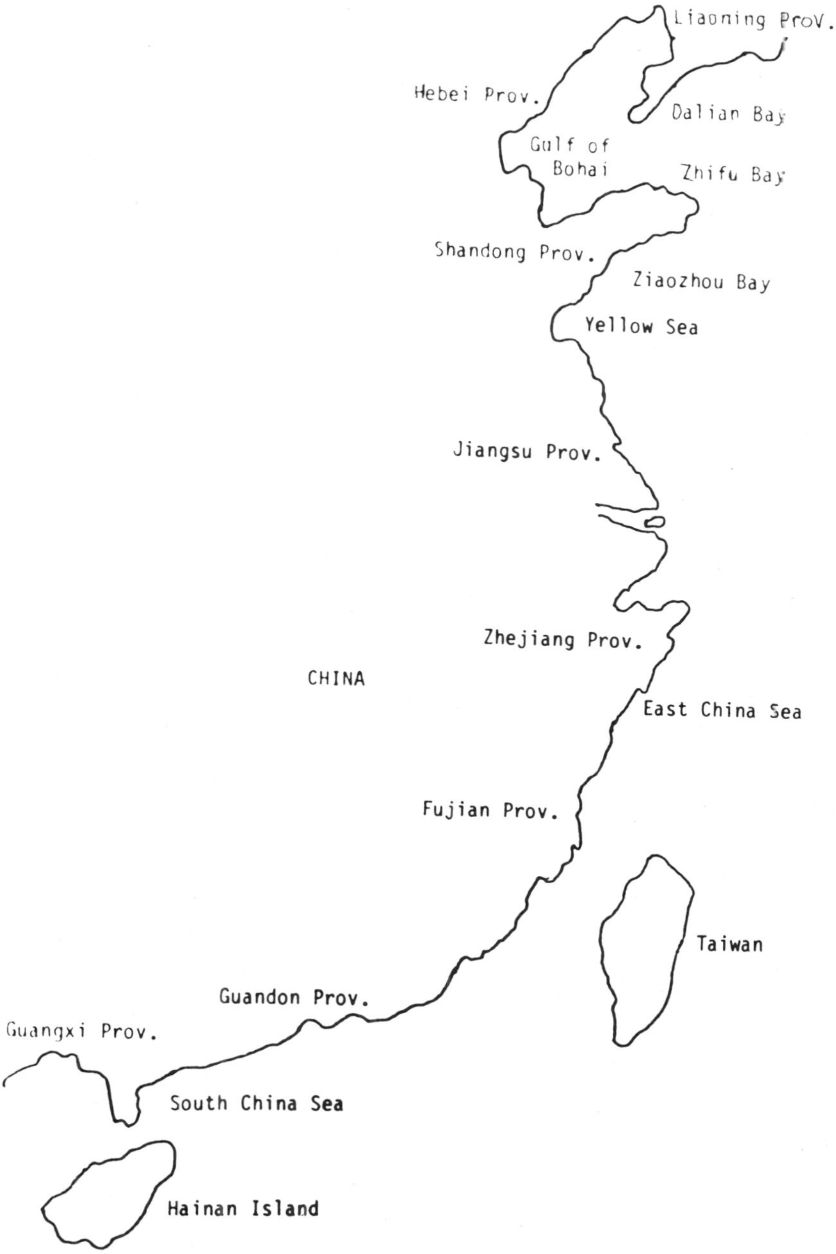

FIGURE 1. Outline map of China Coast.

to the Japanese method (Figure 6). Raft culture gives faster growth and a higher yield. A raft of 84 m^2 will produce in 2 years what 667 m^2 of bottom culture will in 4 years. It has been found that rafts withstand typhoons better than was originally thought.

Crassostrea talienwhenensis (Grosse) (= *C. gigas* Thunberg) is cultivated to some extent.

III. MUSSELS: FAMILY MYTILIDAE

Three commercial species of mussels occur in China: the thick-shell mussel, *Mytilus corucus* (Gould) (salinity 29 to 34‰); the blue or common mussel *M. edulis* Linneaus; and

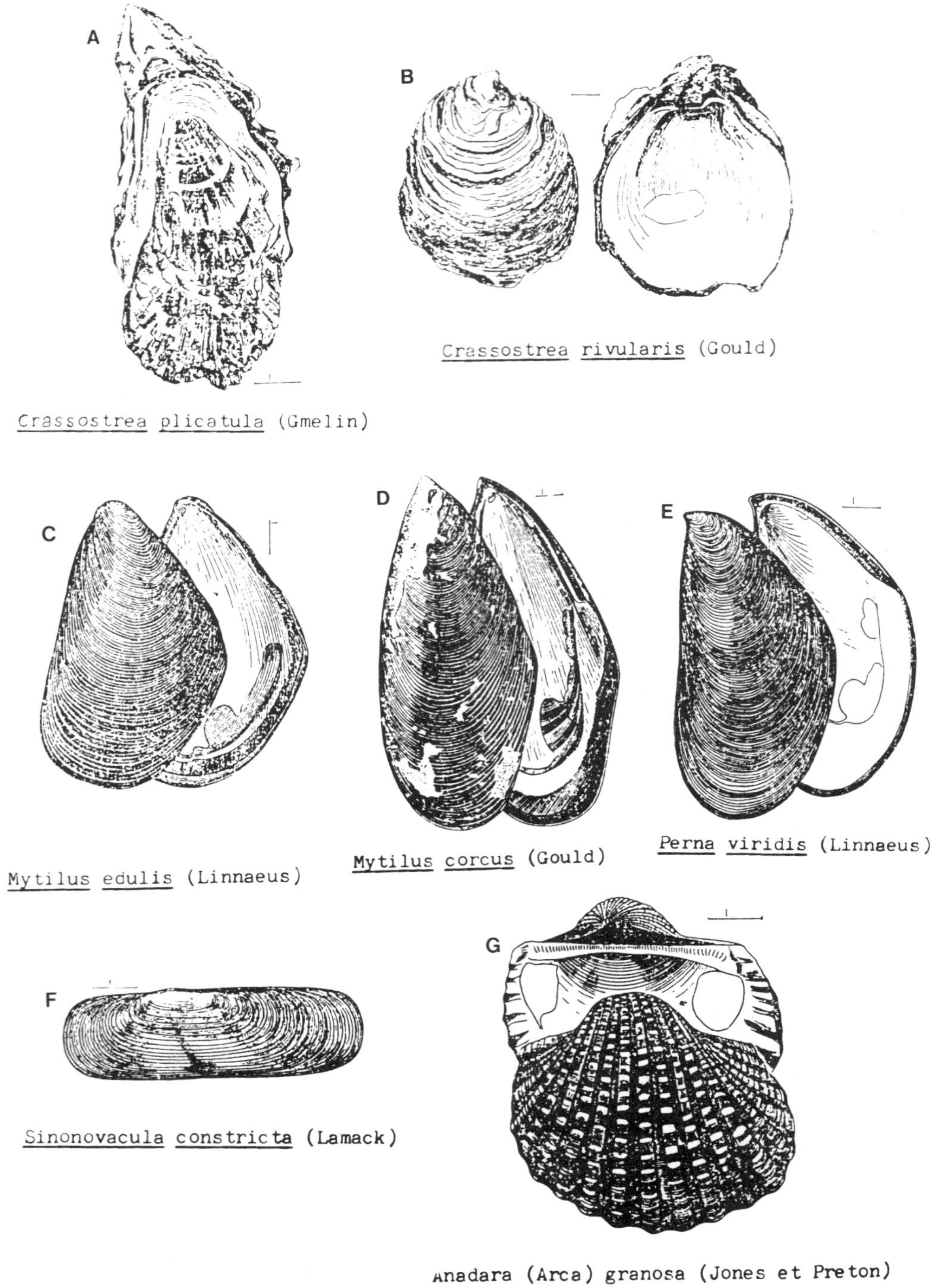

FIGURE 2. Bivalve mollusks cultivated in China. A. *Crassostrea plicatula;* B. *Crassostrea rivularis;* C. *Mytilus edulis;* D. *Mytilus corcus* (not cultivated); E. *Perna viridis;* F. *Sinonovacula constricta;* G. *Anadra granosa;* H. *Tapes philippinarum;* I. Pinctada martensii; J. *Pinctada maxima;* K. *Meretrix meretrix;* L. *Mactra antiquata;* M. *Potamocorbula rubromuscula.*

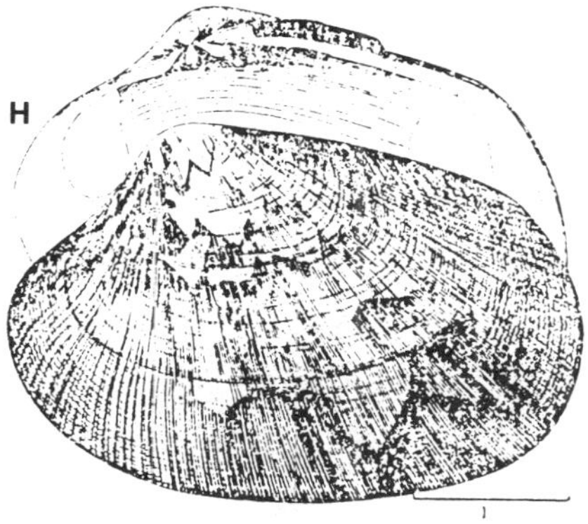

Tapes philippinarum (Adams and Reeve)

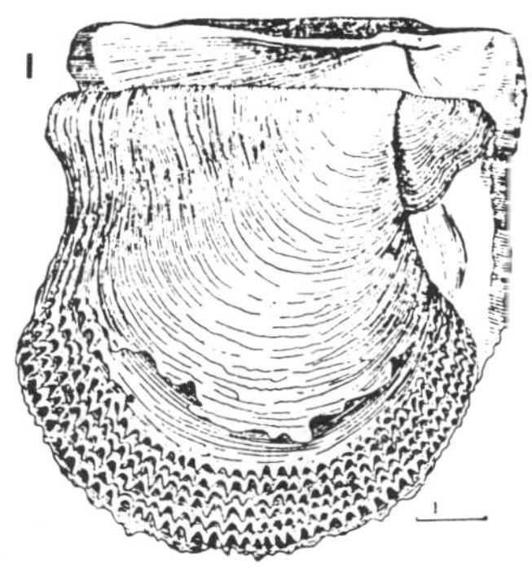

Pinctada (Pteria) martensii (Dunker)

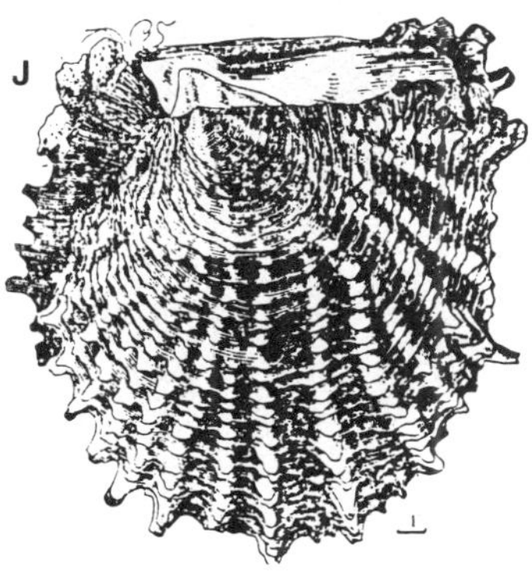

Pinctada (Pterida) maxima (Jameson)

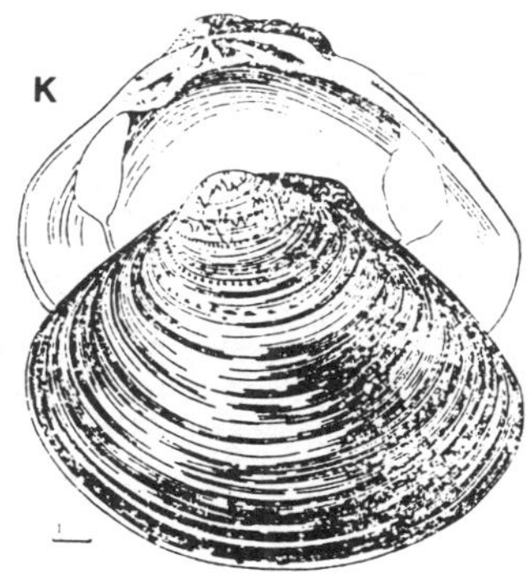

Meretrix meretrix (Linnaeus)

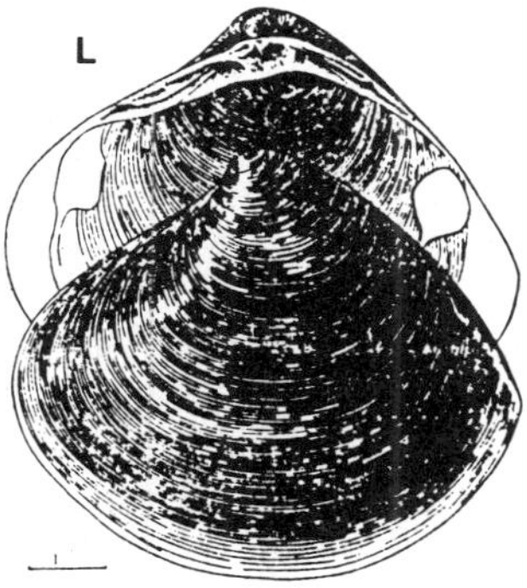

Mactra antiquata Spengler

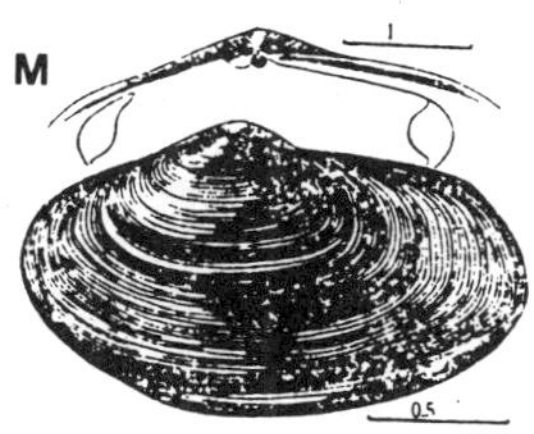

Potamocorbula rubromuscula

FIGURE 2 (continued).

the green mussel, *Perna viridis* (Linnaeus) (Figure 2). *Perna viridis* lives in salinity of 9 to 34‰ with an optimum of 15 to 29‰. *Mytilus corucus* is not cultivated but is harvested on a limited basis where it is abundant on rocks, mostly from Thousand Islands, Zhekiang Province. The following account is mostly of the blue mussel, *Mytilus edulis*. The blue mussel is a cold-water species, circumpolarly distributed, with a relatively broad range of breeding temperatures. (See chapter on blue mussels elsewhere in this volume for details

FIGURE 3. Stone bridge cultch method.

on biology.) The salinity range is 19 to 34‰; optimum is 24 to 30‰. Blue-mussel culture has become a large industry in China. The southernmost range of *M. edulis* is Jiaozhou Bay, where there are two peaks of spawning, spring (March to early July) and fall (early October to mid-December). Further north in Zhifu Bay, most of the spawning occurs in late spring.[12-13]

Experimental studies on hatchery rearing of spat began in the late 1950s, and large-scale hatcheries were initiated in the early 1970s.[14-16] In 1973, a maximum of 2.79 million spat per cubic meter of water were reared (Figure 6).[17] Technical improvements have been made since, including better collectors for the attaching spat and use of drugs to inhibit bacterial contamination. The result has been better survival rates as well as decrease in volume of water needed.

Cultured algae (*Phaeodactylum, Platymonas, Pyramimonas, Chlamydomonas, Dunaliella, Chlorella*) are reared as food for the larvae and added in the proper amounts. *Platymonas* and *Pyramimonas* are the best foods for all larval stages.[13,18,19] (In 1973 no Chrysophyceae had been tried as food.) Palm strings (vascular bundles) are used as cultch for spat attachment (Figure 7). Coarse strings produce toxic effects; thinner strings (with bark removed) are found to be less toxic as well as to provide more surface for attachment.[13,17] Hatchery-reared spat are used primarily in Zhejiang, Fujian, and Guandong.[20-22]

In Zhifu Bay and some other areas, a large raft culture of kelp is practiced. Kelp is harvested in the summer, and the lines are then dried before reuse. It was found that these lines usually have newly attached mussels. Now the lines are relaid in the water, after the kelp is harvested, and by August and September large numbers of mussel seed (5 to 10 mm) are gathered; Zhifu has become an excellent mussel seed ground.

Other areas of kelp growing, especially in Dalian Bay, are utilized in the same manner

FIGURE 4. Bamboo stake cultch method.

FIGURE 5. Oysters prepared for fattening.

FIGURE 6. Raft culture for fattening.

FIGURE 7. Artificial seed attached on palm matting (author second from right).

FIGURE 8. Long-line mussel culture.

as sources for seed mussels. Increases in mussel farming, with increases in larger breeding mussels, have led to an increase in numbers of seed mussels. In Jiaozhou Bay, where about 20,000 to 25,000 rafts of kelp are farmed each year, experimental mussel farming began in 1974. At that time, only 30 to 40 mussel spat attached per meter on the kelp lines, but by 1976, with the increase of mussel farming, the average number of spat had increased up to 1000/m.

Long-line off-bottom culture, the same used for kelp (*Laminaria japonica*) culture, is the current method for mussel culture in China (Figure 8). The long-lines are made of straw or plastic, 50 to 60 m long, and buoyed with 60 to 80 glass or plastic floats (28-cm diameter). To these are attached growing ropes made of palm fiber, each about 1.5 m at densities of about 100 strings per line, each with 1000 to 1500 mussels. Recently, strips of used automobile tires have been used successfully as growing ropes. In the northern regions, only one spatfall occurs, but in the more southern region there are both spring and fall spatfalls, the latter being the most desirable.

One method of attaching the seed mussels to the growing rope is to wrap the seed with netting until they attach with the byssus, usually within 2 d (Figures 9 and 10). Sometimes the seed mussels are inserted between strands of the growing rope. Palm-string cultch with hatchery-produced seed is wrapped around the growing ropes, to which the seed mussels reattach. Sometimes the growing ropes are placed in shallow tanks with seed mussels, and attachment usually occurs in 1 to 2 h.

Mussels on the growing ropes are usually suspended in August or September along the coast (Shandong) and harvested the following March to April at size 4.5 to 5.0 cm. Some are harvested in September to October, a year after planting, when they have grown to 7 to 8 cm. One long-line produces about 100 kg of dried meat.

The green mussel is cultured similarly in the more southern regions such as the north coasts of Fujian Province. Because of temperature, it cannot live through the summer there, so the growers must buy seed from Dalian Bay, Liaoning Province, and then transfer them in refrigerator ships in autumn every year. The harvesting season is in late spring to summer of the next year. The growers prefer transferring seeds from the north to rear them in the locality, because the cost is lower.

There are still many unused areas suitable for mussel culture, and the industry is expanding. The recent surge in shrimp culture in China has also led to an increased interest in culture of mussels as a food for the shrimp.

FIGURE 9. Wrapping seed mussels.

FIGURE 10. Mussel strings wrapped.

IV. RAZOR CLAMS: FAMILY SOLENIDAE

Razor clams, *Sinonovacula constricta* (Lamarck) (Figure 2), occur only in Chinese and Japanese coastal areas in estuaries and bays with little freshwater inflow, at salinities of 4 to 29‰, optimum 7 to 26‰. Over half of the total yield of cultivated species of bivalves in the Provinces of Fujian and Zhejiang is of this species.[5,23]

For a long time the supply of clams was from limited natural recruitment. Observations and experiments by Chinese scientists have led to the development of techniques for large-scale seed production, and intensive culture has reached very high levels in some areas.

The time of spawning is from late September to January, and peaks of pelagic life occur from mid-October to mid-November.[24,25] The water is monitored for the highest numbers of larvae nearly ready to set, and the farmers are informed.[23,26,27]

FIGURE 11. Harvesting razor clams.

Seed beds are prepared by a loosening and smoothing of the bottom, timed to take place 1 or 2 d before the clams set. The fall and winter seed are harvested in January to March and planted on prepared growing beds; the density depends on the condition of the substrate, the size of the seed, and the season. Usually 9 to 18 $\times$ 10^6 clams (size 1 cm) per hectare are planted. By the following August, the clams have grown to 5 cm and can be harvested (Figure 11).[28] Sometimes they are not harvested until the following March to April. The average harvest yield is about 30 ton/ha, but yields up to 82.5 ton/ha have been obtained.[1,23]

V. COCKLES: FAMILY ARCIDAE

Cockles *Anadara (Arca) granosa* (Jones and Preston) (Figure 2) are one of the major cultivated bivalves in Shangdong, Zhejiang, Fujiang, and Guangdong. They occur in the middle to lower intertidal sandy-muddy flats of bays and estuaries with unimpeded tidal and freshwater inflow and weak winds and waves. Their preferred range of salinities is 10 to 33‰, optimum 14 to 19‰. Cockles have very short siphons and are never buried very deep. At low tide, mortality occurs during the winter from freezing temperatures in the north and during summer from high temperatures in the south. Also, there are numerous predators that feed on these bivalves.[1,5,6,12]

Seed is collected and reared in enclosed water pools from the spawning that occurs from July to September (Figure 12).[29,30]As the seed grow, they are thinned several times and also moved to expanded grounds near the low-tide mark, where growth is better. It requires about a year of growth for 800 cockles to weigh 1 kg. By another year they have grown to a marketable size of 2 cm (120 cockles/kg). The harvesting season is usually in the winter, and cultured beds will yield 22.5 to 60 ton/ha.[31,32]

Because of changes in coastal mud flats, cockle culture has decreased in recent years, and traditional seed-rearing areas have been lost, especially in Guangdong, Fujian, and Zhejiang. Attempts are now being made to rear seed in hatcheries.

VI. SMALL-NECK CLAMS: FAMILY VENERIDAE

The small-neck clam *Ruditapes (Tapes) philippinarum* (Adams and Reeve) (= *T. sem-*

FIGURE 12. Collecting and sieving natural cockle seeds.

idecussata Reeve) (Figure 2) is endemic to China, Japan, and the Philippines. In China, it occurs in the muddy beach sand of the intertidal zone and in shallow waters at the mouths of estuaries in salinities of 16 to 36‰, optimum 20 to 26‰. Natural stocks are present and support an important fishery, especially in Fujian.

The spawning season is in October to November in Fujian, and the larval pelagic life is about 2 weeks. Before 1975, all seed was from wild recruitment. Techniques have been developed to increase the seed supply. Shallow ponds of up to several hectares are made in the lower intertidal area from which all enemies and pests, including seaweed, are removed, and the bottom is smoothed. Previous to the introduction of induced spawning clams, the ponds are inoculated with *Chaetoceros,* a good food for both the larvae and newly set clams. Soybean milk is sometimes added as a supplementary food. Each pond can be used to rear 2 or 3 batches a year. About 75 to 150 $\times$ 10^6 seed clams (0.5 cm) can be grown per hectare.

The culture beds should be loosened and smoothed before the spat (both wild recruited and pond-reared) are planted. These clams grow from planting size to 1.5 cm by April and May of the following year. They are then replanted at a density of about 1.8 $\times$ 10^6/ha, and in another year have grown to a marketable size of 3.5 cm, yielding a harvest of 18.7 up to 45 ton/ha.[1,33]

VII. PEARL OYSTERS: FAMILY PTERIIDAE

Pearl oysters are only found in the southern provinces. Two of the most important species are *Pinctada (Pteria) martensii* (Dunker) (Figure 2), which is most abundant, and *Pinctada (Pterida) maxima* (Jameson) (Figure 2) which is the largest, and occurs around Hainan Island. *Pinctada martensii* occurs in salinities of 20 to 36‰, optimum 26 to 33‰.

Although pearls have been used in China for a long time for jewelry and as a medical ingredient, it has only been in recent years that attempts have been made to culture pearls artifically (Figures 13 and 14). Now there are many pearl farms and collectives along the coast of Guangdong and Guangxi, and many phases of culture and of the biology of the oysters are being investigated in fishery laboratories and universities.

FIGURE 13. A raft for pearl oyster culture.

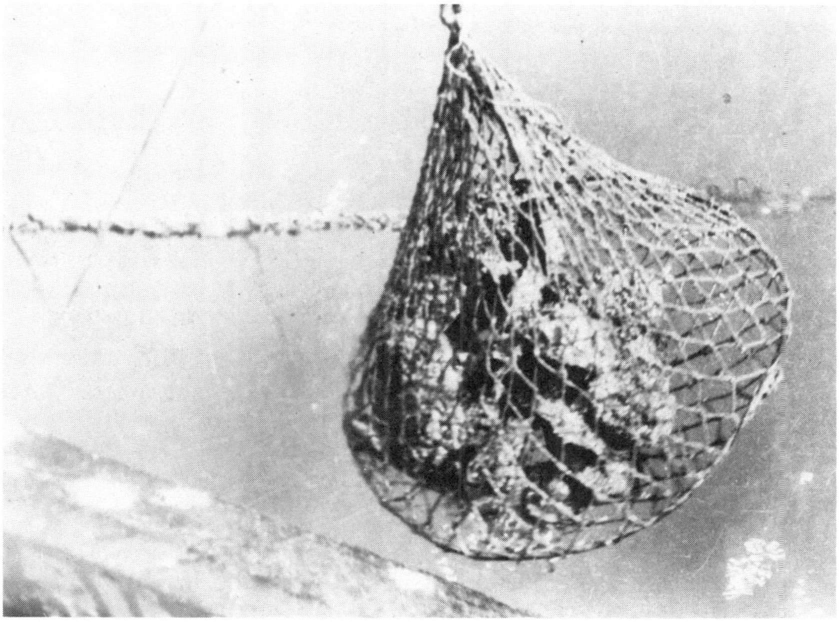

FIGURE 14. Cage for pearl oyster culture.

VIII. MISCELLANEOUS SPECIES

Other species are being cultured on a lesser scale. The clam *Meretrix meretrix* (Linnaeus) (Figure 2) is widely distributed in China, along the coast in fine sand in salinities of 10 to 30‰, optimum 20 to 26‰. Although the wild stock is abundant, there have been some culture efforts in some areas, and these may increase.[34-36] Other species that are receiving some attention are *Mactra antiquata* Spengler (Figure 2),[37] *Brachydontes senhousei* (Benson), and *Potamocorbula rubromuscula* Zhuoang et Cai (Figure 2).

IX. FUTURE PROSPECTS

China is renowned as a leader in aquaculture, especially of freshwater fish, with pond culture dating back to ancient times. The extensive culture of marine algae provides tons of food for its population of over 1 billion.

Remarkable progress has been made in the last several decades in mollusk culture. All of the culture techniques are very labor intensive, but so far this has not been a great deterrent. More advanced technology and mechanization will be used in the future as the culture practices increase. The greatest deterrents probably have been lack of better preservation methods and transportation. Like polyculture in freshwater, so ably practiced by the Chinese, polyculture in the marine habitats is being investigated and is encouraged. The potential is just beginning to be realized, and the future looks very promising.

REFERENCES

1. **Nie, Z. Q.,** Artificial cultivation of bivalves in China. Bivalve Culture in Asia and the Pacific, in Proc. of a Workshop, Singapore, February 16—19, 1982, 21, (English and French).
2. **Zhange, X. and Qi, Z. Y.,** *Essentials of Malcology,* China Science Press, 1961.
3. **Cai, Y. Y., Zhang, Y., and Wei, R. F.,** *Introduction of Malcology,* Shanghai Sci. and Tech. Press, 1979.
4. **Dalian College of Fisheries, (Ed.),** *A Text Book of Molluscan Culture,* Agriculture Press, 1980.
5. The Fisheries Department of Fujian Province (edited), *The Methods of marine Bivalve Culture,* Sci. and Tech. Press, 1958.
6. **Zhang, X. and Xie, Y. K.,** Culture of major bivalves in China, *Bull. Biol.,* 5, 197, 1960.
7. **Zhang, X. and Lou, Z. K.,** Studies on breeding and growth of oyster, *Crassostrea plicatula, Oceanologia et Limnologia Sinica,* 1(1), 123, 1957 (French Abstract).
8. **Zhang, X. and Lou, Z. K.,** *Oyster,* China Science Press, 1959.
9. **Zhang, X. and Lou, Z. K.,** Oyster, *Bull. Biol.,* 2, 27, 1956.
10. **Xie, Y. K.,** Deliberation on oyster food, *Bull. Biol.,* 6, 245, 1959.
11. **Tchang, S., (Zhang, X.), Qi, Z. Y., and Xie, Y. K.,** Moeurs de s'alimenter chez I'*Ostrea rivularis* Gould, *Oceanol. Limnol. Sinica,* 2(3), 163, 1959 (French abstract).
12. The Fisheries Section of Agriculture Department of Shangdong Province, (edited), *Marine Bivalve Culture,* Shangdong People's Press, 1974.
13. **Zhang, F. S.,** Mussel culture in China, *Aquaculture,* 39(1):1, Elsevier Science, Amsterdam, 1984 (English).
14. **Zhang, F. S.,** Observations on living habit of larvae and spats of mussel, *Mytilus edulis, Chinese J. Zool.,* 3, 129, 1963.
15. **Cai, N. E.,** Studies on the life history of *Mytilus edulis* L., *Studia Marina Sinica,* 4, 81, 1963 (English Abstract).
16. **Nie, Z. Q.,** Studies on food of bivalve larvae, *Chin. J. Zool.,* 3, 101, 1964.
17. **Nie, Z. Q., Wang, Z. Y., Niu, X. D., Ji, M. F., Shen, J. F., Chen, W. H., and Liu, L. H.,** Artificial rearing of *Mytilus edulis* larvae on a large scale, *Acta Oceanol. Sinica,* 1(1), 138, 1979 (English Abstract).
18. **Nie, Z. Q., et al.,** Studies on food of straight hinge larvae of mussel, *Mytilus edulis, Chin. J. Zool.,* 2, 1, 1979.
19. **Nie, Z. Q. and Ji, M. F.,** Studies on food for *Mytilus edulis* larvae, *Mar. Fish. Res.,* 1, 85, 1980 (English Abstract).
20. **Chen, X. X., Lin, X. Y., Huang, Z. F., Lieng, L. D., and Xie, Y. K.,** A preliminary experiment on the artificial feeding of the adult mussels of *Perna viridis, Acta Zool. Sinica,* 23(1), 14, 1977 (English Abstract).
21. **Wei, Y. Y., Jin, Q. Z., and Chen, X. X.,** A preliminary study on the food of the larvae of green mussel, *Perna viridis, J. Fish. China,* 6(1), 33, 1982 (English Abstract).
22. **Xu, Z. Z. and Liu, D. J.,** Observations on settlement and growth of green mussel, *Perna viridis* L. in Xiamen Sea, *J. Xiamen Univ.,* 9(2), 1962 (English Abstract).
23. **Shou, Z. H. and Wang, Z. Y.,** The age and growth of *Sinonovacula constricta, J. Agric. China,* 3(1), 72, 1952 (English Abstract).

24. **Wang, Z. Y., Chen, J. L., Yu, Q. S., and Fu, S. B.**, The experimental summation of high production of razor clam, *Sinonovacula constricta,* culture, in *A Corpus of Achievements in Fisheries Sci. and Tech.*, Shanghai Sci. and Tech. Press, 1958.
25. **Wang, Z. Y., Chen, J. L., Yu, Q. S., Fu, S. B., and Chen, W. L.**, *Forecasting of Attaching Time of Pelagic Larvae and Enhancement of Seeds of Sinonovacula constricta,* Shanghai Sci. and Tech. Press, 1958.
26. **Wang, Z. Y., Chen, J. L., Yu, Q. S., Fu, S. B., and Chen, W. L.**, Forcasting of attaching time of pelagic larvae of razor clam, Sinonovacula constricta, in natural sea beach, in *Trans. 5th Meet. West Pacific Fisheries Research Committee,* China Sci. Press, 1962 (English Abstract).
27. **Zhang, J. and Chen, Y. S.**, Experiments on the artificial spawning of *Sinonovacula constricta* (Lamarck), *J. Fish. China,* 1(1—2), 119, 1964 (English Abstract).
28. **Qi, Q. Z. and Yang, M. Y.**, The growth and development of the planktonic larvae, spats and young shells of *Sinonovacula constricta, Taiwan Strait,* 3(1), 90, 1984 (English Abstract).
29. **Chen, W. L., He, L. J., and Liu, D. J.**, Experiments on breeding and rearing of cockle, *Anadara (Arca) granosa* (Jones et Preston) I. Test of inducing spawning, *Fujian Fish. Dev. Sci. Tech.*, 2, 15, 1978.
30. **Chen, W. L. and He, L. J.**, Experiments on breeding and rearing of cockle, *Anadara (Arca) granosa.* II. The rearing of seeds indoors, *Fujian Fish. Sci. Tech.*, 1, 46 1979.
31. The Fisheries Department of Fujian Province (edited), *The Experiences of Anadara (Arca) granosa Culture in Southern Fujian,* Fujian People's Press, 1958.
32. **Yang, J. Y.**, Culture of *Anadara (Arca) granosa* seed, *China Fish.*, 17, 19, 1959.
33. **Liu, Y. F., Liu, Y. X., and Gao, X. S.**, Study on the breeding season of clam, *Ruditapes philippinarum* (Adams et Reeve), *Chin. J. Zool.*, 4, 1, 1979.
34. **Qi, Q. J., Lin, B. S., Wu, T. M., Xu, R. A., and Yang, M. Y.**, Experiments on indoor induced breeding of *Ruditapes philippinarum, J. Fish. China,* 5(3), 235, 1981 (English Abstract).
35. **Lin, B. S., Wu, T. M., and Hang, B. Z.**, The effects of temperature and salinity on the growth and development of the spats of the clam *Ruditapes philippinarum, J. Fish. China,* 7(1), 15, 1983 (English Abstract).
36. **Qui, W. R., Fu, S. B., Zhou, D. T., Zhu, M., and Shi, B. Z.**, Studies on clam, *Ruditapes philippinarum* (Adams et Reeve), in the ponds, Jinjiang, Fujian, *J. Xiamen Univ. Nat. Sci.*, 22(4), 514, 1983 (English Abstract).
37. **Qi, Q. Z.**, The life history of the clam *Ruditapes philippinarum, J. Fish. China,* 11(2), 111, 1987 (English Abstract).
38. **He, J. J., Qi, Q., Wei, X. M., Yang, M. Y., and Xu, Z. C.**, Studies on the food and feeding behaviour of the larvae of *Ruditapes philippinarum, J. Fish. China,* 5(4), 275, 1981 (English Abstract).
39. **Zhang, Y. F.**, Breeding habits and embryo development of clam, *Ruditapes philippinarum, Sci. Tech. Fish.*, 1, 1, 1978.
40. **He, J. J. and Wei, X. M.**, A study on food and feeding habit of the clam spat, *J. Fish. China,* 8(2), 99, 1984, (English Abstract).
41. **Fang, Y. Q.**, A study on the reproductive mechanism of *Ruditapes philippinarum.* II. Ultrastructural study on promoting of egg maturity by NH_4OH sea water, *Taiwan Strait,* 1(2), 99, 1982 (English Abstract).
42. **Shi, B. Z., Fu, S. B., Qui, W. R., Zhu, M., and Zhou, D. T.**, Studies on the spawning Philippine clam, *Ruditapes philippinarum* (Adams et Reeve), in the artificial rearing in earth ponds, *J. Xiamen Univ. (Nat. Sci.,)* 23(2), 211, 1984 (English Abstract).
42. **Zhou, D. T., Zhu, M., Shi, B. Z., Qiu, W. R., and Fu, S. B.**, Studies on the artificial culture of Philippine clam, *Ruditapes philippinarum* (Adams et Reeve), in the ponds, Jinjiang, *Fujian, J. Xiamen Univ. (Nat. Sci.,)* 23(4), 515, 1984 (English Abstract).
44. **Qui, Q. H.**, The morphology, habit and cultivation of clam, *Ruditapes philippinarum, Chin. J. Zool.*, 3, 311, 1959.
45. Guangdong Province Fisheries Research Institute, *Pearl Oyster Culture,* Agriculture Press, 1976.
46. **Xie, Y. K. and Lin, B. P.**, On handling of the pearl oyster after nuclear-insertation, *J. Fish. China,* 7(3), 229, 1983 (English Abstract).
47. **Xie, Y. K.**, *Studies on the South China Sea Pearl,* Tropical Mar. Res., China Ocean Press, 1984, 167 (English Abstract).
48. **Xu, Z. J. and Li, K. K.**, Observations of living habits of *Pinctada maxima* (Jameson), *Chin. J. Zool.*, 1, 30, 1976.
49. **Xu, Z. J. and Li, K. K.**, A preliminary study on the rearing of artificial fertilized larvae of *Pinctada maxima* (Jameson), *J. Fish. China,* 4(3), 295, 1980.
50. **Xie, Y. K. and Xu, Z. J.**, The early feeding habits of the larvae of *Pinctada maxima* (Jameson), *Acta Zool. Sinica,* 26(1), 80, 1980, (English Abstract).
51. **Hu, Y. P.**, *A Comparative Study on Larval Morphology of Pinctada Roding in China,* Trop. Mar. Res., China Ocean Press, 1984, 61 (English Abstract).

52. **Chen, W. L., Liu, D. J., and Xu, W. Z.,** A preliminary study on rearing of spats of *Mactra antiquata, J. Fish. China,* 3(2), 130, 1966 (English Abstract).
53. **Wang, W. D. and Wang, H. C.,** Artificial stimulation and hastening of egg-laying of hard clam, *Meretrix meretrix, Mar. Sci.,* 2, 33, 1983.
54. **Wang, W. D., Zhang, H. R., Wang, H. C., Yu, F. B., and Yu, Y. S.,** Experiments on rearing of seed of *Meretrix meretrix, Chin. J. Zool.,* 4, 1, 1980.
55. **Cui, G. F., Zheng, Q. S., Yu, Y. S., and Xu, S. Y.,** Influence of temperature and specific density of seawater on the development and metamorphosis of clam larvae *(Meretrix meretrix), Mar. Sci.,* 1,, 37, 1981.

Chapter 20

CULTURE OF FARRER'S SCALLOP, *CHLAMYS FARRERI* (JONES AND PRESTON) IN CHINA

Wang Rucai and Shieh Lian-chen

TABLE OF CONTENTS

I. INTRODUCTION

The culture of Farrer's scallop, *Chlamys farreri* (Jones and Preston), has become very important since an abundant and reliable seed source has become available. Changdao County in Shandong Province produced 7.5 million kg, in shell, valued at 1.5 million renminbi each year from 1983 to 1985. In this area, over 80% of the people formerly engaged in seaweed culture have shifted to scallop culture.

II. BIOLOGY

A. MORPHOLOGY

The lateral shell of the scallop is fan-like (Figure 1), and the height is slightly greater than the length.[1,2] The right valve is somewhat flattened and has 20 rather thin ribs, whereas the left valve is more dome-shaped and has about 10 coarse ribs. Both valves have irregularly spaced nodules on the ribs. The ears are unequal, and the anterior ears are larger than the posterior ears. A distinct byssal notch on the anterior ventral surface has 6 to 10 fine catenoid teeth at the edges of the shells. The exterior color of the shell is pale brown, purple brown, or orange yellow to grayish white.

Internally, the nonfused mantle has three folds, with marginal tentacles of different lengths, and well-developed pallial eyes. The single muscle is more or less centrally located. Other parts of the internal anatomy are shown in Figure 2.

B. RANGE

Farrer's scallops are distributed in the Yellow Sea and are found in Japan and Korea. Generally, they are subtidal, occurring in depths up to 10 to 30 m on reefs or on hard, coarse, gravel bottom. They attach with the right valve ventral by the byssus to rocks, shell, etc. and to each other. If disturbed, they break the byssal attachment and move a short distance by rapid closure of the shell before reattaching.[3]

C. TEMPERATURE AND SALINITY TOLERANCES

These scallops are found in full seawater salinity. The optimum temperature for growth is 15 to 20°C. Growth ceases at temperatures below 4°C and above 25°C, and temperatures below −2.0°C and above 35°C are lethal.[3]

D. SPAWNING AND LARVAL DEVELOPMENT

These scallops are dioecious and become sexually mature in 1 year. Mature female gonads are orange red, and those of males are milk white in color. Two spawnings occur, the first in mid-May to mid-July and the second in late August to mid-September. The heavier first spawning commences when the water temperature reaches 15 to 18°C, and the fall spawning occurs when the temperature drops to about 22°C. The gonad comprises about 20% of the body weight in the spring before spawning and about 16% in the fall.

Spawning and fertilization take place in the water column, and the fertilized eggs develop to D-larvae within 24 h at 18 to 22°C. The eyed larvae stage (about 167 μm) occurs in about 13, or as few as 9, d. A foot develops by day 15 (173 to 183 μm), with some crawling, but the velum is still present. A byssal gland develops at the base of the foot within several days; the velum is lost, and byssal attachment occurs. Anterior ears develop when the scallop measures 189 to 200 μm and the posterior ears develop later, after the anterior ones have reached a size of 35 to 70 μm. By the time the scallop measures 780 to 797 μm the mantle edge tentacles extend beyond the shell margins. With further growth, the pallial eyes develop, and the shell surface takes on a rich golden-red color. Two months after spawning, the average size is 4.3 mm in height, but some individuals may reach 6.5 mm.[4,5]

FIGURE 1. Shape of *Chlamys ferreri* shell.

E. GROWTH

Spring-spawned scallops reach an average size of 3 cm by December of the same year, with some over 4.5 cm. Scallops reach the marketable size of about 6 cm by September to October of the following year. The weight of the muscle is closely related to shell size; the relationship can be expressed by the formula: $wt = 2.8995 \times 10^{-3}\ ht^{3.5}$ (wt = weight in grams of the muscle; ht = average shell size). Table 1 gives the average weight of muscle for shell heights and body weights of six sizes of scallops.

III. CULTURE

The development of a seed source, beginning in 1973, has resulted in a large and prosperous scallop culture. The yield of cultured scallops in 1985 was larger than that of the wild fishery in Shandong Province. Seed scallops are obtained by two methods.

A. SEED COLLECTION FROM WILD RECRUITMENT

Wild spat are collected in bags and cages. The bags (30 × 40 cm) are made of polyethylene netting of 1.2 to 1.5 mm mesh, filled with 50 to 100 g of used net of nylon, polyethylene, or other plastics. Collecting cages (60 to 30 cm diameter) are filled with the

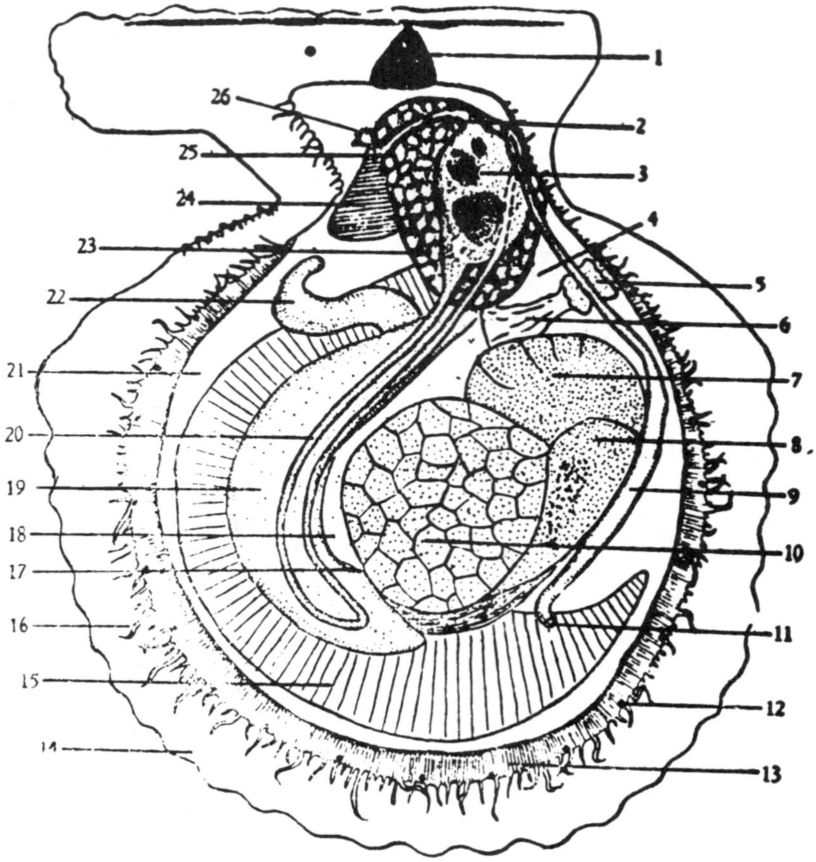

FIGURE 2. Left lateral view of *Chlamys ferreri*. 1. ligament; 2. esophagus; 3. stomach, 4. pericardial cavity; 5. ventricle; 6. auricle; 7. retractor; 8 adductor muscle (smooth); 9. rectum; 10. adductor muscle (striated); 11. anus; 12. pallial eye; 13. inner layer of mantle on the right side; 14. right shell; 15. right side gill; 16. tentacles on mantle margin; 17. nephridipore; 18. nephridium; 19. gonad; 20. intestine; 21. mantle; 22. foot; 23. digestive diverticulum; 24. labial palps; 25. mouth; 26. lip.

TABLE 1
Relationship of Muscle Weight to Shell Height and Body Weight

	Group size					
Av. shell height (cm)	4.1	5.5	6.9	7.7	8.4	8.8
Av. body weight (g)	6.7	17.5	33.9	51.0	56.7	60.6
Av. Compoy weight (g)	0.2	0.6	1.2	2.1	2.3	2.6

same materials as the bags. Cages are partitioned in levels, each about 20 cm thick, and each filled with about 30 g of cultch material. At the proper time during the breeding season, the cultch is placed at the optimum depth of about 2 m. Each will collect several hundred spat and sometimes up to several thousand.

Several factors influence the success of set. Small mesh bags are more effective than large mesh bags. Substratum with a large area for attachment is better than that with a small area. High transparency of the water at the time of breeding seems to give higher numbers of spat collected per unit.[4,6]

B. SEED PRODUCTION BY HATCHERIES

In hatchery operations, a 1 m^2 tank requires about 30 females and 3 males (6 to 7 cm height) to ensure adequate numbers of larvae. The best method for induction of spawning is first to dry the scallops before placing them in the spawning tank and then to raise the temperature 2 to 3°C. After spawning and fertilization, cultured algae (*Isochrysis, Phaeodactylum, Nitzschia, Chaetoceros, Platymonas*) are added in the proper proportions as food for the larvae.[5] Sometimes yeast is added. For best results, the temperature should be kept between 17 and 26°C and not allowed to vary over 2°C within 1 d. The salinity should be between 29 and 38‰, pH between 7.5 and 8.1, NH_4-N below 100 mg/m^3, and illumination between 500 and 1000 lx. The use of antibiotics and increased water circulation decreases mortality.[7] Faster larval development is obtained at the higher temperatures. The normal density of larvae that can be reared is about $1 \times 10^6/m^3$, and up to three times as many have been reared successfully.

C. CULTURE TO ADULT

Small scallops (about 300 μm) can be placed in open water for nursery culture. Both nets and cages are used in nursery culture, as are plastic cylinders (27 to 30 cm diameter, 85 to 90 cm long), closed at each end with small-mesh netting. All of these are suspended. Less mortality occurs in clear water, and survival is about 30% on average but sometimes as high as 50%.[7]

Upon reaching a size of 0.5 to 1.0 cm, the scallops are placed in cages, plastic nets, or cylinders that are suspended from rafts or are planted directly on the bottom.

The 100- to 250-cm-long cages, with plastic netting around metal hoops, are each divided into 5 to 10 compartments. Each compartment of a cage can hold 30 to 50 scallops, and 450 to 600 cages can be cultured in 1 ha of water area. When cylinders are used, the mesh end closures are changed for larger size mesh as the scallops grow. The nets are renewed at regular intervals, and fouling organisms and silt are removed from the scallops. Another method of culture is to bore a 1- to 2-mm hole in the anterior ear of each scallop and to thread it onto nylon string. About five scallops are suspended at intervals of 25 to 30 cm. This method allows about 150,000 spat per hectare, which will yield 4500 to 4750 kg of meat per hectare after a growth period of 12 to 18 months.[3]

A less labor-intensive method is to plant the seed directly on suitable bottom. The bottom plantings, however, are subject to heavy mortality from predation, and the seed may be moved by strong currents.

IV. CONSTRAINTS

Besides being very labor intensive, scallop culture has many hazards that cause mortality or hinder growth. There are many predators, including fish (especially flatfish), starfish, crabs, snails, and octopuses. Shell-boring mollusks and boring sponges (*Cliona*) penetrate the shells, retard growth, and may cause death. A species of crab (*Pinnotheres*) lives inside the mantle cavity and interferes with the feeding of the scallop. Fouling organisms such as barnacles, oysters, mussels, and algae attach and compete with the scallops.

V. HARVESTING AND PROCESSING

Scallops are harvested when they measure about 6 cm. Cages, net bags, and suspended scallops are hand lifted and the mollusks removed. Diving is the usual method of harvest for the bottom plantings.

After shucking, the muscle is separated from the mantle and visceral mass and is often sold fresh. It can be dried, in which case it is first soaked in clean 2% seawater that is sometimes heated first at 80°C and then increased to 90°C. Another method of processing is to steam the scallops before removal of the muscle. Canned meat is sterilized at high temperatures in sealed cans. Frozen meats are cleaned thoroughly, allowed to drip for 5 min, soaked in a lactic acid solution for $^1/_2$ minute, and then quickly frozen on a plate at −20°C.

VI. FUTURE PROSPECTS

Besides the extensive culture of *C. farreri,* China is beginning to culture other species. Seed of *Chlamys nobilis* (Reeve), native to China, is being hatchery reared and cultured to a small extent. *Patinopecten yessoensis* (Jay), introduced from Japan, is spawned and seed is produced. This species has fast growth but can only be grown in the colder waters of northern China. Seed of the introduced *Argopecten irradians* (Lamarck) from the U.S. is now being hatchery produced. Although fast growing, this species suffers heavy mortality during the colder season north of Shandong.

Attempts have been made to hybridize the above species and also to breed new varieties that have fast growth and adaptability to different environmental conditions. Another development is to promote the concurrent scallop culture with the existing extensive kelp culture. The future for scallop culture looks bright in China.

REFERENCES

1. **Zhang Xi and Qi Zhengyan,** *Outlines of Malacology,* Science Press, Beijing, 1961, 1.
2. **Wang Zhenrui,** Studies on Chinese species of the family Pectinidae. III. Chlamydinae (1. *Chlamys) Trans. Chinese Soc. Malacology,* 1983, 1.
3. **Wang, R., Wang, K., Yan, G., Zhong, Y., and Yu, L.,** *Textbook of Mollusk Culture,* Agriculture Publishing House, Beijing, 1980, 1.
4. **Wang Rucai and Gao Jie,** The report of experiment on the artificial rearing of scallop spats, *J. Shandong Fish. Soc.,* No. 2, 1, 1978.
5. **Wang, R. and Gao, J.,** A study on the artificial rearing of scallop spats, *J. Shandong Coll. Oceanol.,* No. 2, 1, 1978.
6. **Wang, R., Wang, K., Yan, G., Zhong, Y., and Yu, L.,** Studies on the techniques of collecting spat of scallop (*Chlamys farreri*) in natural sea area, *J. Shandong Coll. Oceanol.,* No. 3, 1, 1987.
7. **Wang, R., Zhang, L., Wu, A., and Gao, J.,** A study on the artificial rearing of scallop spats in prawn rearing tanks, *QLLU Fisheries,* No. 1, 1, 1985.

Chapter 21

MARICULTURE OF GIANT CLAMS

William K. Fitt

TABLE OF CONTENTS

Seven species of tridacnid clams are found in shallow tropical Indo-Pacific waters, including the "giant clam" *Tridacna gigas,* the largest bivalve ever evolved (Figure 1). These clams are distinct from other species employed in mariculture endeavors in that the juveniles and adults live in symbiosis with, and derive nutrients from, single-celled dinoflagellates of the genus *Symbiodinium,* the same group of zooxanthellae that inhabit corals. The presence of an autotrophic food source, as well as relatively rapid growth rates of the juveniles combined with the high price of the harvested adductor muscle, makes tridacnids one of the most attractive animals for mariculture in the tropics today. This chapter reviews advances in the biological and technical aspects of mariculture of tridacnid clams and does not attempt to address the socioeconomic aspects of mariculture of these animals.

I. RANGE AND HABITAT

Tridacnid clams are found throughout the tropical Indo-Pacific, generally inhabiting the shallow waters of coral reefs. There are currently 7 living species and at least 15 fossil species described.[1,2] Adults of *T. gigas* Linne and *T. derasa* Roding, the largest species (Table 1), live unattached, usually in water less than 10 m deep on forereefs and patchreefs in lagoons with good water circulation. Adults of *Hippopus hippopus* Linne and *H. porcellanus* Rosewater are also found unattached, typically on sandy substrates in water less than 6 m deep. The latter two species will roll in strong ocean waves and currents, a mechanism that may explain their distribution primarily on the leeward sides of reefs.[3] The abyssate adults of the larger species are also able to right themselves by way of strong muscular siphonal action.[4] *T. squamosa* Lamarck attach themselves to rubble with a mucous byssus, usually on backreefs in more protected water. This species exists deeper on the reef than the other species, as *T. squamosa* is sometimes seen in lagoons at depths greater than 20 m. The two smallest species, *T. maxima* Roding and *T. crocea* Lamarck, both live partially embedded in coral while young. *T. maxima* remain attached to hard substrates on the reef crest, but only partially embedded in coral; adult *T. crocea* are almost always found completely embedded in burrows in corals, usually in the lower region of the tidal zone of reef flats and shallow lagoon habitats.

Clams less than 3 months old attach to the substrate via byssal threads and mucus but are also able to move about by means of a muscular foot and reattach to the substrate. Juveniles are difficult to find on the reef,[5-7] leading researchers to hypothesize high fecundity, low larval and juvenile survival, and long adult life as parameters influencing natural tridacnid populations.[8] Exceptions may include *T. maxima* living in the closed lagoon of Takapoto Atoll in French Polynesia, where "trapped" larvae settle on essentially all shallow hard substrates, and densities of clams reach $12/m^2$.[9]

The center of distribution of the tridacnids is the central Indo-Pacific. *T. maxima* has the widest range, extending from the northern Red Sea to eastern French Polynesia. The extent of distribution of tridacnid species is negatively correlated with their maximum shell size, except for *T. crocea* and *H. porcellanus.*[1,2] It has been suggested that the maximum size for each species reflects other physiological factors limiting distribution of the clams. One such factor is the duration of the larval stages: larger tridacnids generally develop more quickly than smaller species.[10,11] Another theory suggests that *T. crocea, T. gigas,* and *H. hippopus,* which are no longer found in the western Indian Ocean, have not recovered their former range after the sealevel and temperature change during the Pleistocene ice ages.[12] Temperature and salinity tolerances, substrate presence, and availability of metamorphic cues are possible factors that may be restricting the range of these species at this time. The limits of distribution of *H. porcellanus,* the newest described species,[2] have yet to be determined. The type specimens of *H. porcellanus* are from the Philippines, and members of this species have recently been found in Palau, Australia, and Indonesia.[13-15]

FIGURE 1. The giant clam *Tridacna gigas* living among corals on a patchreef in Palau.

TABLE 1
Onset of Sexual Maturity in Tridacnid Clams

Species	Maximum size	Smallest size of male phase	Smallest size of hermaphroditic phase	Minimum time to sexual maturity (years)	Ref.
T. crocea	15.0	2	4	n.d.	1,24
T. maxima	35.3	n.d.	11	10[a]	1,10,48
H. porcellanus	21.6	n.d.	n.d.	2	2
H. hippopus	39.7	7	13	n.d.	1,24
T. squamosa	40.8	9	16	n.d.	1,24
T. derasa	60.0	15	26	2—3,[b] 4—5[a]	14,20,24,67
T. gigas	137.0	18.7[c]	n.d.	2—3,[b] 4—5[a]	1,14,20

Note: n.d. = no data.

[a] Hermaphroditic phase.
[b] Male phase.
[c] Mean size.

II. MORPHOLOGY

The tridacnids are a family of the bivalve superfamily Cardiacea and are thought to have evolved from a nonsymbiotic cardid ancestor.[16,17] Yonge[16] has hypothesized that symbiosis with zooxanthellae has been the driving force for evolution of the unique morphology of the tridacnids, which is characterized by a hypertrophied siphon region ("mantle") that extends over the interdigitated margins of the shell in the genus *Tridacna* (Figure 1), but comes only to the margin of the shell in the genus *Hippopus*. The mantle houses 80 to 90%

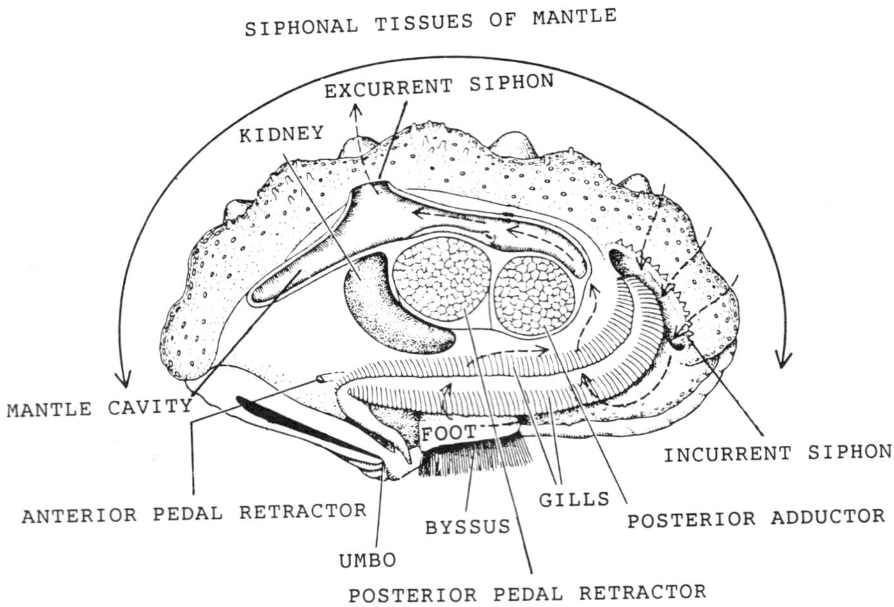

FIGURE 2. General morphology of a tridacnid clam, indicating proposed evolutionary ''rotational'' expansion of the siphonal tissues of the mantle (solid arrows) and ventral position of both umbo and byssus. Dotted arrows indicate direction of water flow. (Adapted from ''Giant Clams'' by C. M. Yonge. © 1975 Scientific American, Inc. All rights reserved.)

of the zooxanthellae in an individual clamp.[18] Evolutionarily, the mantle and shell have extended anterodorsally in relation to the byssus such that the umbo lies ventrally next to the byssus, rather than dorsally 180° from the byssus as in other cardid clams (Figure 2). This evolutionary modification gives more surface area for the mantle on the dorsal (sunward) side of the clam. Indeed, the current tridacnid species are preceded in the geological record by now-extinct species having correspondingly more dorsally placed umbos, indicating less hypertrophy of the siphonal regions.[19]

During the evolution of tridacnids, the anterior adductor muscle was lost, and the anterior pedal retractor displaced posteriorly and ventrally. The two major muscles remaining are the posterior pedal/byssal retractor and the posterior adductor (Figure 2). These are the highly prized edible portions of the clams and are located subcentrally, varying in size depending on the species. The wet weight of edible muscle in *T. gigas* and *T. derasa* is 10 to 15% of the total wet tissue weight[20-22] (Table 2).

The edible mantle makes up an additional 40% of the wet tissue weight.[18] The enlarged kidney in tridacnids contains high concentrations of arsenic compounds[23] and is not eaten.

III. SPAWNING

Tridacnid clams are sequential simultaneous hermaphrodites, developing spermeries at a younger age than ovaries (Table 1). The spawning of eggs usually follows the release of sperm by anywhere from a few minutes to a few hours.[10,20,24,25]

Seasonal patterns of reproduction of adult tridacnids are not always obvious and may depend on latitudinal and local variations in temperature and day length. Natural release of eggs and sperm from *H. hippopus, T. gigas* and *T. derasa* on the Great Barrier Reef and *T. crocea* in Japan occur during the summer months.[7,25-27] *T. squamosa,* on the other hand, has been observed to spawn spontaneously during the winter.[28-30] Seasonal patterns of non-induced spawning of tridacnids in equatorial regions is less clear. For instance, *T. gigas* and *T. derasa* in Palau and *T. gigas* in Papua New Guinea spawn year round.[20,21,31]

TABLE 2
Survival Rates and Tissue Biomass of Different-Aged *Tridacna derasa* in Mariculture in Palau and *Tridacna gigas* in Australia[14,20,21,33,58,59]

	Species	Age	Approx. length (mm)	Total wet wt(g)	Adductor muscle (%)	Survival (%)
			Hatchery Phase			
1. "Extensive" rearing						
Eggs-trochophore-veliger	T.g.	0—72 h	0.1—0.2	n.d.	n.d.	3
Veligers-juvenile	T.g.	3—30 d	0.2—0.4	n.d.	n.d.	20
Juveniles	T.g.	30—150 d	0.4—4.5	n.d.	n.d.	1.1[a]
Veligers-juvenile	T.d.	2—90 d	0.2—6.0	n.d.	n.d.	1
Juveniles	T.d.	5—9 mo	14.5—35.0	n.d.	n.d.	95
2. "Intensive" rearing						
Eggs-trochophore-pediveliger	T.g.	0—9 d	0.1—0.2	n.d.	n.d.	8—40
Pediveliger-juvenile	T.g.	9—16 d	0.2—0.2	n.d.	n.d.	29
Juveniles	T.g.	16—150 d	0.2—4.5	n.d.	n.d.	1.7[a]
			Ocean Nursery and Grow-Out			
Juvenile (protected trays)	T.d.	9—24 mo	35—110	1.5—12.0	6—7	75
Juvenile-adult (unprotected)	T.d.	24—60	110—250	12.0—453.0	7—15	90
Adult (unprotected)	T.d.	72 mo	275	617	13	—
Juvenile						
Protected trays, intertidal	T.g.	11—17 mo	41—93	n.d.	n.d.	100
Protected trays, reef bottom	T.g.	11—17 mo	44—84	n.d.	n.d.	95
Protected trays, rack off bottom	T.g.	11—17 mo	44—99	n.d.	n.d.	90
Protected trays, subsurface floats	T.g.	11—17	42—77	n.d.	n.d.	57

Note: n.d. = no data; T.d. = *Tridacna derasa; T.g.* = *Tridacna gigas.*

[a] Fouling and predation not controlled.[33]

Lunar and diel periodicity of spawning has been suggested based mostly on observations in Palau and Australia. Most spontaneous spawnings of *T. gigas, T. derasa,* and *H. hippopus* have occurred in the late afternoon during the second and fourth quarters of the lunar phase.[7,20,32,33] It has not been possible to predict which individual clams will spawn on a particular lunar or diel cycle.

Tridacnids can be induced to spawn with gonad suspensions,[1,10,28-30,32,34-36] hydrogen peroxide,[30,32,35] serotonin,[15,25,33] and thermal stress.[10,30] However, the most healthy larvae seem to come from spontaneous spawnings,[20,30] probably because spontaneous spawnings occur when eggs are developmentally ready to be released, whereas induced spawnings may force the release of premature eggs. The most promising artificial inducer of spawning is the neurotransmitter serotonin, which is injected directly into the gonad. Though not predictable, this technique is more successful in causing the release of eggs in clams known to have ripe eggs.[15,25,33] Induced spawnings often involve only the release of sperm,[25,30] which appear always ready to be released.[34]

IV. LARVAL DEVELOPMENT

Development of fertilized eggs and larvae of tridacnids follows a course similar to that of other bivalves.[10,29,30,35-37] Free-swimming trochophore larvae typically form about 12 h after fertilization. Veliger larvae develop about 12 h later. Feeding on particulate matter has not been observed during the short trochophore stage, though these larvae may be able to take up dissolved nutrients. Veligers, however, are capable of ingesting phytoplankton within 24 h of their development. Fed veligers have higher growth and survival rates than starved ones. As with other bivalves, optimal concentrations of larvae and food as well as optimal algal food species exist. For instance, larvae of *H. hippopus* survive best at densities less than 5 larvae/ml. Concentrations of *Isochrysis galbana* Parke between 10^4 and 10^5/ml yield optimal growth and survival rates of veligers of *H. hippopus* when they are maintained at 3 larvae/ml.[36]

Veligers take up zooxanthellae of the genus *Symbiodinium* into their stomachs, but digest few if any.[11] However, veligers having zooxanthellae freshly isolated from adult clam mantles in their stomachs grow faster than starved veligers, implying that photosynthetic products may be released from the algae and metabolized by the veligers.[11]

The veliger stage of tridacnids varies in length depending on species of clam, environmental conditions, and batch of larvae. Veligers of the smaller tridacnid species go from pediveligers to metamorphosis in about 2 weeks, whereas veligers of *T. gigas* are typically ready for metamorphosis in less than 1 week at 30°C.[10,20,32,36] Veligers are capable of completing metamorphosis without any feeding, implying that they can obtain all of the nutrients necessary for larval development from the egg. The fact that some batches of larvae have high rates of survival (i.e., 1 to 40%) from early veliger to juvenile, whereas others do not survive at all,[13,20,33,35,36] suggests that it may be possible to select large size classes of eggs or early veligers for mariculture, which are more likely to have higher levels of stored nutrients and higher growth and survival rates through metamorphosis than will smaller eggs or veligers.

Settlement and metamorphosis in tridacnids is a gradual process involving temporary attachment of the pediveliger, degeneration of the velum, and the development of gills. Metamorphosis may take one or more days, and its completion generally correlates with the appearance of a statocyst located at the base of the foot.[10] Juveniles continue to crawl, detaching and reattaching byssal threads when necessary. They attach to many different substrates,[10,33,35] but protection may be more important than substrate composition in choice of attachment sites. Specific cues that trigger settlement and metamorphosis for these bivalves have not yet been found.

V. ESTABLISHMENT OF A SYMBIOSIS

Adult and juvenile tridacnid clams maintain zooxanthellae of the genus *Symbiodinium* in their tissues.[38-40] However, the eggs contain no zooxanthellae, making it necessary for the larvae or juveniles to acquire their own complement of symbionts.[10,29,30] Potential sources of healthy zooxanthellae include those released from symbiotic corals, jellyfish, anemones, and other cnidarians on the reef, as well as from adult clams. Veligers and juveniles are capable of establishing a symbiosis in the laboratory with *Symbiodinium* isolated from a variety of hosts.[11,30] Motile zooxanthellae, which often arise from nonmotile forms after release from an invertebrate host, may facilitate establishment of a symbiosis by swimming toward nonsymbiotic tridacnids.[30,41] This chemosensory behavior may be a response of the motile zooxanthellae to nitrogen (i.e., ammonium) released by the nonsymbiotic clam.[42] No other genera of algae have been found that establish a symbiosis with tridacnid clams.

Shortly after metamorphosis, zooxanthellae in the stomach and digestive gland region

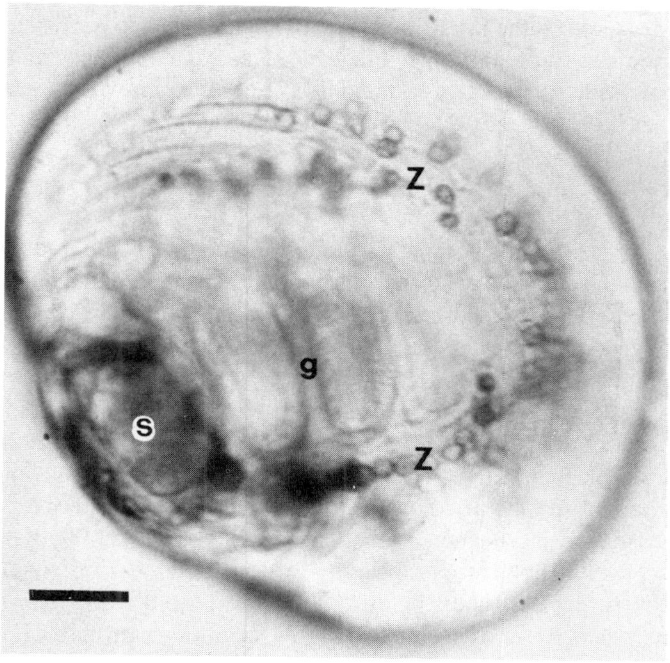

FIGURE 3. Juvenile *Hippopus hippopus,* approximately 40 d old, showing well-developed gills (g) and rows of zooxanthellae (z) extending from the stomach (s) up into the developing hemal sinuses of the mantle. Bar = 50 μm.

of juvenile clams pass by an unknown mechanism into the developing hemal sinuses.[30] The zooxanthellae that grow and divide in the sinuses of the mantle have established a symbiosis (Figure 3). This process can be facilitated in maricultured clams by introduction of freshly isolated or cultured zooxanthellae to veligers such that their stomach contain the future symbionts before metamorphosis.[11,15,33]

VI. GROWTH AND SEXUAL DEVELOPMENT

Zooxanthellae in juvenile and adult tridacnids release photosynthetically fixed carbon to their hosts,[43,44] such that a significant portion of the respiratory requirements of the clam, and potentially many of the metabolic requirements necessary to support the growth and reproduction of the clam, may be met.[11,18,45] Following relatively low larval growth rates (even in fed clams) and virtual cessation of growth during metamorphosis, the growth rates of juveniles after acquisition of zooxanthellae increase markedly, often five to ten times the maximum larval growth rates.[10,11,29,30,32,33,35,37] The maximum rates of growth and survival vary according to species of tridacnid, culture conditions, and the type and numbers of *Symbiodinium* living in each clam.[11,30,46] The relative contribution of external food sources (phytoplankton, dissolved nutrients, particulate organic matter) to growth and survival of juvenile and adults has not been determined.

As with other animals, size-specific growth rates of symbiotic tridacnids decrease with increasing body size.[5,9,32,47] Two factors may be responsible for this trend in tridacnids. First, the number of zooxanthellae per gram of tissue of clam decreases as clam size increases, partially because tissue weight of the clam increases as a cube function while surface area of the mantle increases as a square function.[18] As the clam grows larger, the relative contribution of the zooxanthellae to the nutritional requirements of the metabolic biomass decreases. Second, zooxanthellae in larger clams are shaded to a greater degree than in smaller clams because there are larger numbers of zooxanthellae per unit mantle surface area and mantle tissue is thicker. Additional decreases in growth rate of tridacnids are correlated with the onset of reproduction[14,48,49] (see Table 2).

VII. HISTORY

Clam meat from tridacnids has been a protein source for people in the Indo-Pacific for centuries. All parts of the clam are consumed, with the exception of the bitter-tasting kidney, and shells are used as tools, ceremonial displays, and disk money.[50] Marine "ranching", the collection of clams from the reef and their maintenance in shallow water until ready for use, is still a common village practice. However, as human populations have increased, local extinctions of the larger species of tridacnids have become more common.

Poaching of clams to satisfy the high demand for clam muscle in Asia has also made an impact on tridacnid populations, in spite of the fact that most governments have laws protecting the clams. With the price of dried giant clams adductor muscle up to $120 U.S. per kilogram in Hong Kong, some six-man crews earn over $100,000 per collection trip, making the risk of imprisonment worthwhile.[51] An estimated 450 t of adductor muscle are sold in Hong Kong every year at a wholesale value of $22 million.[51] One boat apprehended on the Great Barrier Reef in 1976 contained 124 bags of clam meat. Each bag held an average of 363 adductor muscles of *T. derasa,* the equivalent of harvesting 1 ha from Swain Reef, the most productive on the Great Barrier Reef.[52] Fortunately, more effective preventative measures have been implemented by the Australian government over recent years.[53,54] However, most island nations do not have the resources even to monitor their clam resources, let alone to police the often remote reefs. For instance, between 1972 and 1975, 98% of all *T. gigas* and 97% of all *T. derasa* from remote Helen Atoll south of Palau were taken.[55]

The International Union for the Conservation of Nature has placed *T. gigas* and *T. derasa* on the "threatened species" list.[56] Nonetheless, over large areas of Micronesia, Indonesia, the Philippines, Papua New Guinea, Fiji, and southern Japan, tridacnids have been so heavily fished by locals and poachers that it is feared that any individuals left occur at densities so low as to be reproductively ineffective in restoring the population.

Tridacnids emerged as a major experimental reef invertebrate during the Great Barrier Reef Expedition led by Yonge in the late 1920s.[16,26] Soon afterward, the Japanese set up a research station on Palau, where Wada[24,34] documented spawning and protandric hermaphroditism in these clams. After LaBarbera[29] and Jameson[10] successfully cultured the larvae in the mid-1970s, rapid progress in developing methods of mariculture occurred. In the last 10 years, the major advancements in both mariculture technology and basic biology of the tridacnids have occurred primarily in Australia,[7,25,33,57] Fiji,[29,31] Palau,[11,14,18,20,21,30,32,36,45] Papau New Guinea,[35,47] and the Philippines.[15,58,59]

VIII. METHODS OF CULTURE

Current mariculture techniques have recently been documented and three major phases are summarized below.[14,20,21,31,33,57,60] The "hatchery phase" of raising tridacnids consists of spawning, larval rearing, settlement and metamorphosis, and juvenile rearing for the first 3 to 8 months. Raising the clams in protected trays in the ocean "nursery" until about 2 years of age follows. At this point the clams are relatively predator resistant and are allowed to "grow-out" unprotected on the reef bottom.

Spawning of eggs in the hatchery may be by either natural or induced release. Natural spawning occurs during the appropriate season and lunar phase and, given a healthy broodstock of 20 or more clams, usually yields millions of healthy eggs. Serotonin-induced spawnings of clams are reasonably successful in producing eggs, as long as only clams with ripe eggs in the gonads (determined by gonad biopsy) are induced. Eggs either are collected from spawning clams in large plastic bags and moved to another container where they are fertilized with sperm from other clams or are spawned directly into the rearing raceways after most of the sperm-laden water is drained. The density of eggs is adjusted to 1 to 10/ml. The larvae may be raised in "intensive" high-density and high-maintenance cultures involving daily feeding with phytoplankton (i.e., *Isochrysis galbana* Tahitian strain) and daily water changes.[57] Alternatively, "extensive" culturing, involving a low-density, low-maintenance regime relying on natural blooms (the Wells-Glancy method) in raceways with minimal water flow, may be used.[20,57] After about 1 week, larvae of *T. derasa* and *T. gigas* begin metamorphosis. If the pediveligers are not already in tanks or raceways exposed to natural light, they are moved to such tanks before metamorphosis. Settlement and metamorphosis of a group of tridacnid larvae may take 2 d or more.

After metamorphosis, juveniles are allowed to grow in outdoor tanks until large enough to be removed safely from the bottom and placed in shellfish rearing trays. During this time, the juveniles establish a symbiosis with zooxanthellae, which may be facilitated by additions of zooxanthellae from the mantles of adult clams. The trays contain small pieces of coral rock or basalt chips as attachment substrate for the juveniles, making it easy to move the clams without injury. The trays are stocked at high densities but thinned regularly and allowed to grow 3 to 6 months in raceways free of most reef predators.

Survival for giant clams in mariculture is generally low (Table 2). However, the number of eggs spawned from a single large *T. gigas* or *T. derasa* is typically in the hundreds of millions, so that even survival rates of a fraction of a percent often yield thousands of clams. The highest mortality of veligers and juveniles during the hatchery phase is usually associated with metamorphosis, when nutrient stores carried over from the egg are apparently exhausted and the transition from velar to gill particulate feeding and establishment of a symbiosis with zooxanthellae takes a few days. Other sources of mortality in the hatchery include overgrowth by benthic algae and predation by pyramidellid gastropods;[61] the latter settle out of the plankton and feed on the soft tissue around the byssus. The problem of algal overgrowth has been solved by coculturing herbivorous gastropods of the genus *Trochus* with juvenile tridacnids.[62] Survival from egg to juvenile in "intensive" hatchery cultures may be as high as 12% but is highly variable, depending on the batch of larvae.[33] Typical survival rates from egg to juvenile in "extensive" hatchery culture conditions are 1%. Survival rates improve markedly, often approaching 100%, after juveniles establish a symbiosis with zooxanthellae, providing fouling and predators are controlled (see Table 1).

Juvenile clams may be placed on the reef at about 8 to 9 months but must be maintained in screened trays for protection from fish and crab predation. During this "ocean nursery" phase, clams in trays placed subtidally on or just off the substrate generally have survival rates of 75 to 100% (Table 2) and fewer problems with pyramidellid gastropods, shading, and crowding from fouling invertebrates and algae, presumably because of the presence of benthic predators of gastropods and fouling organisms. Subtidally raised juveniles have relatively high growth rates during the first year (circa 5 to 10 mm/month).

Giant clams will survive out of water for 4 to 5 h with no adverse effects[57,63] and live intertidally with virtually no mortality.[57] Growth rates of intertidally raised clams were only slightly less than the best growth rates of subtidally raised clams.[57] The advantages of an intertidal habitat are the ease of maintenance and lack of predation by pyramidellid gastropods. Disadvantages include susceptibility to storm and wave disturbance and reduced exposure to light during daytime low tides.

Subsurface floating tray cultures are the least desirable culture system for tridacnids.[57] They are expensive to build and maintain and are subject to high pyramidellid infestations; they foul quickly, and they are easily disturbed by wind and storm waves.

At about 2 years of age, *T. derasa* and *T. gigas* can be placed unprotected on the reef. During this "grow-out" phase, the fresh meat weight of the clams increases dramatically, more than quadrupling in *T. derasa* during the fifth year in Palau.[21] Optimal harvesting time is at the end of this period, because the production of eggs begins at about the fifth year (circa 300 mm length), and growth rates decrease markedly.[14,48,49,64] Survival of *T. derasa* from low-density, low-maintenance cultures in Palau from early veliger to harvestable size at 5 years is about 0.05 to 0.5% (see Table 2).[14,20,21]

IX. USES

Almost all parts of the tridacnid clam can be utilized. Adductor muscle and shell can be exported, with projected earnings of $155,000/ha of 5-year-old *T. derasa* ($3.11 U.S. per 250-mm living clam, 1987 prices).[21] Other meat may be used for local consumption or fertilizer. Total meat production from 5-year-old *T. derasa* in Palau was over 22 ton/ha/year, comparable to temperate mussel and clam production.[21] In addition, seed clams are currently sold for $0.25 (6 months old, 15-mm length) and $1.00 (1 year, 50-mm length) each.[60] Finally, aquarists in the U.S. and Europe pay $5 to $25 per live clam, depending on size and species.

X. FUTURE DEVELOPMENT

The technical and biological advances made in the mariculture of giant clams of the family Tridacnidae have been summarized in a number of reviews over the past few years.[14,20,21,47,60] Besides the ongoing Palauan operation at the Micronesian Mariculture Demonstration Center, experimental and commercial hatcheries have been established in Australia, the Philippines, Papua New Guinea, Fiji, and the Marshall Islands. The increase in giant clam mariculture facilities around the world attests to the vitality of and interest in this new industry.

In spite of the progress made in understanding the biology of tridacnids and developing mariculture techniques, a number of areas of research have potential for increasing yields in the mariculture of these clams. These include the following:

1. First is selection of broodstock from fast-growing F_1 populations. Selection of large eggs may also yield juveniles with high growth rates. Research on nutrient values of eggs[65] in relation to size and subsequent developmental times would be useful in confirming whether egg size can be used as a method of selection.
2. Another area is selection of optimal strains and species of *Symbiodinium* that yield the fastest-growing clams. Preliminary reports show that growth of juveniles may vary by a factor of seven, depending on the type of zooxanthellae living in the clam.[46]
3. Determination of cues that induce settlement and metamorphosis of competent veligers in a few hours rather than a few days would help to increase survival during this critical transition. Chemical control of metamorphosis in oysters and abalone has been successfully exploited in mariculture of these species.[66,67]
4. Another area is increasing survival of trochophore larvae. Survival is extremely variable at present,[15,33,36] but may be increased by judicious use of antibiotics, dissolved nutrients, or vitamins. However, low survival may indicate a natural mechanism of selection of the most hardy and genetically sound larvae for further development.
5. Development of predator-proof subtidal ocean nursery cages for smaller clams would free raceways for more spawnings.
6. Refinement of intertidal culture techniques, including research on long-term growth and survival rates of clams maintained in the intertidal, is needed to establish this as a viable alternative to subtidal culturing.

ACKNOWLEDGMENTS

I thank Drs. A. C. Alcala, S. N. Alcazar, R. D. Braley, E. D. Gomez, G. A. Heslinga, A. D. Lewis, J. S. Lucas, J. L. Munro, J. C. Perretta, and T. D. Sleeter for unpublished data and information concerning tridacnid mariculture and Drs. C. R. Fisher, E. M. Fleischmann, R. W. Menzel, and R. K. Trench for comments on the manuscript.

REFERENCES

1. **Rosewater, J.**, The family Tridacnidae in the Indo-Pacific, *IndoPacific Mollusca,* 1, 347, 1965.
2. **Rosewater, J.**, A new species of *Hippopus* (Bivalvia: Tridacnidae), *The Nautilus,* 96, 25, 1982.
3. **Yonge, C. M.**, Functional morphology and evolution in the Tridacnidae (Mollusca: Bivalvia: Cardiacea), *Rec. Aust. Mus.*, 33, 735, 1980.
4. **Fankboner, P. J.**, Self righting by tridacnid clams, *Nature (London),* 230, 579, 1971.
5. **McMichel, D. F.**, Growth rate, population size and mantle coloration in the small giant clam *Tridacna maxima* (Roding), at One Tree Island, Capricorn Group, Queensland, *Proc. 2nd Int. Coral Reef Symp.*, 1, 241, 1974.
6. **Hirschberger, W. A.**, Tridacnid clam stock on Helen Reef, Palau, western Caroline Islands, *Mar. Fish. Rev.*, 42, 8, 1980.
7. **Braley, R. D.**, Reproduction in the giant clams *Tridacna gigas* and *T. derasa in situ* on the north-central Great Barrier Reef, Australia, and Papua New Guinea, *Coral Reefs,* 3, 221, 1984.
8. **Yamaguchi, M.**, Conservation and cultivation of giant clams in the tropical Pacific, *Biol. Conserv.*, 11, 13, 1977.
9. **Richard, G.**, Quantitative balance and production of *Tridacna maxima* in the Takapoto lagoon, *Proc. 3rd Int. Coral Reef Symp.*, 2, 599, 1977.
10. **Jameson, S. C.**, Early life history of the giant clams *Tridacna crocea* (Lamarck), *T. maxima* (Roding) and *Hippopus hippopus* (Linnaeus), *Pac. Sci.*, 30, 219, 1976.
11. **Fitt, W. K., Fisher, C. R., and Trench, R. K.**, Contribution of the symbiotic dinoflagellate *Symbiodinium microadriaticum* to the nutrition, growth, and survival of larval and juvenile tridacnid clams, *Aquaculture,* 55, 5, 1986.
12. **Taylor, J. D.**, Reef associated molluscan assemblages in the western Indian Ocean, *Symp. Zool. Soc. London,* 28, 501, 1971.
13. **Heslinga, G. A.**, personal communication.
14. **Heslinga, G. A.**, Biology and culture of the giant clam, in *Clam Mariculture in North America,* Manzi, J. and Castagna, M., Eds., Elsevier, Amsterdam, in press.
15. **Alcazar, S. N., Solis, E. P., and Alcala, A. C.**, Serotonin-induced spawning and larval rearing of the China clam *Hippopus porcellanus* Rosewater (Bivalvia: Tridacnidae), *Aquaculture,* in press.
16. **Yonge, C. M.**, Mode of life, feeding, digestion and symbiosis with zooxanthellae in the Tridacnidae, *Sci. Rep. Great Barrier Reef Exped.*, 1, 283, 1936.
17. **Kawaguti, S.**, Bivalve mollusks associated with zooxanthellae, *Bull. Kawasaki Coll. Allied Health Professions,* 3, 77, 1983.
18. **Fisher, C. R., Fitt, W. K., and Trench, R. K.**, Photosynthesis and respiration in *Tridacna gigas* as a function of irradiance and size, *Biol. Bull.*, 169, 230, 1985.
19. **Stasek, C. E.**, The form, growth, and evolution of the Tridacnidae (giant clams), *Arch. Zool. Exp. Gen.*, 100, 1, 1961.
20. **Heslinga, G. A., Perron, F. E., and Orak, O.**, Mass culture of giant clams (f. Tridacnidae) in Palau, *Aquaculture,* 39, 197, 1984.
21. **Heslinga, G. A. and Watson, T. C.**, Recent advances in giant clam mariculture, *Proc. Int. Congr. Coral Reefs,* 5, 531, 1985.
22. **Fisher, C. R., Fitt, W. K., and Trench, R. K.**, unpublished data.
23. **Benson, A. A. and Summons, R. E.**, Arsenic accumulation in Great Barrier Reef invertebrates, *Science,* 211, 482, 1981.
24. **Wada, S. K.**, Protandric functional hermaphroditism in the tridacnid clams, *Oceanogr. Mag. Tokyo,* 4, 23, 1952.
25. **Braley, R. D.**, Serotonin-induced spawning in giant clams (Bivalvia: Tridacnidae), *Aquaculture,* 47, 321, 1985.

26. **Stephenson, A.,** The breeding of reef animals. II. Invertebrates other than coral, *Sci. Rep. Great Barrier Reef Exped.*, 3, 247, 1934.
27. **Murakoshi, M. and Kawaguti, S.,** Seasonal variations in gonad weights of the boring clams and harvest regulation periods, *Bull. Jpn. Soc. Sci. Fish.*, 52, 1709, 1986.
28. **Hardy, J. T. and Hardy, S. A.,** Ecology of Tridacna in Palau, *Pac. Sci.*, 23, 467, 1969.
29. **LaBarbera, M.,** Larval and postlarval development of the giant clams *Tridacna maxima* and *Tridacna squamosa* (Bivalvia: Tridacnidae), *Malacologia*, 15, 69, 1975.
30. **Fitt, W. K. and Trench, R. K.,** Spawning, development, and acquisition of zooxanthellae by *Tridacna squamosa* (Mollusca, Bivalvia), *Biol. Bull.*, 161, 213, 1981.
31. **Lucas, J. S.,** personal communication.
32. **Beckvar, N.,** Cultivation, spawning and growth in the giant clams *Tridacna gigas, T. derasa* and *T. squamosa* in Palau, Carolina Islands, *Aquaculture*, 24, 11, 1981.
33. **Crawford, C. M., Nash, W. J., and Lucas, J. S.,** Spawning induction, and larval and juvenile rearing of the giant clam, *Tridacna gigas, Aquaculture*, 58, 281, 1986.
34. **Wada, S. K.,** Spawning in the tridacnid clams, *Jpn. J. Zool.*, 11, 273, 1954.
35. **Gwyther, J. and Munro, J. L.,** Spawning induction and rearing of larvae of tridacnid clams (Bivalvia: Tridacnidae), *Aquaculture*, 24, 197, 1981.
36. **Fitt, W. K., Fisher, C. R., and Trench, R. K.,** Larval biology of tridacnid clams, *Aquaculture*, 39, 181, 1984.
37. **Murakoshi, M.,** Larval and post-larval development of a boring clam, *Tridacna crocea* Lamarck, *Biol. Mag. Okinawa*, 16, 29, 1978.
38. **Taylor, D. L.,** Identity of zooxanthellae isolated from Pacific Tridacnidae, *J. Phycol.*, 5, 336, 1969.
39. **Blank, R. and Trench, R. K.,** Speciation in symbiotic dinoflagellates, *Science*, 229, 656, 1985.
40. **Trench, R. K. and Blank, R.,** *Symbiodinium microadriaticum* Freudenthal, *S. goreauii* sp. nov., *S. kawagutti* sp. nov., and *S. pilosum* sp. nov., gymnodinioid dinoflagellate symbionts of marine invertebrates, *J. Phycol.*, in press.
41. **Fitt, W. K.,** The role of chemosensory behavior of *Symbiodinium microadriaticum,* intermediate hosts, and host behavior in the infection of coelenterates and molluscs with zooxanthellae, *Mar. Biol.*, 81, 9, 1984.
42. **Fitt, W. K.,** Chemosensory response of the symbiotic dinoflagellate *Symbiodinium microadriaticum, J. Phycol.*, 21, 62, 1985.
43. **Muscatine, L.,** Glycerol excretion by symbiotic algae from corals and *Tridacna* and its control by the host, *Science*, 156, 516, 1969.
44. **Goreau, T., Goreau, N. I., and Yonge, C. M.,** On the utilization of photosynthetic products from zooxanthellae and of a dissolved amino acid in *Tridacna maxima* f. *elongata* (Mollusca: Bivalvia), *J. Zool. London*, 169, 417, 1973.
45. **Trench, R. K., Wethey, D. S., and Porter, J. W.,** Some observations on the symbiosis with zooxanthellae among the Tridacnidae (Mollusca: Bivalvia), *Biol. Bull.*, 161, 180, 1981.
46. **Fitt, W. K.,** The effect of different strains of the zooxanthellae *Symbiodinium microadriaticum* of growth and survival of their coelenterate and molluscan hosts, *Proc. Int. Congr. Coral Reefs*, 6, 131, 1985.
47. **Munro, J. L. and Heslinga, G. A.,** Prospects for the commercial cultivation of giant clams (Bivalvia: Tridacnidae), *Proc. Gulf Caribb. Fish. Inst.*, 35, 112, 1983.
48. **Jones, D. S., Williams, D. F., and Romanek, C. S.,** Life history of symbiont-bearing giant clams from stable isotope profiles, *Science*, 231, 46, 1986.
49. **Romanek, C. S., Jones, D. S., Williams, D. F., Krantz, D. E., and Radtke, R.,** Stable isotopic investigation of physiological and environmental changes recorded in shell carbonate from the giant clam *Tridacna maxima, Mar. Biol.*, 94, 385, 1987.
50. **Weingarten, R. and Tuxson, R.,** The giant clam, *Mar. Aquarist*, 8, 32, 1977.
51. **Munro, J. L.,** Giant clams—food for the future?, *ICLARM Newsl.*, 6, 3, 1983.
52. **Pearson, R. G.,** Impact of foreign vessels poaching giant clams, *Aust. Fish.*, 36, 8, 1977.
53. **Dawson, R. F.,** Can Australia expect a resurgence of Taiwanese clam boat activity?, *Aust. Fish.*, 43, 20, 1984.
54. **Dawson, R. F.,** Report on a Study of the Market for Giant Clam Products in Taiwan, Japan, Hong Kong and Singapore, South Pacific Forum Fisheries Agency, 86/37, 1986.
55. **Bryan, P. G. and McConnell, D. G.,** Status of giant clam stocks (Tridacnidae) on Helen Reef, Palau, Western Caroline Islands, *Mar. Fish. Rev.*, 38, 15, 1975.
56. **Wells, S. M., Pyle, R. M., and Collins, N.,** Giant clams, in *The IUCN Invertebrate Red Data Book,* International Union for the Conservation of Nature, Gland, Switzerland, 1983, 97.
57. **Crawford, C. M., Lucas, J. S., and Nash, W. J.,** Growth and survival during the ocean-nursery rearing of giant clams, *Tridacna gigas* 1. Assessment of four culture methods, *Aquaculture*, in press.
58. **Alcala, A. C., Solis, E. P., and Alcazar, S. N.,** Spawning, larval rearing and early growth in *Hippopus hippopus* (Linn.) (Bivalvia: Tridacnidae) *Silliman J.*, in press.,

59. **Alcazar, S. N. and Solis, E. P.,** Spawning and larval development and growth of *Tridacna maxima* (Roding) (Bivalvia: Tridacnidae) *Silliman J.*, in press.
60. **Heslinga, G. A. and Fitt, W. K.,** Domestication of reef-dwelling clams, *BioScience,* 37, 332, 1987.
61. **Perron, F. E., Heslinga, G. A., and Fagolimul, J. O.,** The gastropod *Cymatium muricunum*, a predator on juvenile tridacnid clams *Aquaculture,* 36, 35, 1981.
62. **Heslinga, G. A. and Hillman, A.,** Hatchery culture of the commercial top snail *Trochus niloticus* in Palau, Caroline Islands, *Aquaculture,* 36, 35, 1981.
63. **Lopez M. D. G. and Heslinga, G. A.,** Effect of desiccation on *Tridacna derasa* seed: implications for long distance transport, *Aquaculture,* 49, 363, 1985.
64. **Watson, T. C. and Heslinga, G. A.,** Optimal harvest age for *Tridacna derasa:* maximizing biological production, Publ. Int. Giant Clam Workshop, Aust. Ctr. for Intl. Agr. Res., in press.
65. **Southgate, P. C.,** Biochemical development and energetics of *Hippopus hippopus* larvae, Aust. Ctr. for Intl. Agr. Res., 1988.
66. **Coon, S. L., Bonar, D. B., and Weiner, R. M.,** Induction of settlement and metamorphosis of the Pacific oyster, *Crassostrea gigas,* by L-DOPA and catecholamines, *J. Exp. Mar. Biol. Ecol.,* 94, 211, 1985.
67. **Yonge, C. M.,** Giant clams, *Sci. Am.,* 232, 96, 1975.

Chapter 22

THE GREEN MUSSEL (*PERNA VIRIDIS*) IN THE PHILIPPINES*

Neon C. Rosell

TABLE OF CONTENTS

I. BIOLOGY

A. MORPHOLOGY

The shell is equivalval and wedge-shaped like *Mytilus* but without the anterior adductor, except in young specimens; the lunule is bent inward, forming one or two tooth-like ridges, with both surface and margin smooth.[1] The periostracum is anteriorly dark brownish-green, becoming olive green to light green posteriorly. The posterior adductor muscle is confluent with the posterior byssal retractor muscle scar II; the posterior byssal retractor muscle scar I is widely separated from the latter (Figure 1).

The illustrated specimen, a female with length and height of 113.5 mm and 45.1 mm, respectively, exhibits a small adductor muscle scar on its anterior end.

B. GEOGRAPHIC RANGE

P. viridis, formerly known as *Mytilus smaragdinus* Chemnitz, 1785, and *M. viridis* (L.), 1758, is a Southeast Asian species. Its present known range of distribution is the northeastern Indian Ocean, i.e., from the coasts of India, peninsular Malaysia and Singapore, to the South Chine Sea in Thailand and the Philippines.

In the Philippines, the species is found over almost all of the archipelago, in bays or coves, estuarine mangrove areas from Ilocos Norte in northern Luzon Island to Panguil Bay in Mindanao Island in the South. The biggest mussel-producing areas are in Bacoor Bay within Manila Bay, Sapian Bay in Capiz, Panay Island, and Maqueda Bay in Samar Island.[2]

C. SALINITY AND TEMPERATURE TOLERANCES

Salinity and temperature requirements of adult and larval stages do not differ much. Adults thrive best in 25 to 30°C and 27 to 35‰.[3-4] Larval stages from trochophore to spat (settlement stage) survived and metamorphosed in 26 to 30.5°C and 29 to 30‰. Young[5] reported that according to Tan and Rao, in Singapore and India larvae of *M. viridis* (= *P. viridis*) survived and metamorphosed in 23 to 25°C and 25 to 27.4°C, respectively.

D. TYPES OF SUBSTRATE

P. viridis is an epifaunal species. Larvae are pelagic, while adults are sessile; however, adults do not live on or near the bottom. They live in abundance along the upper water column, settling on various kinds of submerged substrates—wharf pilings, protruding portions of driftwood, bamboo poles, or other material stuck in the mud—attached by byssal threads.

The species prefer hard to semihard and sticky mud or firm and nonshifting bottoms. Highly productive grounds are usually in muddy or sandy substrata.[4]

E. SEXUAL DEVELOPMENT AND SPAWNING

Sexes are separate but not externally dimorphic. Gonads are not discrete organs but rather diffuse gonadal tissues in the mantle. The color of ripe female gonadal tissue is orange red, while that of the male is dirty white to creamy. *P. viridis* becomes sexually mature on reaching a size of 18.5 to 30 mm.[2,3,5] Fertilization is external: the males spawn first with sperm released as dense white streams; eggs are discharged as disconnected orange rods or cords 1 to 2 mm in length, averaging 3.8 to 12 million eggs in number.[5-6]

F. LARVAL DEVELOPMENT

Pelagic larval stages last for about 3 weeks, after which spat start settling, henceforth leading a sessile mode of existence. Ciliated blastula are formed in 3 to 4 h following fertilization. In 29 to 30.5°C and 30‰, succeeding pelagic larval stages are attained: as trochophore in 5 to 6 h, as veliger in 24 to 36 h, as pediveliger in 14 to 15 d, and as ready-to-settle spat on the 17th d.

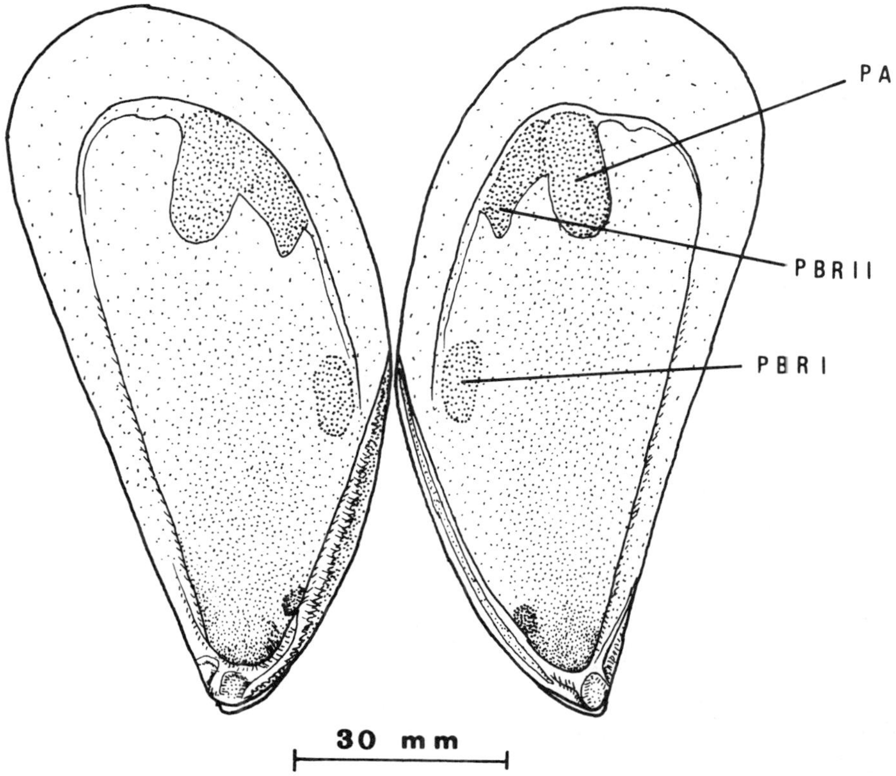

FIGURE 1. *Perna viridis* shells, inside view. Posterior adductor muscle scar (PA); posterior byssal retractor muscle scar (PBR I, II).

When temperature and salinity were lowered to 27.2°C and 28 to 30‰ respectively,[5] pediveliger stage was reached in 22 d. In *Mytilus edulis* (L.), 1758, temperature and salinity reportedly influence larval growth rates as well.[7]

G. GROWTH

The estimated rate of growth in *P. viridis* is 0.5 to 1.0 cm, averaging 0.88 ± 0.04 cm/month.[3] Growth rate decelerates as the animal gets larger and older.

Growth varies with availability of food, salinity, population, location along the water column, and culture method used. Mussels attain marketable size in 6 to 10 months, or when they are 5 to 10 cm in length.[2]

II. FISHERY

A. HISTORY

Before 1950, oyster farmers regarded green mussels only as a nuisance, inimical to oyster cultivation because they competed with oysters for food and space.[8] In early 1950, personnel of the Bureau of Fisheries at Binakayan, Cavite recognized the importance of mussels as a primary crop and as a valuable food source for the increasing human population. Therefore, in 1955, a 300-m^2 mussel farm was established. This is the earliest record of green mussel commercial production in the Philippines. Since then, mussel farming has gone rapidly from a fad to a countrywide commercial endeavor.[2]

B. IMPORTANCE

Green mussels nowadays are widely cultivated and are a popular food item because they have high nutritive value, being rich in protein, fat, carbohydrates, and vitamins and minerals. The shells are used also as shellcraft ornaments and decors.

The production cost of a family-size, stake-method farm of 1000 m^2 is U.S. $986.58*. With two croppings a year, a farmer could derive a gross income of U.S. $1592.10**, yielding a net income of U.S. $605.60. Estimated return per dollar invested is U.S. $1.63.

For a 1-ha (10,000 m^2) stake-method mussel farm, the production cost is U.S. $7939.00*. With two croppings a year, one would have a gross income of U.S. $15,687.80** and a net income of U.S. $7748.78. Estimated return per dollar invested is U.S. $1.92.[2]

Manpower requirement depends on the size of the farm. Several persons are needed in the preparation of materials and in putting up the farm. Once the farm is established, only a caretaker is engaged to inspect the plot periodically. This duty is usually assigned to a family member.

Harvesting and marketing are left to the buyer if the farm owner sells the crop *in situ* on a contractual basis.

C. RECRUITMENT AND METHODS OF CULTURE

There are several kinds of cultches or spat collectors employed for recruitment of mussel spat. Bamboo poles staked in mud and polypropylene or cabonegro ropes with coconut husks inserted in rope laid 5 to 6 cm apart are both regularly used. These collectors are submerged in natural mussel beds hanging from a raft or tied to horizontal bamboo bars anchored to bamboo poles staked in mud.

Clusters of mussel spat 10 to 15 mm in length are transplanted to growing ropes using abaca cloth as binders. More often, transplanting is no longer resorted to, and spat are allowed to grow on the cultches or collectors where they initially settled.

Commonly practiced effective and profitable culture methods are

Staked method—Bamboo poles or other appropriate materials are staked in mud, 1 to 2 m apart along rows and 2 m apart between rows (Figure 2). Prior to staking, bamboo poles are sharpened at their basal end by making a diagonal cut. Holes are punched on the upper end of internodes so that the hollow portion inside will fill with sea water and reduce buoyancy. The poles are then staked in mud at about 0.5 m deep.

Rope-web method—This method makes use of webbed synthetic ropes and bamboo poles (Figure 3). Two 6-m-long synthetic ropes positioned 2 m apart are tied to bamboo poles staked in mud 5 m apart. Another 40-m rope (polypropylene, 12-mm i.d.) is webbed between the two 5-m parallel baseline ropes in zigzag fashion at 40-cm intervals. Pegs are inserted at contact points with baseline ropes to preclude sliding of zigzag lines when top heavy with attached mussels.[3]

Bamboo raft method—Once again, bamboo poles are used as substrate materials (Figure 4). Here, 10 to 12 poles lying side by side are bundled together into a single layer by tying or punching bore holes 5 m apart. Long wooden pegs are driven through these holes, and the outermost poles are tied to the end of the raft to secure the poles. Holes are punched in some of the poles to reduce buoyancy and hold the bundle submerged 1 to 1.5 m below the surface. If the above procedure is not sufficient to submerge the bundle to the desired depth, additional weights (such as large rocks) are placed on the raft.

Raft hanging method—This consists of a platform raft mounted on buoys made of metal drums, plastic drums, Styrofoam blocks, or ferroconcrete buoys (Figure 5).[3] Culture ropes or other appropriate materials are hung from the platform 0.5 m apart. Culture ropes must reach 4 m below the surface, in waters 5 to 8 m deep, and are weighted at the end to

* *Perna viridis* (Linne), 1758; family Mytilidae; common name, green mussel; Philippine name, "tahong".

* Based on an exchange rate of P20.50 to U.S. $1.00.

** Based on 1981 figures.

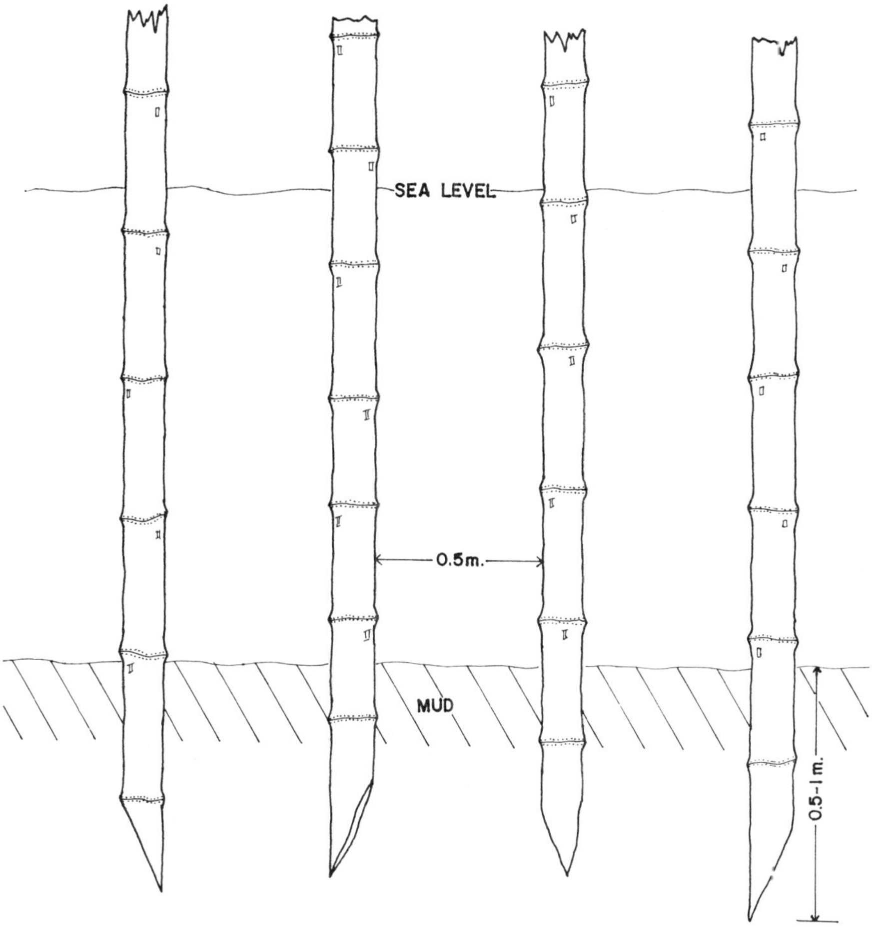

FIGURE 2. Staked method of mussel culture.

maintain the ropes in vertical positions. Grow-out could either be from spat recruited *in situ* or spat transplanted from highly productive and crowded areas. Rafts are set in place by rope cables anchored to heavy or fixed objects on the seabed.

The most productive portion of the water column in estuaries or mangrove inlets is 2 to 4 m below the surface. Test panels showed highest biomass production of sessile organisms in this section of the water column. Rafts go up and down with tidal changes; therefore, hanging culture ropes are maintained, at all times, in a fixed location within the most productive zone of the water column. In this regard, raft method of culture is most advantageous in terms of optimal food abundance conditions; besides, mussels are not within reach of or exposed to benthonic predators. Furthermore, this method prevents the aggravation of bottom siltation that can lead to rapid deterioration of natural mussel beds.

D. METHODS OF HARVESTING

Mussle-laden stakes or rope-web culture ropes are uprooted or untied, respectively, from anchor bamboo poles and loaded on a boat or raft. Clusters of mussels are stripped from bamboo poles with a spade or any hard object. Those on rope-web culture ropes are removed by cutting off byssal threads with sharp-bladed instruments.

In other areas, harvesting is done underwater by free diving using mask and snorkel or air compressor. Mussels are selectively harvested by cutting off their byssal threads. Only

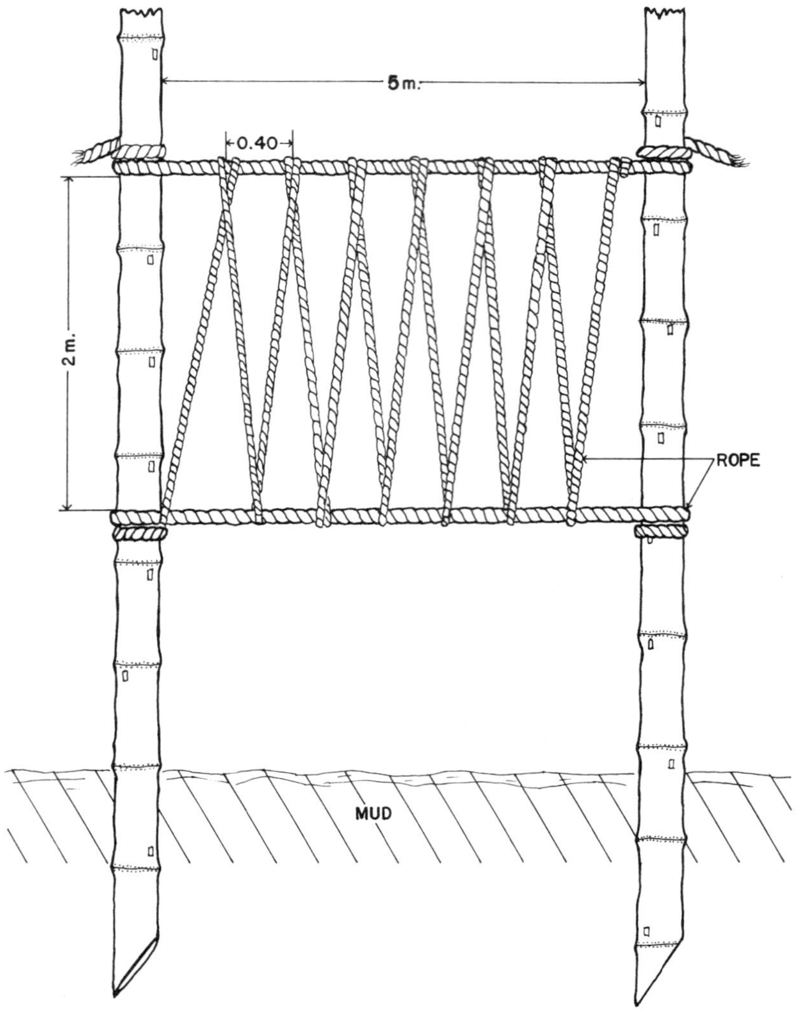

FIGURE 3. Rope-web method of mussel culture. (Redrawn from The Philippine Recommends for Mussels and Oysters, PCARRD Tech. Bull. 26A, 1983, 1.)

the bigger mussels are taken, leaving undersized ones for the next cropping. Harvest from a single bamboo stake averages 6 to 8 kg in the first cropping and 2 to 4.4 kg in the next.[2]

Detaching mussels by cutting off byssal threads is a better method of harvesting them than stripping mussel clusters with hard objects. Cutting the byssal threads avoids damage to the byssal gland and prolongs the life of harvested mussels.

E. USES AND PROCESSING

Green mussels are harvested for their meat and shells. Shells are used as raw materials in shellcraft industries, where they are made into ornaments and other industrial products. Since the shells contain 90% calcium, they may also be ground into fine powder and used as an additive to poultry feed.

Mussel meat is used as food and prepared in various delectable recipes. It may be processed by freezing, smoking, canning, bottling, drying, or pickling.[2]

F. CONSTRAINTS

Potential mussel farming areas are too far from population centers. Roads and transportation from farm to market are also inadequate.

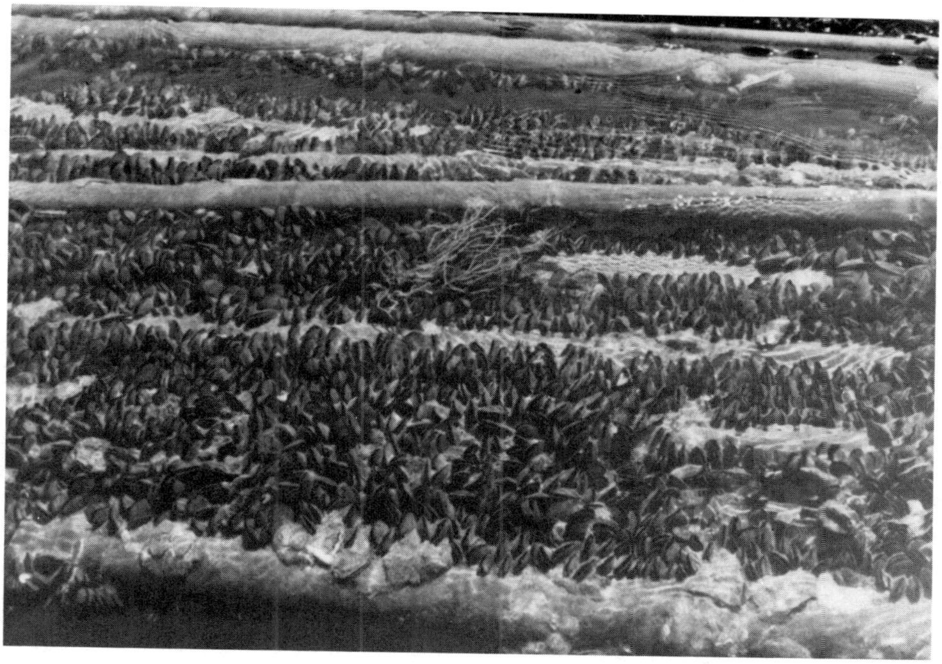

FIGURE 4. Bamboo raft method of mussel culture. Densely recruited mussels are shown on submerged bamboo raft in Aguisan River, Himamaylan, Negros Occidental, July 22, 1981.

FIGURE 5. Raft hanging method of mussel culture. Clusters of green mussels attached to rope suspended from a platform raft on SEAFDEC Mussel Farm, Himamaylan river, Himamaylan, Negros Occidental, July 22, 1981.

Another problem is the speedy disintegration and collapse of the bamboo poles that are used as substrates for settlement on account of mollusk-boring organisms such as teredinids (*Teredo bartschii* Clapp, 1923, and *Lyrodus pedicellatus* (Quatrefages), 1849), and pholads (*Martesia fragilis* Verill and Bush, 1890 and *M. striata* (Linne), 1758).

Existing and potential farming areas also suffer deterioration since bays and estuaries serve as sinks for harmful effluents discharged by industries and agricultural activities. There has been a reduction or elimination of natural mussel grounds as well, due to reclamation of mangrove areas for fishponds and human settlements.

G. MANAGEMENT

Restrictions on gathering, such as a minimum allowable size and a quota on volume collected, are not imposed inasmuch as the species is cultured in farms and not obtained from wildstock populations. Permits are required when putting up a culture farm to avoid conflict of interest.

Depuration to rid mussels of harmful microorganisms is desirable but not practiced. Heavy metal pollution, like that from mercury, has not reached critical levels. Metal body burden of those tested from culture farms is still within permissible levels; however, *P. viridis* is known to bioaccumulate rapidly when exposed to 10 ppb mercury concentration: metal body burden can increase by as much as 3426% wet weight in 15 d of exposure.[8]

III. RESEARCH AND DEVELOPMENT

A. BIOLOGICAL TECHNIQUES

The life history of *P. viridis* has been documented. Seed production under laboratory conditions has been successfully demonstrated, but the technology for mass production of spat in hatcheries is not yet available. At the moment, establishment of large-scale hatcheries is not essential because spat from natural beds are still plentiful and sufficient to meet existing demands. Nonetheless, development of a hatchery technology for mass production of mussel spat must be seriously evaluated.

Spatfall prediction requires further investigation and refinement of technique in order to determine the appropriate timing in placing spat collectors to maximize recruitment. Culture methods must be developed that will forestall environmental degradation of natural mussel grounds.

B. FUTURE DEVELOPMENT

Research efforts are to be directed to (1) expanding existing productive natural grounds by transplanting breeding stocks or spat to new suitable areas, (2) transplanting spat in cultch from densely populated areas to areas with scanty spat, and (3) determining effective population size to be set aside as breeding stocks or a gene pool from the existing crop to ensure a continuous supply of spat for succeeding cropping periods.

REFERENCES

1. **Soot-Ryen, T.,** Mollusca (Bivalvia) in *Treatise on Invertebrate Paleontology,* Moore, R. C., Ed., Geol. Soc. Am. and University of Kansas, 1, N271, 1969.
2. Mussel and Oyster Committee, The Philippine Recommends for Mussels and Oysters, PCARRD Tech. Bull. 26A, 1983, 1.
3. **Walter, C.,** Reproduction and growth in the tropical mussel *Perna viridis* (Bivalvia: Mytilidae), Kalikasan, Philipp. *J. Biol.,* 11, 83, 1982.

4. **Mussel and Oyster Committee,** The Philippine Recommends for Mussels and Oysters, PCARRD, Mimeographed, 1977, 36.
5. **Young, A. L.,** Larval and post-larval biology of the tropical green mussel, *Mytilus smaragdinus* Chemnitz, M. S. thesis, University of the Philippines, Diliman, Quezon City, 1978.
6. **Escritor, G. L.,** *Tahong,* Phil. Fish. Yearbook, 1966, 35.
7. **Hrs-Brenko, M. and Calabrese, A.,** Combined effects of temperature and salinity on survival and growth of larvae of *Mytilus edulis, Mar. Biol.*, 4, 224, 1969.
8. **Rosell, N. C.,** Uptake and depuration of mercury in the green mussel, *Perna viridis* Linnaeus (Bivalvia: Mytilidae), *Phil. J. Sci.*, 114, 1, 1985.

Chapter 23

THE SLIPPER-SHAPED OYSTER (*CRASSOSTREA IREDALEI*) IN THE PHILIPPINES

Neon C. Rosell

TABLE OF CONTENTS

Crassostrea iredalei (Faustino), 1932; family Ostreidae; common name, slipper-shaped oyster; Philippine name, ''talabang tsinelas''.

I. BIOLOGY

A. MORPHOLOGY

The shell is inequivalval, as the upper valve is small, thin, flat, and scaly at the margins, whereas the lower or left valve is slightly excavated, the hinge toothless, and the inside margin smooth. The upper valve is usually yellowish brown, sometimes with fainter radiating rays of purple, and the lower valve generally has a purple tint. The interior of the shell is chalky white and the adductor muscle scar, purple. The shell shape, although variable, is largely dependent on the surface of the substrate and the oyster's mode of attachment (Figure 1). The length varies from 30 to 80 mm.[1,2]

This is one of the most common and widely cultivated species found near Manila and other parts of the archipelago. Other species of oysters cultivated together with *C. iredalei* are *C. malabonensis* (Faustino), 1932, the subtrigonal-oblong oysters, ''talabang kukong kabayo''; *C. palmipes* (Sowerby), 1867, the palm-rooted oysters, ''talabang pulid-pulid''; and *Saccostrea cucullata* (Born), 1871, the curly oysters, ''talabang kulot''.

Oysters, regardless of species, are generally referred to as ''talaba'' (Tagalog), ''sisi'' (Visayan), or ''tirem'' (Ilocano).

B. GEOGRAPHIC RANGE

C. iredalei, so far, has not been reported to occur outside the Philippine archipelago. Based on available information, the species is indigenous and endemic to the country.

The species is a familiar inhabitant of tidal rivers, channels, gulfs and land-locked bays. These areas receive enough freshwater drainage to carry the nutrient-rich terrigenous deposits that fertilize these areas and enhance primary productivity. All other ecological factors being favorable, *C. iredalei,* as a filter-feeding organism, would thrive best in areas of high primary production.

C. iredalei is widely distributed in the Philippines. Although it is particularly abundant in certain localities—notably, Northern Luzon, Lingayen Gulf, Manila Bay, Batangas Bay, and the coasts of Quezon and Sorsogon, Catbalogan, Northern Leyte, and the Negros, Panay, and Palawan islands—the entire coast serves as its natural habitat.

C. SALINITY AND TEMPERATURE TOLERANCES

Freshets periodically cause wide salinity fluctuations of the coastal zones that are customary habitats of *C. iredalei.* In areas where wild and cultivated populations occur, ambient salinities range from 17.0 to 29.0‰ and temperatures from 18.0 to 28.0°C. However, under laboratory conditions, early embryonic and larval stages survived and metamorphosed in salinities of up to 32.0‰ and temperatures to 30.0°C.[3,4]

D. TYPE OF SUBSTRATES

C. iredalei is an epifaunal species: larval stages are pelagic whereas adults are sessile. They attach by their lower left valve, which partly cements to substrates on the seabottom or on a wall of submarine rocks and other structures sticking out of the seabed. Attachment is made possible by byssal gland secretions deposited on the substrate by settling spat. The left valve is then pressed against the secretions, which harden rapidly. Vertical surfaces of fixed submarine objects are the most desirable habitat for attachment, as there the animals are not exposed to heavy siltation, which could be detrimental to their survival. Once oysters are dislodged, they are not capable of reattachment, because the byssal gland degenerates after effecting settlement.

FIGURE 1. *Crassostrea iredalei* shells.

The species is frequently encountered in sheltered bays, coves, or estuaries with hard and sticky blue mud, or semisandy or hard nonshifting eroding bottoms. Furthermore, the species thrives best with adequate water circulation along a water column rich in both phyto- and zooplankton and organic detritus.[3,5,6]

E. SEXUAL DEVELOPMENT AND SPAWNING

C. iredalei are protandric hermaphrodites. However, they not only change from male to female, as in most true protandric hermaphrodites, but also change back from female to male.[7]

Individuals measuring 50 to 80 mm are sexually mature. Ripe gonads are creamy white in appearance. The species was observed to spawn at temperature and salinity ranging from 18 to 27°C and 17 to 26‰, respectively. However, under laboratory conditions, spawning took place at temperatures and salinities up to 28°C and 32‰, respectively.

Spawning is seasonal but varies from locality to locality; nevertheless, in general it falls within the months of May to September.[4,5,7,8] Spat start settling in August and early September, and ready-to-settle spat are attained in 20 to 22 d following fertilization.[4] Considering the length of the pelagic larval stage, first settlement would be in early or late June, depending on the date in May that oysters started spawning.

Fertilization is external. During spawning, eggs are discharged aided by rhythmic contractions of the adductor muscles. Unfertilized eggs are teardrop shaped, ranging in size from 47 to 50 μm; however, following fertilization, they assume a spherical shape. A spawner can release millions of eggs every spawning. Sperm is shed continuously as dense white streams lasting anywhere from a few minutes to 1 h.[4,5,7]

F. LARVAL DEVELOPMENT

C. iredalei has pelagic larval stages which last for approximately 20 to 22 d. Under controlled conditions, larvae develop at water temperatures and salinities of 26.5 to 30°C and 30 to 32‰, respectively.[4] Blanco and Montalban,[5] however, reported that larvae develop at a temperature of 15°C and salinities of 17 to 26‰.

The ciliated gastrula stage is attained in 5 to 6 h following fertilization. The trochophore stage is reached in 12 to 15 h, the straight-hinge veliger stage in 22 to 26 h, and the pediveliger with lengths of 216 to 275 μm in 16 to 21 d. Ready-to-settle spat have lengths greater than 270 μm.[4]

G. GROWTH

Oyster spat start settling when their size is greater than 270 μm. Spat settle in abundance at 84 cm from datum level, and rates of growth are more or less uniform at this level up to 2 m below.

Rates of growth depend on several factors, including sufficiency of food organisms, size of population, location within the water column, and method of culture used. Generally, oysters grow faster when not exposed to the harmful effects of heavy siltation.[9]

In fairly favorable environments, spat grow to a marketable size of 6.5 to 12.5 cm in about 7 to 12 months. Bacoor Bay spat planted in May, June, July, and August are harvested in December, January, February, and March, respectively.[5,6,7]

II. FISHERY

A. HISTORY

The earliest available record on oyster culture in the Philippines was in 1931. In this year, an oyster farm was established at Hinigaran, Negros Occidental, employing the broadcast method of culture. In this, the oldest traditional method, old disarticulated oyster shells are scattered on solid nonshifting seabed where oyster spat are known to occur. It is the simplest and most economical method, but it is less productive than other methods.[5]

Modern methods were introduced in May 1935 by the then-Bureau of Science at Binakayan, Cavite.[7] Henceforth, the bureau initiated research and development to evolve more efficient, economical, and productive culture methods to "help oyster farmers develop their oyster industry along scientific methods". Since then, the oyster industry has undergone tremendous expansion in terms of farm size and has transformed areas of potential cultivation into productive sites.

B. IMPORTANCE

Commercial cultivation of oysters, nowadays, is a highly profitable industry. It caters to local markets and recently has begun to cater to foreign markets as well.

Oyster farming has augmented the incomes of finfish farmers. Because of this additional income, rural communities have received a "shot in the arm", raising the living standards of rural coastal communities.

Culturing oysters, in most instances, is only a family enterprise. Maintenance and harvesting of crops requires the labor of very few people. Inasmuch as oysters are generally sold fresh in their shells, shucking, which is a labor intensive operation, is seldom or rarely undertaken.

Oyster production in 1977 was only 33 t. This has been consistently increasing ever since, and by 1981 production rose to 579 t.[10] Current (1988) cost of a kg of fresh unshucked oysters is U.S. $0.68,* while shucked oyster meat costs U.S. $2.44.

Production cost of a 5000-m^2 oyster farm, stake method, is U.S. $949.52*. Estimated gross income in 1 year is U.S. $1158.54, giving farmers a yearly net income of U.S. $209.02. Return per dollar invested is U.S. $1.22. With a farm of the same size using the hanging method, total production cost is U.S. $2341.47 and the gross income U.S. $6690.74. This gives a farmer a yearly net income of U.S. $3349.27, and a return per dollar invested of U.S. $2.43.[3]

* Based on P20.50 to U.S. $1.00 foreign exchange rate.

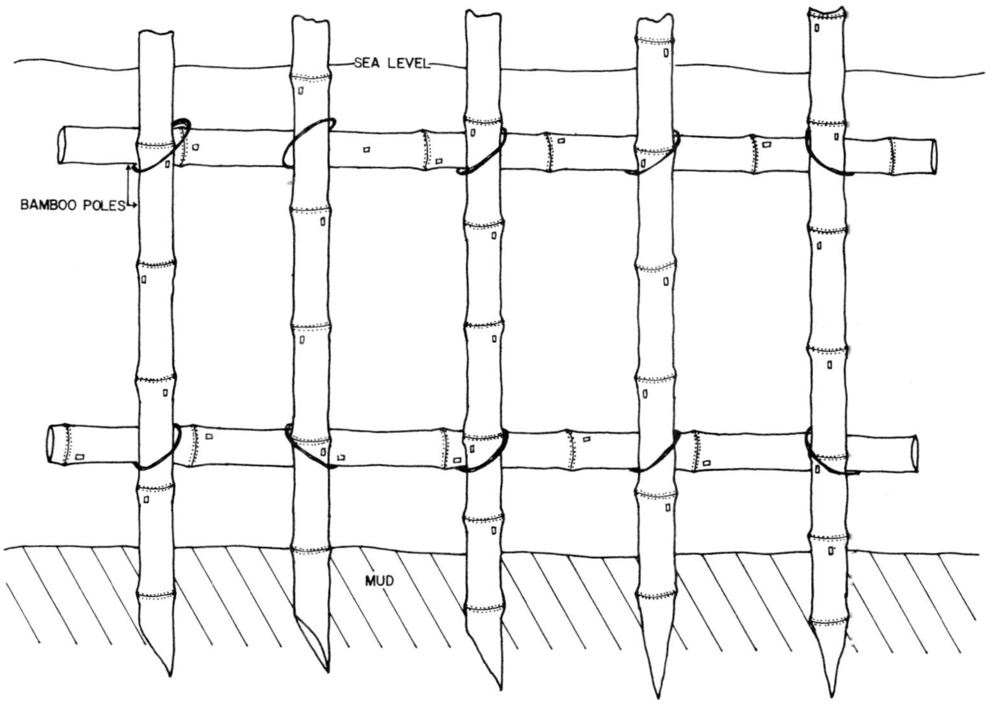

FIGURE 2. Schema of stake method of oyster culture using bamboo poles as substrates.

C. RECRUITMENT AND METHODS OF CULTURE

Oyster spat are recruited from the wild. These spat are either raised to marketable size in the same locality or transplanted to other suitable sites for grow-out.

Cultch are made of different materials. Most common are old uncooked empty oyster, scallop, and clam shells placed in wire bags or baskets or threaded on wire, monofilament nylon cords, cabonegro ropes, or small round rattan stems, and bamboo poles staked in mud.[3,5,6-9]

Recommended culture methods are[3]

Stake method—This is the most common, often using bamboo poles staked in mud 0.5 m deep and 2 m apart (Figure 2).

Hanging method—This is considered both effective and practical (Figure 3). Oyster or coconut shells are threaded on synthetic ropes 10 cm apart separated by knots, loops, or bamboo tubes. Lines with threaded shells are tied at 25- to 30-cm intervals to bamboo rafts or to horizontal bamboo pole bars attached to a series of bamboo posts staked in mud. Length of each line depends on depth of water, but it should not touch the bottom during low tides, particularly if the culture module is a floating raft.

Broadcast method—This is the traditional method. It requires a substratum of hard, semihard, sandy, or eroding nonshifting bottom. Old oyster, scallop, and clam shells, stones, or other solid objects are scattered in areas where spatfall is known to occur.

D. METHODS OF HARVESTING

The manner of harvesting varies with the type of culture method used.[3,5,6] For the stake method, bamboo pole stakes are pulled up and loaded in a small boat or banca. Oysters are detached by knocking them off with an iron rod or any hard object. Bamboo poles not seriously damaged by borers may be reused in another cropping; those that have deteriorated and are weakened are discarded and replaced.

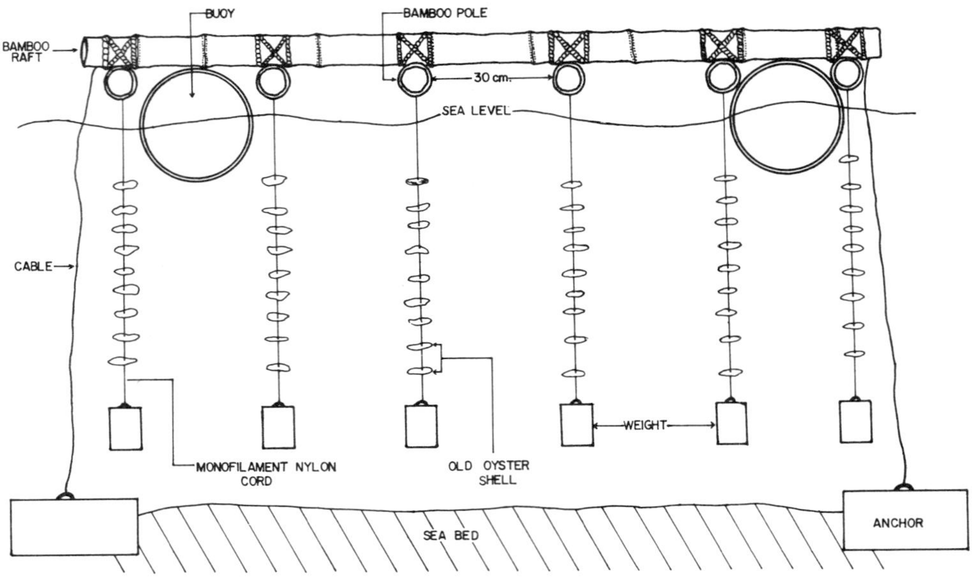

FIGURE 3. Schema of hanging method of oyster culture using threaded old oyster shells on monofilament nylon cord tied to a floating bamboo raft mounted on buoys.

In the hanging method, harvesting is quite easy. Strings with threaded shells are pulled up and loaded into a raft or boat. With a pair of shears, oyster clusters are unfastened, together with the substrate to which they are adhering.

In the broadcast method, oysters are raked or dredged with a wire scoop net. Selective harvesting is practiced: smaller ones are left behind and allowed to grow until the next cropping.

Whatever the method of harvest, the harvested oysters are then put in baskets, dipped in water, and vigorously shaken to remove silt and epizoic organisms.

E. USES AND PROCESSING

Shells and meat of oysters are both economically important. The shells are sources of raw materials for the manufacture of poultry grit, paint, and lime for construction and other industrial purposes. Disarticulated shells are either employed as oyster cultches or processed as fertilizer.

Oyster meat is a valuable food item, a good source of protein, minerals, and vitamins. It is in great demand either fresh or preserved. Fresh oyster meat is incorporated in delicious items like stews, chowders, omelets, or baked oyster dishes. It is processed by icing, freezing, salting, and canning.[11]

F. CONSTRAINTS

One has to contend with several factors when putting up an oyster farm. Borers may attack bamboo poles, and clusters of oysters readily fall off from rotten bamboo pole stakes. Fouling organisms, siltation, and sedimentation are serious problems with which oyster farmers are confronted.

Foulers are nuisance organisms on culture modules. Sponges, hydroids, barnacles, other mollusks (such as *Pteria* spp.), carnivorous animals such as oyster drill (*Thias* sp.), conch (*Melongena* sp.), and starfish are pests to oyster beds.

Moreover, there are the increasing pollution problems from oil spills and other heavy metals from industries. Oysters are known to bioaccumulate rapidly trace metals that could be passed on to higher trophic levels.[12] In addition, typhoons or very strong winds destroy oyster farms by wave action.

G. MANAGEMENT

Oyster seed depletion in the wild has not yet reached critical levels. Enough spat are still available in localities with large populations. It is heartening to see oyster farm operators, in spite of the abundance of wild spat, setting aside a portion of their farm unharvested as broodstock or a gene pool for succeeding croppings.

Establishment of an oyster farm requires licensing from local government units to avoid conflicts among oyster farmers. Culture methods which do not promote sedimentation are highly recommended.

III. RESEARCH AND DEVELOPMENT

A. BIOLOGICAL TECHNIQUES

Documentation of the early developmental stages up to spat settlement in *C. iredalei* has been achieved. As of the moment, there is no imperative need for the establishment of oyster hatcheries. Seedlings from the wild are still sufficient to supply the needs of existing oyster farms. Research is geared toward discovering better cultch materials from the standpoint of spat preferences, durability, and economy.

Spatfall forecasting methods need improvement to maximize recruitment. The introduction of spat on cultches or mature breeders to potential sites devoid of natural populations should be stimulated and promoted. An appropriate size for the broodstock or gene pool relative to the farm size needs to be determined. Finally, depuration procedures for microorganisms and heavy trace metals must be seriously considered and developed.

REFERENCES

1. **Faustino, L. A.,** Recent and fossil shells from the Philippine Islands, *Philipp. J. Sci.*, 49, 543, 1932.
2. **Talavera, F. and Faustino, L. A.,** Edible mollusks of Manila, *Philipp. J. Sci.*, 50, 1, 1933.
3. Mussel and Oyster Committee, The Philippine Recommends for Mussels and Oysters, PCARRD Tech. Bull. (26A), 1983, 1.
4. **Ver, L. M. M.,** Early development of *Crassostrea iredalei* (Faustino, 1932) (Bivalvia: Ostreidae), with notes on the structure of the larval hinge, M. S. thesis, University of the Philippines, Diliman, Quezon City, 1981.
5. **Blanco, G. J. and Montalban, H. R.,** Scheme of oyster farming in the Philippines, *AIL* 17, 8, 1955.
6. **Villaluz, D. K.,** Oyster farming, *Philipp. J. Sci.*, 65, 303, 1938.
7. **Blanco, G. J., Villaluz, D. K., and Montalban, H. R.,** The cultivation and biology of oysters at Bacoor Bay, Luzon, *Philipp. J. Fish.*, 1, 35, 1951.
8. **Young, A. L. and Travena, R.,** On predicting the setting season of the oyster, *Crassostrea iredalei* in Himamaylan River, Negros Occidental, Kalikasan, *Philipp. J. Biol.*, 12, 107, 1983.
9. **Villaluz, D. K.,** Vertical distribution of oyster spat in Bacoor Bay, Cavite Province, *Philipp. J. Sci.*, 70, 375, 1939.
10. **Flores, E. C.,** Harvesting of marine molluscs, in *Status of Mollusc Resources and Prospects for Development*, Proc. PCARRD, Vol. 52, 1987, 68.
11. **Guevara, G.,** Post harvest technology for molluscs, in *Status of Mollusc Resources and Prospects for Development*, Proc. PCARRD, Vol. 52, 1987, 102.
12. **Cunningham, P. A. and Tripp, M. R.,** Accumulation and depuration of mercury in the American oyster *Crassostrea virginica*, *Mar. Biol.*, 20, 14, 1973.

Chapter 24

THE PACIFIC OYSTER, *CRASSOSTREA GIGAS*, IN AUSTRALIA

Trevor G. Dix

TABLE OF CONTENTS

I. BIOLOGY

The general biology of the Pacific oyster, *Crassostrea gigas* (Thunberg) (family Ostreidae), is described elsewhere in this volume. The growth of Pacific oysters in Tasmania, Australia, has been studied,[1,3] as have growth rates in a hypersaline pond in South Australia.[4] Commercial size (about 75-mm shell length) is reached at an age of near 18 months.

II. FISHERY

A. HISTORY

Pacific oysters were introduced into Australia by the Division of Fisheries, CSIRO. Sea shipments in 1947 and 1948 were followed by air consignments in 1951 and 1952.[5,6] The oysters established breeding populations in the north of Tasmania, large colonies becoming established in the Tamar River. Early farming simply involved collection of naturally set oysters, but during the 1960s industry began to collect oysters on sticks[7] using methods similar to those used for Sydney rock oysters. Brief interest was shown in collection of spat on scallop shell rens,[8] but sticks were employed for the bulk of catch. These were set out each summer and transported during autumn and early winter to growing areas throughout Tasmania.[1] Small quantities were sent to South Australia.

Spatfalls proved irregular in the Tamar River, and in 1978 the writer established a pilot-scale oyster hatchery at the Fisheries Research Laboratory, Taroona, Tasmania. After 2 years of operation, oyster farmers established their own hatchery in 1980, and two further hatcheries began operating in 1985. Virtually all seed is now supplied to industry by the hatcheries.

B. IMPORTANCE

Pacific oysters are produced mainly in Tasmania, Australia, with limited production on several leases in South Australia. Tasmanian production has shown steady growth (Table 1) and is restricted primarily by shortages of seed stock, first from the Tamar and in the early 1980s from the hatchery.

The industry is most important aquaculture activity in Tasmania, and about 100 farms are leased to produce.

C. RECRUITMENT AND METHODS OF CULTURE

Almost all culture is with single microcultch (''cultchless'') spat set in the hatchery on ground scallop-shell particles. Spat are raised in upwell nursery systems until 3 to 4 mm in length. They are then grown to 10 to 15 mm either in larger floating upwell nurseries or in fine mesh seed boxes.

The 10- to 15-mm seed continue their growth in a variety of containers including soft mesh polyethylene bags (similar to the French pouches). Larger oysters (>40 to 50 mm) continue growth in similar but larger containers or in open baskets or trays either on wooden intertidal racks or suspended below long-lines.

D. METHODS OF HARVESTING

Oysters in their growing containers are brought ashore by boat and graded at the oyster farmer's shed (depot). Market-size oysters are usually packed in labeled bags for shipment to buyers.

E. USES AND PROCESSING

Almost all Pacific oysters are eaten in Australian restaurants, where they are supplied live or more frequently ''opened on the half-shell''. Opened oysters are either fresh or fresh frozen. Small quantities are processed into seafood chowders.

TABLE 1
Production and Value of Pacific Oysters in Tasmania, Australia

Year	Production[a] (t, live weight)	Value (thousand Australian dollars)
1974/75	105	60
1975/76	94	50
1976/77	149	92
1977/78	138	106
1978/79	116	110
1979/80	105	109
1980/81	190	237
1981/82	353	483
1982/83	569	853
1983/84	645	1066
1984/85	846	1608

[a] Production is often expressed in million dozen oysters. An approximate conversion is 1 million dozen = 900 t.

F. CONSTRAINTS

Until recently, shortages of seed have been the major constraint to industry development. With three hatcheries now producing, this constraint has been overcome.

Diseases have not been found in Pacific oysters in Tasmania, and it is rare to find a dead oyster on an oyster farm. Minor losses have been caused by predators, including flatworms and fish. Mudworms are generally not a problem as oysters are not grown on the bottom.

In the summer of 1985 to 1986, the dinoflagellate *Gymnodinium catenatum* Graham caused toxin accumulation in oysters and mussels in parts of southeast Tasmania. Harvesting was restricted until toxin levels decreased.

G. MANAGEMENT

Oyster growing and harvesting is permitted in Tasmania only in areas with a lease or permit (Fisheries Amendment [Marine Farming] Act 1982). Areas are tested to establish a classification under the Tasmanian Shellfish control Programme (TSSCP, very similar to the U.S. N.S.S.P.). Handling of shellstock including the standard offshore facilities is subject to codes under the control program.

Growers must possess licenses, and their vessels (punts, barges) are licensed. Processors are licensed, and those exporting product are subject to federal regulations.

Leases are not granted in areas known to be subject to pollution by human or industrial wastes.[9] There has been no attempt to limit the number of leases, although public opinion is resulting in restrictions on further leasing. A business plan must be submitted with each new application.

III. RESEARCH AND DEVELOPMENT

A. BIOLOGICAL TECHNIQUES

Industry development has been characterized by the combined efforts of government and industry in developing techniques for growing single oysters, which were unknown to Tasmanian growers prior to 1980. Evolution of techniques for growing and handling is now done by industry and includes mechanized system development, e.g., graders

Hatchery techniques to produce triploid spat are under current investigation, as are systems for setting "eyed larvae" and production of native oysters, *Ostrea angasi* Sowerby.

Biological investigations by the government currently center on monitoring of occurrence and the life cycle of dinoflagellates (Department of Sea Fisheries and CSIRO).

B. FUTURE DEVELOPMENT

With adequate supplies of spat, growing of Pacific oysters in Tasmania is set to expand,[10] and it is considered that production will be about 50 million per year by 1990 (at current prices this represents nearly $10 million first-sale value). Domestic markets should absorb most of the production, although export is being considered.

Although the current farmers employ modern culture techniques, it is certain that methods that have evolved in the last 7 years will continue to be changed as experience increases.

REFERENCES

1. **Sumner, C. E.,** Growth of Pacific oysters, *Crassostrea gigas* Thunberg, cultivated in Tasmania. I. Intertidal stick culture, *Aust. J. Mar. Freshwater Res.,* 31, 129, 1980.
2. **Sumner, C. E.,** Growth of Pacific oysters, *Crassostrea gigas* Thunberg, cultivated in Tasmania. II. Subtidal culture, *Aust. J. Mar. Freshwater Res.,* 32, 411, 1981.
3. **Dix, T. G.,** Growth and survival of hatchery produced oysters, *Crassostrea gigas* in Tasmania, in preparation.
4. **King, M. G.,** Cultivation of the Pacific oyster *Crassostrea gigas* in a non-tidal hypersaline pond, *Aquaculture,* 11, 123 1977.
5. **Thomson, J. M.,** The acclimatization and growth of the Pacific oyster *Gryphaea gigas* in Australia, *Aust. J. Mar. Freshwater Res.,* 3, 64, 1952.
6. **Thomson, J. M.,** The naturalisation of the Pacific oyster in Australia, *Aust. J. Mar. Freshwater Res.,* 10, 144, 1959.
7. **Wilson, R.,** Oyster farming spreads to Tasmania, *Aust. Fish.,* 29(1), 7, 1970.
8. **Dix, T. G.,** Farming the sea, in *Resources of the Sea,* Banks, M. R. and Dix, T. G., Eds., Royal Society of Tasmania Symposium, November 1974, 1975, 93.
9. **Thrower, S. J. and Eustace, I. J.,** Heavy metals in Tasmanian oysters in 1972, *Aust. Fish.,* 32(10), 7, 1973.
10. **Dix, T. G.,** Tasmanian oyster farming finally set to grow, *Aust. Fish.,* 45(11), 21, 1986.

Chapter 25

NEW ZEALAND SCALLOPS, *PECTEN NOVAEZELANDIAE* REEVE

M. F. Bull

TABLE OF CONTENTS

I. BIOLOGY

Pecten novaezelandiae (common names: scallop, New Zealand scallop, queen scallop, tipa [Maori]) is endemic to New Zealand and is the largest pectinid in the country. Apart from some recent experimental fishing of the deep water fan scallop (*Chlamys delacatula*), it is also the only commercially exploited scallop species in New Zealand. It has the typical asymmetric shape of its genus, with convex right valve and flat or slightly concave left valve, and is also typical in being a functional hermaphrodite with separate male and female sections of the gonad. The shell is usually white, sometimes pink or brown, and both valves are strongly ribbed. Rib numbers range from 17 to 22 per valve. The species occurs sporadically around the whole New Zealand coastline including Stewart Island and the Chatham Islands and, in three areas of the country (Nelson/Marlborough, East Coast Auckland, and East Coast Northland), is abundant enough to support moderate-sized commercial fisheries (Figure 1).

Largely on the basis of shell characteristics such as number and shape of ribs, *P. novaezelandiae* has been subdivided into a northern and a southern subspecies *P. n. novaezelandiae* and *P. n. rakiura,*[1] but there is no published evidence of any physiological or ecological differences between the two. The species occurs on a wide variety of substrates in semiestuarine and coastal waters from low tide level to at least 90 m and is not associated exclusively with any single bottom community.[2] The main commercial beds of the Nelson/Marlborough fishery occur in an open bay situation in 10 to 25 m of water on a soft muddy substrate, whereas on the North Island beds the substrate is predominantly hard sand.

The life cycle of *P. novaezelandiae* is similar to that of other members of the genus. Juveniles usually become mature at the end of their first year but probably contribute little to the spawning pool until the end of their second year. Observation of recently spent adults, presence of *Pecten* larvae in plankton samples, and records of spat settlement on artificial collectors indicate that in the Nelson/Marlborough area some scallops spawn from at least August to April. Spat settlement in this area is, however, usually highest from November to January. The data available for other areas of the country suggest a peak of settlement in early summer is probably widespread.[3,4] At metamorphosis, late stage larvae attach themselves by byssus threads to filamentous material such as *Zostera* debris, algae, and hydroids on the seabed and can also be caught on artificial collectors such as synthetic rope and mesh. This attachment is usually retained until the spat are at least 5 mm in shell length, although smaller spat are sometimes found free living on the seabed.

Studies of size frequencies and measurement of tagged scallops indicate growth rates differ widely between populations in different areas. In the Nelson/Marlborough area, it is estimated that the majority of scallops on the commercial beds reach harvestable size (100 mm in length) about $2^1/_2$ years after settlement, but instances have been recorded in which this size was reached in 18 months, and in other areas scallops have reached only 95 mm after $3^1/_2$ years. The longevity of the species in the wild is uncertain as determination of aging is not possible from the shells.[2] In the Nelson/Marlborough area, strong year classes are usually only traceable for 3 or 4 years, and it seems probable that few scallops reach more than 4 or 5 years of age. Large scallops in excess of 160 mm are occasionally taken, but these are probably the result of especially rapid growth rather than extreme old age.

In many areas harvesting by amateur and commercial fishermen is the main cause of mortality. Other important predators include fish, octopus, and starfish.

II. WILD FISHERY

A commercial fishery for scallops, based mainly in the Nelson/Marlborough area at the top of the South Island, has been in existence for more than 25 years. This fishery reached

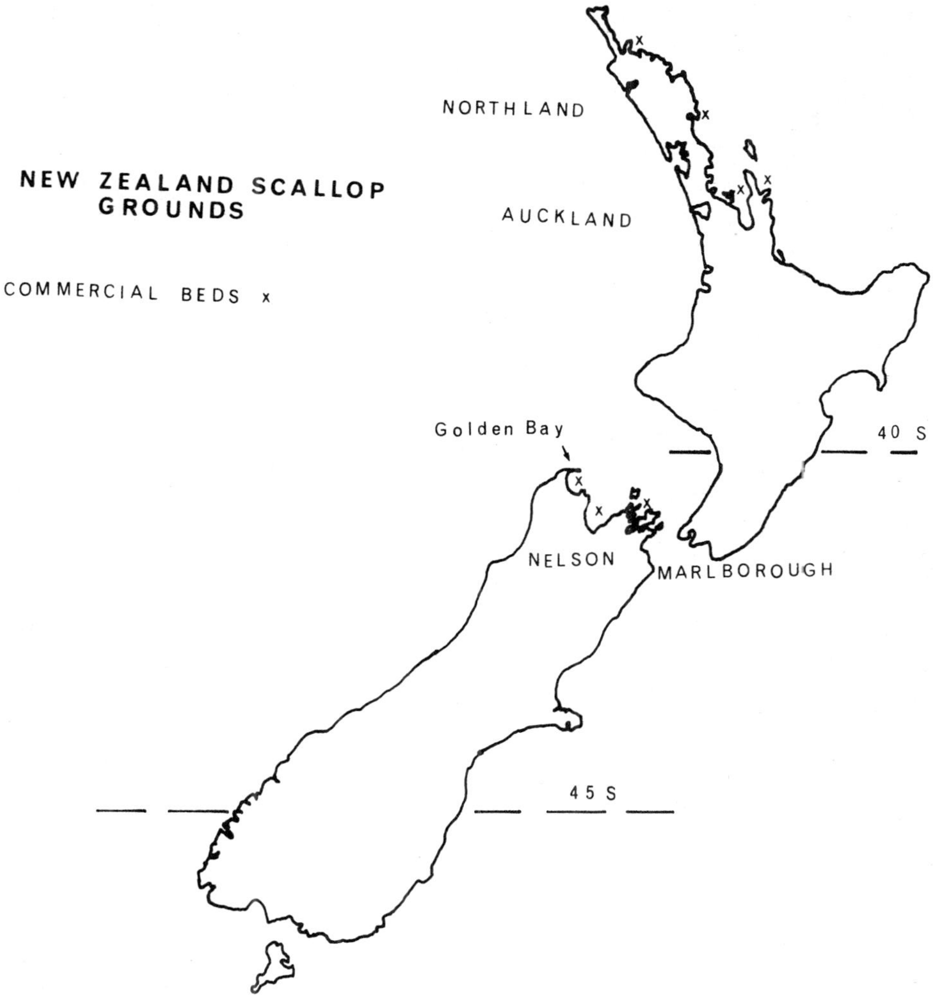

FIGURE 1. Location of commercial scallop grounds in New Zealand.

its peak in 1975, when approximately 10,000 t, live weight, was taken by a fleet of 190 dredge boats.[5] This level of effort was not sustainable, however, and although grounds along the east coast of northern New Zealand have since been developed, the national annual catch is now only 3000 to 4000 t and is taken by a fleet of about 80 vessels. Approximately 70% of the catch comes from the Nelson/Marlborough beds.

The entire commercial catch is taken by dredging, mostly by vessels in the 10- to 15-m range. On areas of soft substrate, such as occur in the Nelson/Marlborough fishery, ring-bag dredges with chain tickler are used, whereas box dredges, with rigid tooth bar, are more common on the North Island grounds, where hard sand is the predominant substrate.

Following the decline in catches in the late 1970s, a controlled licensing system was introduced, and there is currently only very limited provision for access of new entrants into the fishery. With this limitation, daily and seasonal quotas, and controls on type of gear (2 × 2.5-m dredges) and fishing period, fishing effort is now tightly regulated.

There is also a size limit of 100 mm, which is aimed at protecting a portion of the spawning stock and giving some stability to the fishery by spreading the harvest of strong year classes over more than 1 year. A closed season extending from February to July prevents harvesting of scallops during that part of the year when meat yields are usually lowest, and

FIGURE 2. The shell of a scallop recovered from the seabed 20 months after release from a spat-collector bag. The size of the scallop at release (14 mm) is clearly shown by the prominent disturbance ring.

in recent years commercial dredging on the Nelson/Marlborough beds has been further limited to avoid disturbance of the seabed environment during the peak December to January spat settlement period.

III. PROCESSING AND MARKETING

Scallops are generally landed live in either 25-kg sacks or 35-kg cases and taken to onshore processing facilities, where they are hand shucked. Only the adductor muscle and gonad are retained, and the shell and viscera are usually dumped. Meat weight recoveries are generally 12 to 20% of total weight; individual meats weigh 10 to 25 g, of which the gonad may account for 30% or more. After shucking, the meats are usually frozen in layer packs or free-flow form, although some are further processed to a crumbed or battered form.

The local market currently consumes approximately half the catch. The remainder goes to a variety of markets, including Australia, the U.S., and France. The quantity and destination of exports is very dependent on availability and demand, and at the time of the collapse of the resource in the early 1980s, New Zealand became a net importer of scallops from Australia.

IV. FARMING TRIALS

Since the late 1970s, there has been considerable local interest in farming scallops with Japanese hanging-culture techniques, and a number of small-scale trials have been carried out, mainly around the Nelson/Marlborough area. Japanese-type scallop spat collectors consisting of fine mesh bags containing either used monofilament gill net or a special polyethylene mesh have been successfully used and catches of up to 5000 spat per 5-l bag made. However, attempts at growing scallops using Japanese pearl nets and lantern-cage systems have proved discouraging, mainly because of problems with fouling of mesh cages, especially by mussels, *Perna canaliculus* (Gmelin). Because of these fouling problems and relatively high costs of equipment and labor in New Zealand, it has been concluded that scallop-farming operations based on Japanese hanging-culture techniques are unlikely to be economical there.[6] This situation is unlikely to change unless an efficient and low-cost method of attaching juvenile scallops to ropes or tapes can be developed.

In addition to the hanging-culture, trials of the feasibility of enhancing scallop production by seeding spat caught in collector bags onto the seabed were begun in 1982. In 1983, the New Zealand Ministry of Agriculture and Fisheries in conjunction with the Overseas Fisheries Cooperation Foundation of Japan expanded this work and embarked on a 3-year pilot-scale trial in Golden Bay in the Nelson province. These trials were aimed at assessing the suitability of seeding methodology developed in Japan, the economics of such an operation, and the modifications that would be necessary to suit local conditions. Spat were collected in Japanese spat-collector bags hung from subsurface long-lines. The catch in the 44,000 collectors set in the first year resulted in a release of approximately 35 million spat 3 to 4 months after settlement. An operation of similar scale was carried out in the second and third years.

Follow-up work on the seeding trials indicated that, provided the newly released spat were not disturbed by activities of trawlers in the area, rates of survival to harvest could be in excess of 20% and were comparable to those achieved with spat that had been grown for a further 3 to 6 months in pearl nets before release. It also became apparent that survival of spat that settled on the outside of the collector bags and eventually released themselves naturally onto the seabed was quite high and that a simplified enhancement operation consisting of provision of suitable settlement material followed by natural release might be possible. Further work has shown that this method followed by dredging and redistribution of spat to suitable growing areas is likely to be at least as attractive as the method of direct release from spat bags, and both methods are currently being proposed for development of a commercial operation.

V. FUTURE DEVELOPMENTS

Provided the level of success with spat collecting and survival of released scallops achieved during the recent trials can be continued, prospects for future expansion of production of scallops by enhancement operations seem extremely bright. The 1984 seeding trials produced a harvest of 170 t live weight in 1986, and scallops seeded in 1986 could yield up to 1000 t in 1988. In the past, allocation problems associated with alienation of large areas of seabed, which is traditionally a common-property resource, and interference with historical fishing rights have been major obstacles to this type of development. However, with the potential of such operations now clearly indicated by the trials, considerable effort is now being put into finding a solution. In the interim, it is planned that the trial operations in the Golden Bay area will now move to a full commercial phase under a joint arrangement between the Ministry of Agriculture and Fisheries and local fishing interests. If this type of operation is to spread to other suitable areas of the country in the future, identification of local spat-catching sites will probably be necessary. If such sites cannot be found, the possibility of transporting spat from the Golden Bay area or hatchery rearing of spat could be contemplated. However, both methods may well be uneconomical given the relatively low survival rates to be expected, and the hatchery-rearing option is some way off, as to date all attempts at rearing the larvae of *P. novaezelandiae* to settlement have failed.

REFERENCES

1. **Fleming, C. A.,** The Genus *Pecten* in New Zealand, New Zealand Geological Survey Paleontological Bulletin, 1957, 26.
2. **Bull, M. F.,** Aspects of the Biology of the New Zealand Scallop *Pecten novaezelandiae* Reeve 1853 in the Marlborough Sounds, Ph.D. thesis, Victoria University, Wellington, 1976.
3. **Bull, M. F.,** Scallop farming studies, in Proc. Aquaculture Conf., Fisheries Research Division Occasional Publication, No. 27, 1980.
4. **Nicholson, J.,** Feeding and Reproduction of the New Zealand Scallop, *Pecten novaezelandiae,* M.Sc. thesis, University of Auckland, 1978.
5. **King, M. R. and McKoy, J. L.,** Scallop Landings in the Nelson-Marlborough Dredge Fishery, 1959—80, Fisheries Research Division Occasional Publication, Data Series No. 14, 1984.
6. **Cameron, M. L.,** A preliminary investigation of the economic viability of scallop farming in New Zealand, Economic Division discussion paper 15/83, Ministry of Agriculture and Fisheries,1983.

Chapter 26

PERNA CANALICULUS (GMELIN) IN NEW ZEALAND

Robert W. Hickman

TABLE OF CONTENTS

I. BIOLOGY

Perna canaliculus (Gmelin) (family Mytilidae; common names: green-lipped mussel, green mussel, *kutai*), which is indigenous to New Zealand, is the largest of the twelve species of mytilids in the country. It can grow to a length in excess of 230 mm but is more typically 100 to 170 mm long. It can be distinguished from the largest of the other common mussels, *Mytilus edulis aoteanus* (Powell), by the distinct bright green rim around the interior of the shell and by the presence of two posterior mussel scars; *Mytilus* has only one. These are the only two mussel species of economic importance, and both are widely distributed throughout the country. *Perna* is more abundant in central and northern New Zealand and is found predominantly subtidally or in the low intertidal. Typically, *Perna* occurs in crowded patches attached to any firm substrate, frequently mingling with the holdfasts of large red and brown algae. The patches generally comprise mussels of similar size, although the jumble of byssus threads attaching the mussels to the substrate and to each other may contain numerous small juveniles, which, when recently settled, have shells with distinct red zigzag markings. The mussels themselves often provide the substrate for a variety of epifaunal organisms, which may include barnacles, tubeworms, tunicates, hydroids, and even the large brown kelps.

Beds of green-lipped mussels are also found in deeper water (to 30 m or more) on sandy or muddy bottom where initial attachment may have been to gravel or shell fragments and subsequently to other mussels. These beds have provided the basis for dredge fisheries in several parts of the country.

Mussels have long been a significant item in the diet of New Zealanders, as shown by the quantities of shells in Maori middens and the strong demand for mussels from the dredge fisheries. However, it was not until the potential of the New Zealand species for aquaculture was realized in the 1960s that their basic biology began to be studied. Much of the biological data concerning *P. canaliculus* consequently relates to its growth and physiology in suspended cultivation. The green-lipped mussel can become mature in its first year. The sexes are separate; ripe males are creamy white in color, ripe females reddish apricot. Spawning can occur throughout the year. *Perna* larvae have been recorded in the plankton[1] and spat settlements observed on natural[2] and artificial substrates[3-5] during most months. There are, however, two spawning peaks, one in spring to early summer and the other in autumn when water temperatures are about 18°C.[6] The spring peak is heavier and more consistent in the warmer northern parts of the country,[7] and the autumn peak more prominent in central areas.[4,8] The planktonic larval life was thought to be of 3- to 6-week duration,[9] but hatchery-reared larvae have reached settlement size (250 to 300 μm) in 14 to 21 d at 18°C.[10]

Natural settlement of *Perna* spat frequently occurs on the epifaunal hydroids and algae or the byssus threads of adult mussels,[2] but dense settlements of seed mussels have also been recorded on beach-stranded seaweed[11] and on various natural and synthetic materials offering a finely divided branching network of niches for initial attachment.[6] On commercial spat collectors, *Perna* settles predominantly below 2 m, down to at least 20 m. The juveniles are highly mobile,[12] and postsettlement behavior together with predation from starfish, crabs, and fish can radically alter initial settlement patterns.

The growth and allometry of *P. canaliculus* differ markedly between wild mussels grown intertidally (which reach 30 to 50 mm in length in 12 months, 50 to 70 mm in 24 months) and those grown in suspended cultivation (70 to 115 mm and 113 to 165 mm, after 1 and 2 years, respectively).[2,3,7,13,14]

Growth continues throughout the year but is correlated directly with water temperature (and food availability) and therefore shows seasonal variation. The longevity of the species in the wild is unknown because calculating age is not possible from the shells. Fish, starfish, and humans are the principal causes of mortality, although wave scour and freshwater floods may intermittently cause mass mortalities in intertidal populations.

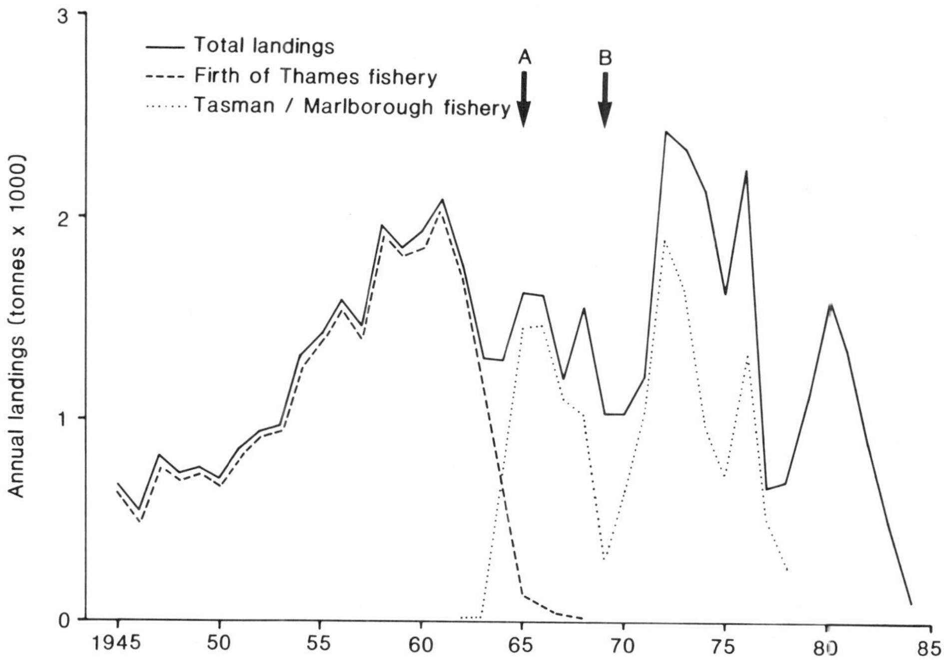

FIGURE 1. Annual landings of dredged green-lipped mussels *Perna canaliculus* in New Zealand. Arrows indicate the establishment of the first commercial mussel rafts in the Hauraki Gulf, North Island (A) and the Marlborough Sounds, South Island (B). (Redrawn from Hickman, R. W., *N. Z. Vet. J.*, 28, 226, 1980.)

II. WILD FISHERY

Until the early 1960s the demand for mussels in New Zealand was met by the dredge fishery for green-lipped mussels and a small amount of hand-picking of blue mussels.

The Firth of Thames (North Island) dredge fishery showed a steadily increasing yield over 40 years to peak landings in 1961 of just over 2000 t. Up to five boats operated in the Firth, towing 2- to 3-m-wide heavy dredges. There were no attempts to manage the fishery. No minimum size limit was imposed, and the dredging process was, by its very nature, destructive of the beds, removing at the same time the adult stock and the potential settlement surfaces for the spat. The rapid and complete collapse of this fishery in the mid-1960s was compensated for by the opening up of new mussel beds in the South Island,[15] and annual production was maintained at around 1500 to 2500 t until the late 1970s, when farmed mussels started to become available (Figure 1).

III. FARMING

A. HISTORY

The collapse of the dredge fishery provided the impetus for the development of mussel farming in New Zealand, and the first commercial rafts were established in the late 1960s (Figure 1). Farming was initially styled on Spanish raft cultivation techniques. However, the already established local market for large (>100 mm) mussels, combined with medicinal properties supposedly unique to *P. canaliculus,*[16] focused all farming development on the green-lipped mussel, as opposed to the blue mussel, *M. edulis aoteanus*.

From its outset the industry was hampered by a variety of socioeconomic constraints mainly focusing on its acceptability and the ''alternative use potential'' of the coastal areas suited to mussel farming.[17] Potential for conflict was seen to exist with other water uses

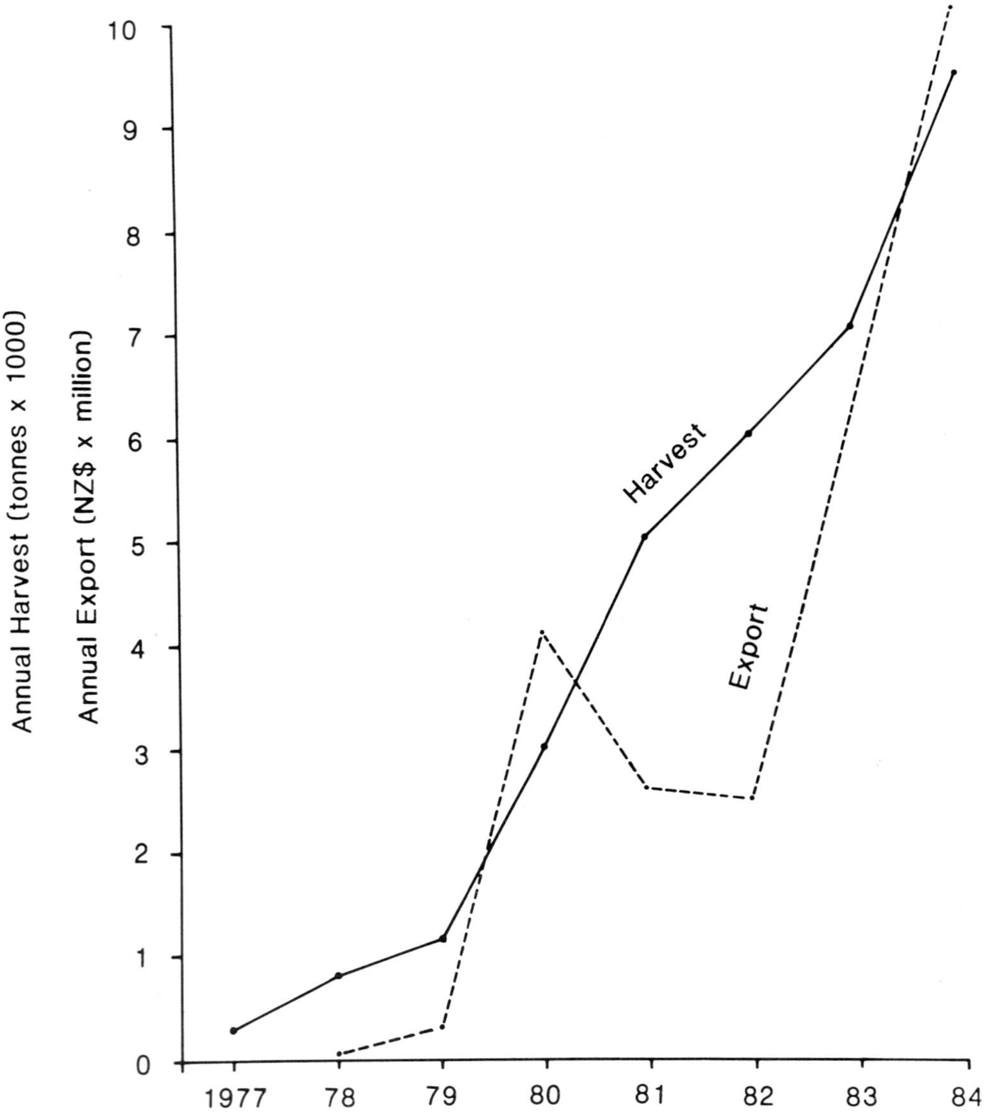

FIGURE 2. Annual harvest and export of farmed green-lipped mussel *Perna canaliculus* in New Zealand. (The 1980 export figure was boosted by high-value mussel powder.)

including fishing, navigation, transportation, recreation, and sewage disposal, as well as adjacent land uses such as agriculture, forestry, horticulture, industry, and housing. These conflicts, actual and imagined, supported by a widely-held belief in every New Zealander's inalienable right to free and unrestricted use of the coastal waters, necessitated extensive advocacy and planning before the industry was able to establish itself in the Marlborough Sounds[18] and to achieve its first significant crop of 300 t in 1977. The rapid expansion in production and particularly in exports from 1977 onward (Figure 2) was largely facilitated by three factors: an efficient long-line culture system suited to New Zealand conditions, a fast-growing mussel that produces a marketable crop within 18 months, and a major export marketing drive by individuals and groups within the industry.

B. CURRENT STATUS

Mussel farming in New Zealand is centered on the Marlborough Sounds at the northern

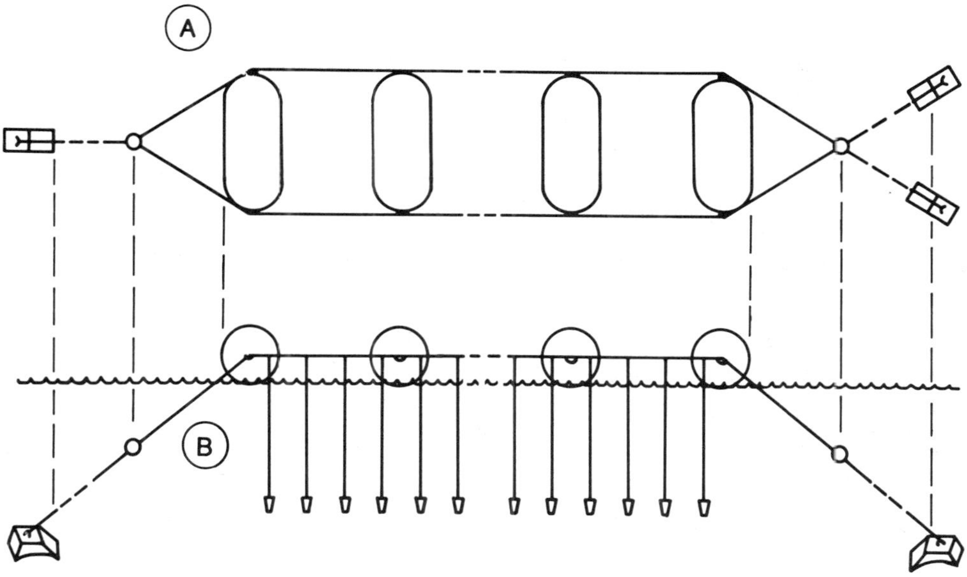

FIGURE 3. Diagrammatic representation of a mussel long-line. (A) Surface view, showing floats, paired horizontal lines, and two alternative mooring arrangements; (B) side view, showing weighted vertical growing ropes at 0.5-m spacing. The typical long-line is 110 m in length with 50 floats of up to 400-kg displacement each, 5-m-long growing ropes, and 2.5-ton anchors; the length of mooring line is 3 × the depth. (Redrawn from Hickman, R. W., *N. Z. Vet. J.*, 28, 226, 1980.)

tip of the South Island. There are presently 305 license holders (average license has 3 ha), and annual production is about 12,000 t, although not all production is currently harvested. Most licenses are held by individuals or family partnerships. Only recently has there been any major company involvement through the purchase of existing licenses. The majority of farms are owner-operated on a part-time basis, but with an increasing trend toward full-time farm managers, each operating several farms. Both casual and contract labor, particularly for harvesting, are significant elements in the operation of a farm. Up to ten contract harvesters provide full-time employment for seven or eight persons each, and the mussel processing sector directly employs over 100 persons.

Investment in farming equipment probably exceeds NZ $16 million with at least a further NZ $1 million each in harvesting and processing equipment. The first-sale price of mussels to the farmer is about $2.00 to $2.30/kg meat or $10 to $12/25-kg sack of live mussels, which puts the farmgate value of the current annual production at about NZ $6 million. In 1984, mussel exports returned NZ $10.5 million FOB to New Zealand, and the wholesale value of mussels on the domestic market was estimated at NZ $9 million.

C. METHODS OF FARMING

Mussel farming in New Zealand uses a simple, low-technology system suited to relatively sheltered nearshore situations. The basic unit is the long-line (Figure 3), a system of ropes and plastic floats that has a low capital cost, is easy to establish and maintain, and is generally considered to be more aesthetically acceptable than a large raft. Ten or more long-lines can be accommodated in the typical 3-ha (200 × 150 m) license area, the lines usually being oriented parallel to the shore. There is a minimum spacing of 50 m between farms and between a farm and the shoreline (Figure 4).

The industry relies entirely on natural supplies of spat for stocking the farms. Spat caught locally within the Marlborough Sounds provide about half the current supply, but a majority of the farmers obtain their spat from seed-mussel-encrusted seaweed (Figure 5), irregularly

FIGURE 4. Mussel farming in the Marlborough Sounds, New Zealand.

washed up on a beach in the north of the North Island.[11] *In situ* spat catching is concentrated on communal long-lines established at two sites in the sounds that have, over a number of years, been found to accumulate commercial quantities of spat seasonally. "Christmas tree" rope, a black polypropylene rope incorporating sacking offcuts to increase its surface area, is the preferred spat-catching medium. Bundles of 5-m lengths of the rope are hung on the communal long-lines, at 2- to 5-m depth during spring and autumn, and transferred to farms throughout the sounds some 6 to 8 weeks after a successful catch of spat.

Seed mussels from the seaweed source are hand-gathered from the beach, together with the weed, and rapidly air-freighted to the sounds, where they are seeded onto ropes. Cotton mesh tubing that slowly rots away is used to retain the seed mussels until they have attached to the rope.

Because it is not possible to control the density of settlement from either source of spat precisely, the mussel ropes are generally stripped and the juveniles (<20 mm) reseeded onto the definitive growing ropes at densities of 200 to 250/m. Ropes of juvenile mussels can be spaced more closely on the long-line and subsequently spread out to the optimum 0.5-m intervals as the mussels grow to maturity. The spacing and number of floats can also be adjusted to the differing crop weight throughout the growing cycle.

The green-lipped mussel grows to harvestable size (80- to 100-mm length) in 13 to 18 months, by which time each 5-m-long rope of mussels (at optimum density) weighs approximately 100 kg in air or 25 kg in the water. A well-managed farm will produce 25 to 40 ton/long-line per growing cycle. By utilizing spat from different sources and different seasons and by manipulating stocking densities, it is possible to have mussels of harvestable size and condition available throughout the year.

FIGURE 5. Piece of drifted seaweed from Ninety Mile Beach encrusted with several thousand seed mussels (0.5- to 5-mm length).

D. HARVESTING

Almost all harvesting is done by a few contracting companies using highly mechanized equipment capable of harvesting up to 20 to 30 ton/d. The expendable trace used to attach the culture ropes to the headline is cut, and the ropes are fed through a notch that strips off the mussels and allows them to drop into a revolving washer and declumper (Figure 6 A and B). Conveyor belts subsequently permit hand sorting, including removal of dead shell and unwanted *Mytilus,* before the mussels are packed into 25-kg sacks or 250-kg bulk bags or bins. The mussels are landed daily, mostly at the small town of Havelock, for processing in one of its two mussel factories or for road transport to other processors and distributors throughout the country.

E. PROCESSING

A proportion of the harvested mussels are sold live on the domestic market or chilled and air-freighted to the U.S., but the majority are processed by steam cooking, shucking, and debearding before being assigned to one of several product forms. Marinated meats are supplied extensively for the domestic market in plastic bottles (375 g net, 250 g meat), whereas mussels frozen on the half-shell are the major export form. Other lines include "instant quick freeze" meat; bulk frozen meat; smoked, crumbed, or battered mussels; and a small quantity of freeze-dried powder. The principal overseas markets are Japan, the U.S., and Australia. No use has yet been found for the large quantities of shell, which are presently dumped.

IV. CONSTRAINTS

No significant disease or parasite problems have affected the mussel farming industry of New Zealand Trematode (*Cercaria haswelli* Dollfus) and pea crab (*Pinnotheres novae-*

A

B

FIGURE 6. Mussel harvesting. (A) Section of the long-line is lifted and mussel-ropes detached; (B) ropes are fed into the stripping, declumping, and washing apparatus.

zelandiae Filhol) infestation is not uncommon in shore-grown mussels but rare in farmed mussels.[19] Suspended cultivation eliminates predation from bottom-dwelling starfish and crabs but may accentuate fish predation. Spotty (*Pseudolabrus celidotus* Bloch and Schneider), leatherjackets (*Patika scaber* Bloch and Schneider), and snapper (*Chrysophrys auratus* Bloch and Schneider) can cause serious stock losses, particularly of spat or reseeded juveniles. Management of the farm to reduce disturbance of the ropes (during spatting or reseeding) or to hold the young mussels closer to the surface, combined with frequent trapping of the fish, can minimize fish predation.

The industry is well aware that one of its main advantages in the world market is the pollution-free high quality of the growing waters. In the long term, increased tourism and boating, and possible logging of forests, may adversely affect water quality, but at present the major impact on water quality is from periodic freshwater floods. These affect mussel farms, in the inner sounds only, by restricting harvesting for a set period following the flood—the number of days' restriction depends on the quantity of rainfall over a 24-h period.

V. FUTURE DEVELOPMENTS

New Zealand has developed a mussel farming industry capable of producing high-quality mussels in quantities frequently exceeding market demand. Annual production from existing licenses, if all were fully developed, could reach 50,000 ton. To achieve this expansion several problems will have to be resolved.

A reliable supply of spat will be essential for an expanded industry. Location of alternative sources within the sounds and around the North Island, together with the improvements in catching and maintaining spat that will result from further research into settlement behavior, should permit continued reliance on natural spat supplies. Hatchery rearing of *Perna canaliculus* has proved difficult but might provide back-up to natural spatfall.

Improvements in the efficiency of the farming operation may result from increased mechanization, but optimization of production will require experimentation in farm layout, stocking densities, and prevention of predation, as well as further research on natural productivity, hydrology, and food availability in mussel-farming areas.

Maintainance of water-quality standards will be of increasing importance as the industry expands. The potential for diseases will increase, and the impact of mussel farming itself (particularly in combination with other forms of aquaculture, such as cage farming of salmon) on the water quality will need to be monitored.

Marketing is the area of most immediate concern to the industry. There is a dichotomy of opinion between those who see *P. canaliculus* as a premium-quality upscale market product and those who see it as a high-volume, low-price commodity. If a greater degree of integration and trust between the various industry sectors could be achieved and equitable pricing structures, quality standards, and marketing strategies firmly established, it should be possible for the mussel-farming industry to expand to its full potential.

REFERENCES

1. **Booth, J. D.,** Common bivalve larvae of New Zealand: Mytilacea, *N. Z. J. Mar. Freshwater Res.*, 11, 407, 1977.
2. **Luckens, P. A.,** Settlement and Succession on Rocky Shores at Auckland, North Island, New Zealand, No. 70, Mem. N. Z. Oceanogr. Inst., 1976, 64.
3. **Greenway, J. P. C.,** Settlement and Growth of a Colony of the Large Green Mussel *(Perna canaliculus,* Gmelin) from a Pontoon in the Kouma Harbour, Coromandel, Fisheries Tech. Rep., No. 43, N. Z. Marine Department, 1969, 14.

4. **Meredyth-Young, J. L. and Jenkins, R. J.,** Depth of settlement of two mussel species on suspended collectors in Marlborough Sounds, New Zealand (note), *N. Z. J. Mar. Freshwater Res.,* 12, 83, 1978.
5. **Johns, T. G. and Hickman, R. W.,** A Manual for Mussel Farming in Semi-Exposed Coastal Waters; with a Report on the Mussel Research at The Kaha, Eastern Bay of Plenty, New Zealand, 1977—82, Fisheries Research Division Occasional Publication No. 50, 1985, 28.
6. **Tortell, P.,** Investigations into Settlement Behaviour of Mussels and the Hydrology of a Selected Area with a View to Farming the Shellfish, Ph.D. thesis, Victoria University, Wellington, 1976.
7. **Greenway, J. P. C.,** Development of a colony of green mussels, *Perna canaliculus* in Coromandel Harbour, 1971—72, Fisheries Tech. Rep. No. 141, N. Z. Ministry of Agriculture and Fisheries, 1975, 22.
8. **Hickman, R. W. and Illingworth, J.,** Condition cycle of the green-lipped mussel *Perna canaliculus* in New Zealand, *Mar. Biol.,* 60, 27, 1980.
9. **Jenkins, R. J.,** Mussel Cultivation in the Marlborough Sounds (New Zealand), N. Z. Fishing Industry Board, Wellington, 1979, 75.
10. **Redfearn, P., Chanley, P., and Chanley, M.,** Larval shell development of four species of New Zealand mussels (Bivalvia, Mytilacea), *N. Z. J. Mar. Freshwater Res.,* 20, 157, 1986.
11. **Hickman, R. W.,** Potential for the use of stranded seed mussels in mussel farming, *Aquaculture,* 9, 287, 1976.
12. **Kennedy, V. S.,** Crawling and clustering behaviour of three species of intertidal mussels in New Zealand, *N. Z. J. Mar. Freshwater Res.,* 18, 417, 1984.
13. **Hickman, R. W..,** Allometry and growth of the green-lipped mussel *Perna canaliculus* in New Zealand, *Mar. Biol.,* 51, 311, 1979.
14. **Flaws, D. E.,** Aspects of the Biology of Mussels in the Cook Strait Area, Ph.D. thesis, Victoria University, Wellington, 1975.
15. **Hickman, R. W.,** Management problems in mussel farming, *N. Z. Vet. J.,* 28, 226, 1980.
16. **Croft, J. E.,** *Relief from Arthritis: A Safe and Effective Treatment from the Ocean,* Thorsons Publ. Ltd., Wellingborough, Northamptonshire, 1979, 128.
17. **Hickman, R. W.,** Marine Farming and Coastal Water Use Conflicts, in Proc. 11th N. Z. Geography Conf., 1981, 159.
18. Anon., Declaring areas in the Marlborough Sounds not available for marine farming licenses or leases (Notice No. 2211, Ag. 54/7/1), *N. Z. Gazette,* August 16, 1979, No. 76, 2435, 1979.
19. **Hickman, R. W.,** Incidence of a pea crab and a trematode in cultivated and natural green-lipped mussels, *N. Z. J. Mar. Freshwater Res.,* 12, 211, 1978.

Chapter 27

THE NORTHERN ROCK OYSTER, *SACCOSTREA GLOMERATA* (GOULD, 1850), IN NEW ZEALAND

P. Dinamani

TABLE OF CONTENTS

I. BIOLOGY

A. MORPHOLOGY

This species produces small- to medium-size oysters, up to 75 to 100 mm (Figure 1). The shell outline is slightly elongate and crescentic, and the shell is dense and solid, subequivalve with a large attachment area. The left valve is deeply cupped and has a pronounced umbonal cavity. The right is opercular, slightly domed, and bears lamellar squamae near the shell margin. The commissural shelf is narrow and the chomata well developed, usually along the dorsal third of the valves on either side of the hinge. The hinge plate is fairly broad; the external ligament is blue black to blackish brown and the internal ligament light brown. The muscle scar is reniform, placed close to the posteroventral margin of the valves. The gills are accommodated in a recess of the shell marked by a line of indentations made by the mantle retractor muscles. The tough periostracum covers the right valve when formed but is eroded away unevenly, resulting in pittings and uneven layering on the shell. The shell margin is brownish purple or violet and the interior of the valves solid white, with occasional brownish blotches of conchiolin in the right valve. The mantle edges are usually light gray with flecks of yellow, particularly over the bases of the inner row of papillae. The rectum bears no anal collar or flange.

B. RANGE

*Saccostrea glomerata** ranges along both coasts of the peninsular part of the North Island, north of 38°, and extends to the coast of the Coromandel peninsula and tapers out along the Bay of Plenty up to East Cape. In the days before the advent of the Pacific oyster, the rock oyster formed very thick beds, 10 to 12 cm deep and about 1 m wide, in the intertidal zone along the coast, around the numerous islands off the coast, and among the mangroves. Populations have declined since the 1970s and occur only sparsely now, in isolated pockets in Kaipara Harbor, the Bay of Islands, and the Coromandel. The oysters prefer rocky substrates and can tolerate exposures of up to 60%.

C. SALINITY AND TEMPERATURE

The rock oyster is essentially subtropical in character and can withstand temperature maxima of 25°C in the summer, and winter minima of 12°C. It prefers more saline areas but ranges from open coast (salinity 35‰) to estuarine conditions (20 to 33‰) and has been found in the upper reaches of estuaries and rivers (12 to 16‰).[1] The larval distribution and spat settlement follow the adult distribution. Larvae and spat are found in large numbers toward the mouth of the estuaries at salinities of 33‰ at periods of high tide[1] and probably follow the salt wedge up the estuaries.

D. ASSOCIATED FAUNA

The small barnacle *Elminius modestus* Darwin settles densely on the oysters in the top third of the intertidal zone, and the black mussel, *Xenostrobus pulex* (Lamarck), often forms a dense cover in the spring. In the lower zone, the blue mussel, *Mytilus edulis aoteanus* Powell, the serpulid *Pomatoceros cariniferus,* and the gastropod *Littorina unifasciata antipodum* Philippi occur. The oyster borer *Lepsiella scobina* (Quoy and Gaimard) is a very conspicuous predator on both barnacles and oysters.

E. SEXUAL DEVELOPMENT

Gametogenesis begins in July or August, and further development and maturation follow in the spring.[2] Most oysters are fully ripe by about the end of November, and the first spawning usually takes place in December, though some trickle spawning may occur earlier.

* Editor's note: Some authorities think this species is conspecific with the Sydney rock oyster, *S. commercialis.*

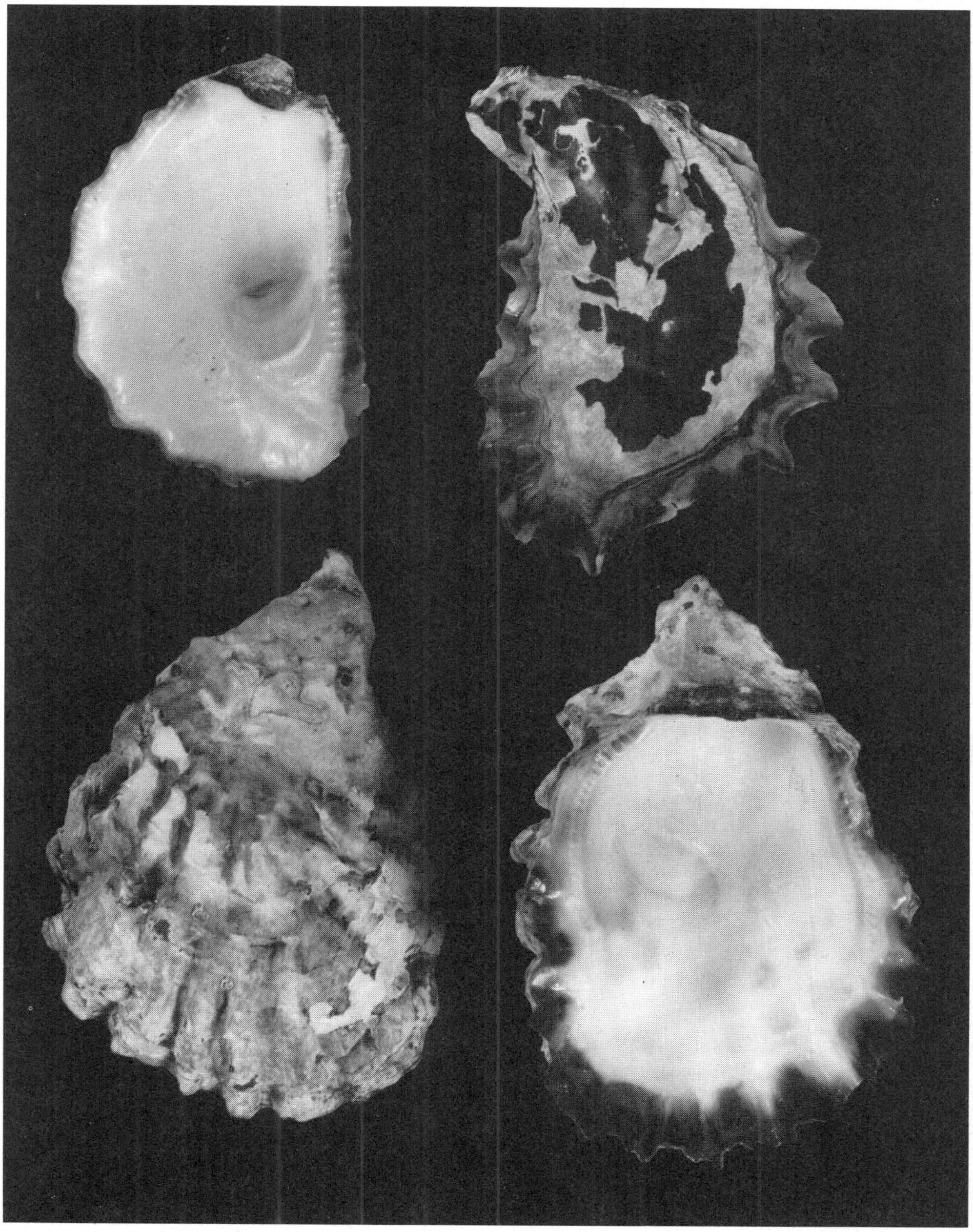

FIGURE 1. Views of the shell of *Saccostrea glomerata*.

Major spawnings take place in January and February, depending on water temperature. Spawning may continue right up to the end of March or the beginning of April. After spawning, the gonads go through a resorptive phase, when leukocyte infiltration and phagocytosis take place. The gonad passes through a distinct resting phase between May and July.[2]

Gonial cells proliferate in 2- to 3-month-old oysters, but sexual differentiation occurs when the oysters are about 6 months old. In young-of-the-year oysters, the percentage of males ranges from 60 to 90%, but 2- to 5-year-old oysters are predominantly (to 70%)

female.[2] The occurrence of an indeterminate phase during the resorptive cycle accounts for wide fluctuations of the sex ratio over this period. The reproductive cycle is closely related to the condition cycle,[3] and both are related to the seasonal water-temperature pattern—winter low of 14°C to summer high of 22 to 24°C.

F. LARVAL DEVELOPMENT

Larvae are planktonic for 18 to 24 d at temperatures around 20°C.[2] The pediveliger has a taller form than crassostreid larvae, measuring 320 to 330 × 350 to 360 μm.[4] Larvae prefer to settle on undersurfaces of substrates but avoid densely settled surfaces and move onto upper surfaces progressively.[5] The larval shell shows well-developed umbones on both valves, a strongly defined posterior sulcus,[6] and symmetrical hinge teeth. Under hatchery conditions, larval development is completed in 14 d at 27°C.[7]

G. GROWTH

Spat growth is influenced by position on shore in relation to tidal height and amount of exposure and shows great variation. Maximum growth occurs in the spring at mean temperatures of 17 to 20°C in northern harbors. Spat grow 2 to 3 mm/week soon after settlement, but the rate falls off after summer, and at the end of the first year, oysters are usually up to 25 mm in size and weigh 15 g. Moderate growth occurs during the second and third year, at the end of which oysters measure about 65 mm and weigh 60 g (meat wt. 10 g). The condition index shows peaks just before the summer breeding period.[3]

II. FISHERY

A. HISTORICAL

Exploitation of rock oyster beds under government control began in 1877 in the northern parts of the North Island, and licenses for ''picking'' natural beds were issued between May and September every year. Departmental records show that on average 20,000 to 24,000 bags (1 bag = 50 dozen oysters) were picked between 1914 and 1925, and production dropped to about 8000 bags in 1930. Production averaged 12,500 bags between 1940 and 1960. Some attempts were made to follow Australian methods of intertidal rack-type cultivation at the turn of the century, but preliminary results were recorded as disappointing.[8] However, between 1908 and 1960, considerable effort was made by the government not only to provide additional settlement surfaces (rock boulders, stakes, renewal of old cultch surfaces), to thin out overcrowded beds, and to transplant the culls to new areas, but also to provide for the manual removal of the oyster borer and weeds from the beds.

In 1962, as a result of the findings of a select committee investigating the fishing industry in the country and a recommendation that rock oyster farming be pursued in the Auckland province,[4] licenses were issues in 1963 for experimental culture in the Kaipara Harbor. Then in 1964 the Rock Oyster Farming Act was passed, specifically for the purpose of establishing farming on leased areas of the foreshore. Leases were granted for periods of 14 years with a preferred right of renewal. Each lease was divided into lots of 3 ha, and the lessee paid an annual rent. In 1967, the government established two demonstration farms, and in the following year three more farms and a field area for investigation of spatfall were set up.

The Marine Farming Act of 1971 superseded the Rock Oyster Farming Act of 1964, being broader in scope to provide for the farming of any marine species under leasing or licensing. In 1975, the responsibility for marine farming passed to the Ministry of Agriculture and Fisheries, and by then 121 oyster leases covering an area of 650 ha had been established.

B. OYSTER INDUSTRY

Rock oyster production reached more than 500 t in the middle 1970s, over half of which was exported at a value close to NZ $0.5 million. However, during this period the rock

FIGURE 2. "Rock boulder" cultivation of oysters in the Kaipara Harbor.

oyster began to be overrun by the Pacific oyster, and by 1978 it had been virtually eliminated in the farms. Most of the culture methods described for the Pacific oyster were, in fact, developed for the rock oyster.[9] In earlier years, rock cultivation was the most successful mode, particularly for west coast harbors, where large boulders were placed on the foreshore (Figure 2) of coastal inlets. When proper farming began about 1967, intertidal racks of timber were built, and oyster cultivation on sticks and trays (Figure 3) was adopted everywhere. Controlled harvesting of natural beds by farmers, on the basis of permits, was allowed occasionally, but the oysters had to be grown or conditioned in trays in the farm for 3 months before sale.

Rock oysters were exported live in the shell after cleaning and grading and had a long shelf life of about 10 to 14 d. Domestic markets were also supplied with whole oysters.

C. CONSTRAINTS

In natural beds the most serious predator is *Lepsiella scobina,* a small muriid borer that attacks spat and adults, especially in areas of higher salinity, but is less common in estuarine conditions. A few other carnivorous gastropods (*Haustrum haustorium* [Gmelin], *Cominella maculosa* [Martyn], *Morula [Neothais] smithi* [Brazier]) that live near the oyster zone occasionally prey on the rock oyster. *Polydora websteri* Hartman attacks the oyster shell at lower levels but does not invade farm oysters growing at mean tidal level. Fouling by the bryozoan *Watersipora cucullata* (Busk) occurs on oysters growing in shaded and overcrowded situations. Another seasonal fouling organism is the black mussel, *Xenostrobus pulex* Lamarck, which settles densely on oyster sticks in the spring. The copepod *Pseudomyicola spinosus* Raffele and Monticelli is the most persistent associated organism within the mantle cavity of the oyster, and the copepod has been found to attack the gut and

FIGURE 3. Intertidal rack cultivation of rock oysters in one of the northern harbors.

complete its life cycle inside the oyster.[10] There have been records of what is known as "winter mortality"[9] in some areas, where oysters have shown symptoms of ulceration of the viscera, particularly of gonads, palps, and inner surfaces of mantle. The oysters die in some localities after a dry winter, but the problem can be avoided if oysters are moved upstream to higher levels before the winter season.

D. RESEARCH AND DEVELOPMENT

First came the development of the technology of intertidal farming and evolution of practical methods for growing, maintaining, harvesting, and spat collection.[9] Scientific studies on growth, reproduction, larval development, settlement, and spatfall[2-5,11] were undertaken at the same time. It became apparent that natural spatfall was quite irregular and that cultching material had to be placed at the right level in relation to the tide to avoid competitors, particularly a noncommercial species of oyster[12] that confused the early farming efforts. Farming suffered a setback again in the 1970s[13] when competition from Pacific oysters introduced an element of uncertainty: it was not clear in 1974 to 1976 which species of oyster was a better candidate for farming or whether it was possible to grow the native species in preference to *C. gigas*. Poor and uncertain spatfall added to the downturn in farming, so hatchery methods were investigated. It was demonstrated[7] that suitable methods of larval rearing and settlement on selected substrates were commercially feasible.

Heavy spatfalls of *C. gigas* between 1977 and 1982 in the Mahurangi, and the demonstrated advantage of growing a market-size oyster in half the time taken by *S. glomerata*, led to the end of rock oyster farming by 1979 and the dominance of the Pacific oyster.[14]

Interest in looking at hybridization of the two commercial species arose when farmers began to report occurrences of odd-looking oysters in the farm, which they saw as proof that hybridization was occurring in the wild. Preliminary experiments designed to verify whether it was possible to cross the two species and whether they would produce viable hybrids were carried out in 1978. These indicated that fertilization and development to the D stage was possible every time. So a detailed series was carried out in 1982, when the larvae of *S. glomerata* × *C. gigas* were taken through full development and settlement.[14] However, high spat mortality was noticed, and it was not possible to substantiate the results through genetic analysis, but further work is planned.

REFERENCES

1. **Dinamani, P.,** unpublished data, 1974-1977.
2. **Dinamani, P.,** Reproductive cycle and gonadal changes in the New Zealand rock oyster, *Crassostrea glomerata, N. Z. J. Mar. Freshwater Res.,* 8, 39, 1974.
3. **Greenway, J. P. C.,** Seasonal condition of rock oysters, *Crassostrea glomerata,* in North Auckland, *Fish. Tech. Rep. Minist. Agric. Fish.,* 142, 11, 1975.
4. **Dinamani, P.,** Embryonic and larval development in the New Zealand rock oyster, *Crassostrea glomerata* (Gould, 1850,) *Veliger,* 13, 295, 1973.
5. **Dinamani, P.,** Some aspects of the spatfall of the New Zealand Rock oyster during 1974, *Veliger,* 20, 17, 1977.
6. **Chanley, P. and Dinamani, P.,** Comparative descriptions of some oyster larvae from New Zealand and Chile, and a description of a new genus of oyster, *Tiostrea, N. Z. J. Mar. Freshwater Res.,* 14, 103, 1980.
7. **Curtin, L.,** Oyster hatchery pilot scheme, *Fish. Tech. Rep. Minist. Agric. Fish.,* 155, 16, 1979.
8. Marine Department, New Zealand, Annual Reports, 1908-1960.
9. **Dinamani, P. and Gordon, D. B.,** On the habits and nature of the association of the copepod, *Pseudomyicola spinosus,* with the rock oyster, *Crassostrea glomerata* in New Zealand, *J. Invertebr. Pathol.,* 24, 305, 1974.
10. **Curtin, L.,** Oyster farming in New Zealand, *Fish. Tech. Rep. N. Z. Mar. Dep.,* 72, 99, 1971.
11. **Rainer, S. F.,** The Ecology of the Auckland Rock Oyster, *Crassostrea glomerata,* M.Sc. thesis, University of Auckland, 1964.
12. **Dinamani, P.,** Identification of oyster species competing with rock oyster for settlement space, Fisheries Research Division, New Zealand, *Inf. Leaf.,* 1, 9, 1971.
13. **Dinamani, P.,** Pacific oyster may pose threat to rock oyster, *Catch '74 Minist. Agric. Fish. N. Z.,* 1, 5, 1974.
14. **Dinamani, P.,** Wary researchers watching alien's rapid rise, *Catch '78 Minist. Agric. Fish. N. Z.,* 5, 24, 1978.
15. **Dinamani, P.,** A world's first in oyster mating, *Catch '82, Minist. Agric. Fish. N. Z.,* 9, 11, 1982.

Chapter 28

THE PACIFIC OYSTER, *CRASSOSTREA GIGAS* (THUNBERG, 1793), IN NEW ZEALAND

P. Dinamani

TABLE OF CONTENTS

I. BIOLOGY

A. MORPHOLOGY

The Japanese or Pacific oyster, *Crassostrea gigas* (Thunberg), grows to large size, and specimens up to 300 mm (all measurements are umbo to valve margin) have been recorded (Figure 1). Growth squamae are conspicuous in the right ventricle, particularly in spat and young oysters. Color consists of shades of light purple and pink on a whitish background, occasionally with wide radial bands of deep purple. The mantle fold is usually edged with black, but occasionally with golden brown or light gray. The rectum ends slightly below the curve of the muscle, and the anus is flanged by a short collar.

B. RANGE

Crassostrea gigas is farmed in all the east coast and a few west coast harbors of the northern peninsula of the North Island and in Coromandel and Ohiwa Harbor to the east (Figure 2). It forms intertidal beds on the foreshore rocks and grows among the mangroves in all the above areas. Particularly dense wild populations are found in some west coast harbors, such as the Kaipara Harbor. Pacific oysters have been recorded at some sites at the top of the South Island.

C. TEMPERATURE AND SALINITY

Sea surface temperatures around the northern peninsula range from a mean of 14°C during the winter months to above 20°C in the summer,[1] though mean temperatures of 22 to 24°C have been recorded in the summer in coastal inlets and harbors.[2,3] Salinity of inshore waters, unmodified by coastal effects, shows value above 35.5‰,[4] and high salinity values are recorded within harbors in the summer.[1] Salinity falls to 20 to 22‰ in the spring and winter in some east coast harbors,[5] but is particularly noticeable in the Kaipara Harbor, where salinity values of 16 to 23‰ occur in the upper reaches of inlets.[6] As the oysters are normally distributed between mean-high-water neaps and mean-low-water springs, periods of exposure to low salinity, which coincide with low water and freshets in the inlets, are brief.

D. SUBSTRATE TYPES

The shoreline of inlets and coast where oysters occur intertidally is hard bottom, generally consisting of "greywacke" formations (a type of stratified sandstone) on the east coast and of typically hydraulic limestone made up of calcareous mud and sand on the west coast. Oysters also settle on bits of shell or grit on muddy bottoms, and these grow continually over the mudline into spatulate shapes. In farming areas, oysters are spatted on fiber-cement or wooden battens and grown to market size.

The most common epifaunal organisms in oyster beds are the barnacle *Elminius modestus* Darwin, three species of mytilids (the blue mussel, *Mytilus edulis aoteanus* Powell; the black mussel, *Xenostrobus pulex* Lamarck; and the brown mussel, *Modiolus areolatus* [Gould]), and two species of polychaetes (*Polydora websteri* Hartman and *Pomatoceros cariniferus*). Oysters suspended from long-lines are fouled by red sponges, tunicates (*Microcosmus australis* Herdman), and the large barnacle *Balanus decorus* Darwin. Farmed oysters on intertidal racks are fouled by barnacles and black mussels particularly in the spring months. In west coast harbors, fouling by *M. areolatus* is common, particularly under estuarine conditions and at lower salinities.

E. SEXUAL DEVELOPMENT

The establishment and spread of *C. gigas* in the north has been mainly due to its spectacular breeding success. The breeding period is extensive, and the gonad becomes

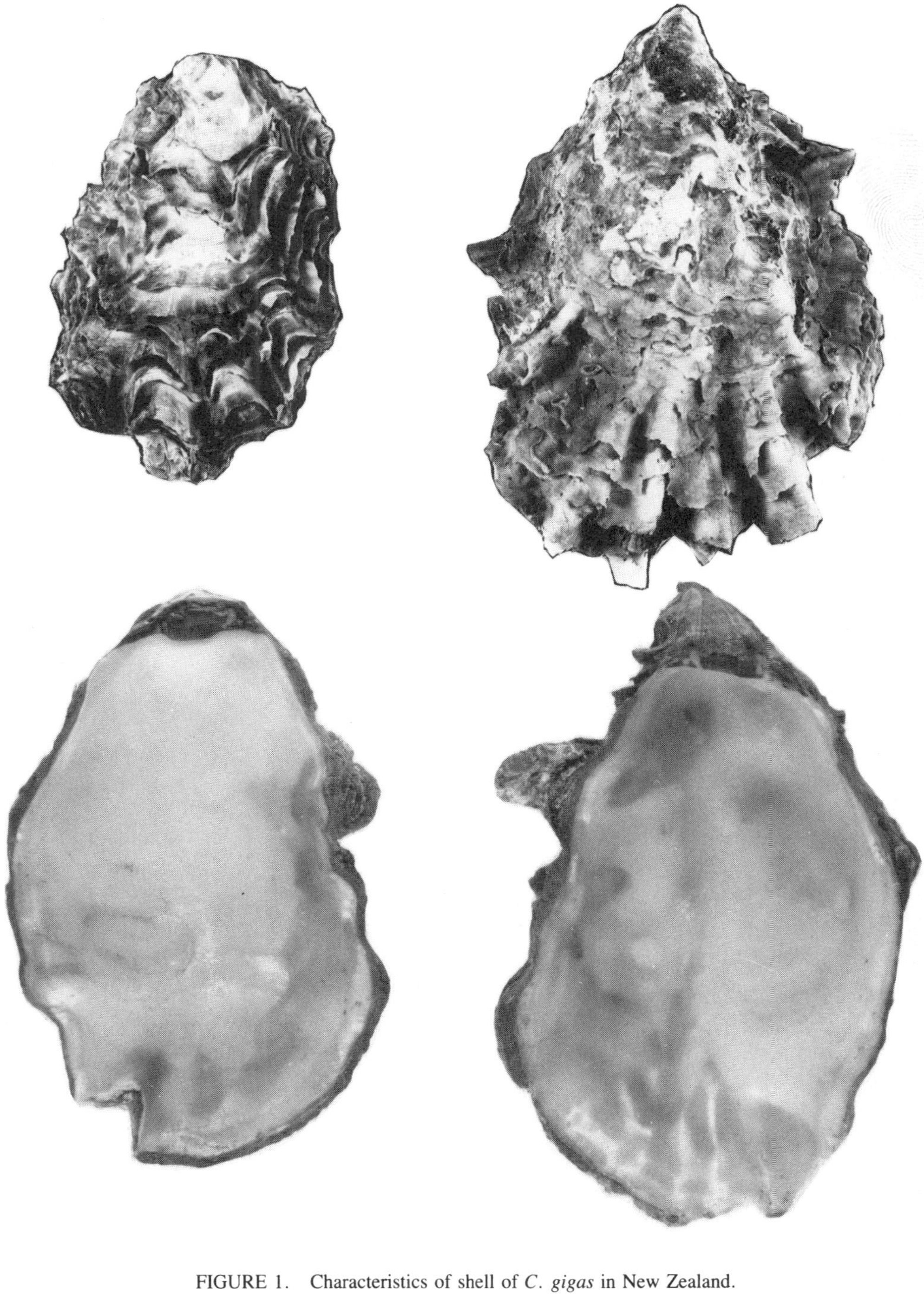

FIGURE 1. Characteristics of shell of *C. gigas* in New Zealand.

active in August, just before the southern spring. Follicles proliferate in 3-month-old (about 35 mm) spat, but sexual differentiation and enlargement of follicles takes place when the spat are about 6 months old (about 50 mm), though the gonad may have some undifferentiated gonial cells in a small percent of the oysters. Full ripeness has been observed in 9- to 10-month-old oysters, and at that age the larger percentage of the population consists of males (~60%). The sex ratio changes significantly after the first spawning, and the proportion of

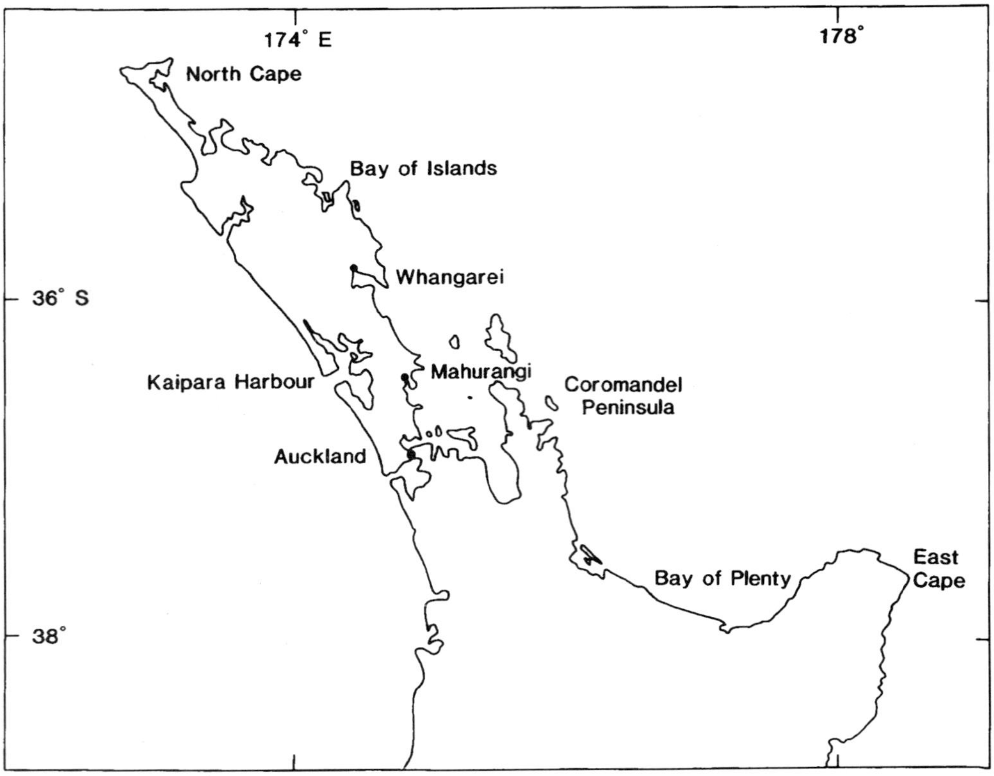

FIGURE 2. The northern part of the North Island of New Zealand, above 38° 30′S, to show locations of places mentioned in the text.

females increases appreciably in oysters more than 1 year old. This trend continues in older oysters, and 2- to 3-year-old oysters are predominantly female, though sex changes can occur immediately after spawning in this population.[7]

Some spawning activity has been observed in October and November, but the first main spawning takes place in December, when mean temperatures of around 20°C occur. The breeding period continues through summer and autumn, and several spawning may occur, but the major events coincide with the spring-summer temperature rise and the summer-autumn drop in temperature. A percentage of oysters appear to be active throughout the breeding period.[7]

F. LARVAE

The duration of larval life varies with the temperature regime at the time of development but is usually 21 to 24 d at a mean temperature of 20°C. At the lower temperatures occurring during March and April, larvae may take 30 d or more to complete their development, and late stages have been recorded in the plankton at the end of April. These larvae tend to be of a darker color and grow to over 320 μm. Under hatchery conditions, larval development takes 13 or 14 d at 26 to 27°C and 17 or 18 d at 22°C.[7]

G. GROWTH

Spat grow at different rates during the year, depending upon the time of settlement. Normally those that set in December on fiber-cement cultch grow to an average size of 35 to 40 mm by the end of April, but spat set in February or March reach only 10 to 15 mm by the end of the season. Growth slows down between May and July (southern winter) but speeds up toward September; at the end of the first year, early-set oysters have grown to 60 mm and weights of 50 g.[8]

Oysters are marketed when they are about 15 to 18 months old, at an average size of 70 mm and weight of 90 to 100 g. They remain in good condition through most of the year, except from February through April or May, and continue to grow in the second and third years. They reach over 120 mm and 200 g at the end of the second year.

II. FISHERY

A. HISTORY

The Pacific oyster was introduced accidentally, and though the actual time of introduction remains uncertain, the first accurate collection and identification was in 1971,[9] when six specimens were discovered in the Mahurangi Harbor on the east coast, among cultch put out by farmers for collecting rock oyster, *Saccostrea glomerata* (Gould), spat. These oysters measured 60 to 80 mm and were over a year old, obviously the result of a larval settlement in the harbor in the summer of 1969 to 1970. Unconfirmed reports in old Marine Department files mention large and odd-looking oysters, which may have been *C. gigas*, turning up in the Kaipara in the late 1950s. Now, vast areas of the foreshore in this harbor show a heavy settlement of Pacific oyster occurring side by side with the rock oyster; the two species appear morphologically very similar.

Since its accidental arrival in New Zealand, the Pacific oyster has made rapid gains, invading most coastal inlets and mangrove areas. In the farming areas, its progress has been equally spectacular,[8-10] and within about 7 years of the first sighting,[9] the alien species had completely taken over as the sole commercial oyster in the north.

B. OYSTER INDUSTRY

A small but efficient industry has developed in the north of the North Island and consists of a core of about 50 farmers who have developed about 350 ha of foreshore for mostly intertidal cultivation. There are approximately 35 establishment and ancillary units employing about 150 people in the industry, which produces (1984 figures) about 2000 t of oysters valued at about NZ $5 to $6 million. There are seven main processing units, most of which are certified for export to the U.S., though only a small fraction of the oysters go to that market. A few processors have depuration facilities to meet import requirements of certain countries. The majority of the oyster export is in the form of "oysters in the half shell", with the right valve removed. The cleaned and washed oysters are packed in layers in polystyrene boxes with ice and chilled. Oysters are alternatively packaged in wetlock boxes in layers and frozen. Some are marketed whole, alive, or frozen, or as potted meats. Harvesting starts when the crop is 15 to 18 months old and normally continues for 8 months of the year; most of the crop is sold before the oysters are 2 years old, though some markets prefer jumbo-sized oysters.

Harvesting is carried out between tides, and a flat-decked, flat-bottom punt is used to work between the racks. Fiber-cement battens are loaded into small cradles (Figure 3) and taken ashore, where the oysters are chiseled off, usually with a pneumatic tool or special culling chisel, and dropped onto a conveyor belt. The oysters enter a washer unit, which uses high-pressure water jets to remove mud and adherent organisms. The clumps of oysters are broken apart into singles by hand, graded, and passed on to the processing unit. All shucking (mostly involving the removal of the right valve) is done by hand by skilled openers. Smaller-sized oysters are packed into mesh bags or trays and grown on the farm to market size (Figure 4).

C. RECRUITMENT

Oyster farming is based wholly on the collection of wild spat. There are three main

FIGURE 3. Harvested oyster sticks loaded into cradles for transport to processing units.

spatting areas, two on the east coast and one on the west coast. In the east coast harbors, bundles of fiber-cement or wooden battens (collectively called "sticks") are set out on wooden racks (Figure 5) at about mean tidal level (about 30 cm above ELWN) just before the main breeding season. Each bundle is made of 30 sticks, arranged in 10 layers of 3 each, with 2 slivers of pine wood (12 mm thick) separating the layers. Plastic binding tape is used to hold the sticks together. Wooden sticks are also made into similar bundles of 30 to 40; all the bundles are placed in an interlocking pattern on supporting racks, which are built parallel with the tidal flow. The collectors are left on the racks for a period of about 4 months.

On the west coast, where sandstone strata form level platforms, bundles are placed directly on the rocky bottom or just off-bottom on rails or pallets at midtide level. In recent times, some interest has developed among farmers in producing single oysters, and collectors from which spat can be stripped easily have been tried. The most promising of these have been 50 × 1200-mm polyvinyl chloride (PVC) plastic strips, assembled together like fiber-cement bundles, and exposed for 4 weeks. Spat densities of over 5000 spat per strip were obtained in the 1984 to 1985 season. The spat, when they measure about 10 mm, are easily removed as singles when the strip is flexed. Interest has also been shown by the industry in finding an alternate cultch to the fiber-cement sticks which, because of their brittleness, are easily broken and cannot be reused. Successful trials have been carried out with 20-mm-diameter PVC tubes, imported from Canada and also manufactured locally, as cultch. Whether the tubes can be used to grow oysters to market size remains to be seen, because in some trials the oysters have been easily dislodged by wave action.

D. METHODS OF CULTURE

Growing methods were first developed for the intertidal culture of the native rock oyster, where Australian techniques of erecting timber racks at optimal level were followed. This method makes efficient use of coastal inlets and bays, which have gently sloping muddy

FIGURE 4. Pacific oysters being grown in plastic mesh bags, placed in intertidal racks. (Photo by L. Curtin.)

bottoms with a tidal range of about 3 m. The racks are built at least 450 mm off the bottom and extend offshore to where the depth of water is approximately 2 m at LWS. Racks and posts are built of marine treated timber; posts, 1.2 to 1.8 m tall, are spaced 1.5 m apart, and rough-sawn timber rails (75 × 25 mm) are fixed to the posts. Sets of racks are built 5 m apart, to allow working space for punt and equipment, and battens are placed across the rails at intervals to yield about 7000 sticks per hectare.

The spat bundles are opened up and the sticks set out individually on the growing racks, usually in May. Some farmers may keep the bundles intact on the farm until spring (a technique locally known as "depot") to give the spat sheltered conditions and to stagger their growth.

E. HARVESTING METHODS

Mechanization is evident only in the loading and offloading operations from the punt onto a landing or transporter. Most harvesting operations are carried out manually after the flat-deck punt is worked alongside the racks.

Some harvesting from natural beds is allowed under a system of permits, which requires the farmer to retain the oysters in his farm for a period of at least 3 months before sale and to set out a prescribed number of additional sticks for spat collection to make up for the

FIGURE 5. Fiber-cement battens, made up into bundles, for spat collection in Mahurangi Harbor.

natural harvest. Harvesting is carried out for about 8 months of the year and is tailored to meet market requirements; restrictions are placed on harvesting under export certification and shellfish sanitation criteria, which have to be strictly observed.

Some of the larger farms operate floating barges anchored in deeper water off the farm, where the oysters are stripped from the sticks, washed, and culled into bags.

F. PROCESSING

The type of processing is very much governed by preferences of the export market, because more than 50% of the production is exported. Regulations require that oysters be alive on arrival at the processing plant, and the farmer therefore ensures that fresh oysters are delivered immediately after harvest. Some oysters are exported alive in the shell, and these are kept in trays or mesh bags on the farm for at least 2 d before delivery to the packing house. The majority of the oysters are processed in the half-shell, chilled or frozen (see above).

G. CONSTRAINTS

The oyster farming industry has been fortunate that, from its inception in the early 1900s, there have been no mass mortalities, disease, or pollution problems; the country continues to enjoy a high reputation for quality oysters produced in fairly clean waters. However, oyster farmers intending to export their crop must meet export certification requirements. The U.S. Shellfish Sanitation Standards have been adopted, and growing areas have been classified as approved, conditionally approved, or nonapproved. Oysters from nonapproved areas have to be relaid in clean waters. In other areas, harvesting criteria have been laid for each locality and, in some cases, for individual farms, depending on the amount of rain in the previous 24 or 48 h and the amount of runoff from the land, periods when they are subject to closure. Regulations also extend to handling, transport, equipment, packing, storage, and labeling, all of which must meet export standards.

Among predators and organisms attacking the oyster, the major pest is the mudworm, *Polydora websteri* Hartman, which in certain years infests an appreciable percentage of oysters. Because most oysters are sold in the half-shell, the presence of mudworm trails and occasional staining of the meat renders the product unfit for sale. Of parasites and associate organisms, the copepod *Pseudomyicola spinosus* Raffele and Monticelli and the pea crab *Pinnotheres novaezealandiae* Filhol are the major species. The copepod usually occurs around the labial palps and the gill surfaces, and nauplii and copepodid stages are often seen within the mantle cavity and occasionally invade the gut. Histopathological surveys carried out as part of the disease-monitoring work have failed to show symptoms such as digestive diverticula epithelial atrophy, leukocyte infiltration, and disorganization of vesicular tissue. There has been no outbreak of disease in either farmed or wild stock.

H. MANAGEMENT

Farming operations are carried out in areas of the foreshore that are leased from the Crown and come under the Marine Farming Act 1971. Applications for leases are made to the corresponding district fisheries office, and after a detailed procedure during which the County Council, the Harbor Board, the Ministry of Transport, adjacent holders of land, and other users of the water are consulted, a lease for an initial term of 14 years is given to the farmer. This term may be renewed for one or more further terms. There are moves at present to revise the farming legislation so that there is more security of tenure and the farmer can make long-term plans and investments to utilize his farm fully.

In many suitable areas, farming interests vie with other user groups, particularly recreational interests, and usually farming gets lower priority. This situation has placed limits on further expansion, and any increase in production can only come with fuller and better utilization of existing farms.

Depuration is an import requirement for sale of oysters to New South Wales in Australia, and some processing units in the north are equipped to carry it out, using UV for disinfection, usually for 24 h.[11] There is no closed season as such, though between the end of December and March the oysters are usually partly or fully spawned out and are not salable, although the market continues to be buoyant for oysters and there is a good demand. In certain harbors, the beginning of the spring season marks the arrival of dense blooms of a ciliate, *Mesodinium rubrum,* which produces a condition known as "bleeding oysters" when the stomach and the digestive diverticula of the oysters are stained red and appear to bleed when cut. The ciliate contains chromatophores full of red pigment; the chromatophores break down in the stomach of the oyster, and the pigment is absorbed by the gut. Sometimes the bloom lasts for a week or 10 d before it is flushed out by the tides. There has been no known or recorded occurrence of any toxic dinoflagellates in the oyster-growing area.

I. RESEARCH AND DEVELOPMENT

Conditions for spawning adults and rearing of larvae were studied in 1977-1978 for all commercial species of oysters, and detailed descriptions of larval development and morphology were produced.[11,12] A pilot-scale hatchery was established by the fisheries in the north to serve as a demonstration unit for farmers to produce larvae or spat on a small commercial scale.[13] The government also carried out a few trials with "eyed larvae" and remote setting. At the same time, other natural spatting areas were explored, and new material was tested as cultch. When demand for growing single oysters was high, spat were collected on thin plastic strips in large quantities, then stripped from the collectors and grown in trays. This development, particularly in its effectiveness and economy compared with hatchery-based systems, resulted in loss of interest in "eyed larvae" and remote setting.

Attention was also directed toward investigating the causes of spatfall decline, and one of the aspects that was examined was whether the cause was genetic. In particular, it was feared a single small founder stock had given rise to the entire Pacific oyster stock in New Zealand and that inbreeding depression was leading to the decline. Tests carried out in New Zealand and Japan showed, however, that there was no evidence for lowered genetic variation in the local stock.[15]

REFERENCES

1. **Garner, D. M.,** The seasonal range of sea temperature on the New Zealand shelf, *N. Z. J. Mar. Freshwater Res.*, 3, 201, 1969.
2. **Skerman, T. M.,** Seasonal variations in sea water surface temperatures within New Zealand harbors, *N. Z. J. Geol. Geophys.*, 1, 197, 1958.
3. **Booth, J. D.,** Observations on the hydrology of Bay of Islands, New Zealand, *N. Z. J. Mar. Freshwater Res.*, 8, 671, 1974.
4. **Garner, D. M.,** Hydrology of New Zealand coastal waters 1955, *N. Z. J. Dept. Sci. Ind. Res. Bull.*, 138, 85, 1961.
5. Department of Health, personal communication, 1985.
6. **Rainer, S. F.,** The Ecology of the Auckland Rock Oyster, *Crassostrea glomerata,* M.Sc. thesis, University of Auckland, 1964.
7. **Dinamani, P.,** Gametogenic patterns in populations of Pacific oyster, *Crassostrea gigas* in Northland, New Zealand, *Aquaculture,* 64, 65, 1987.
8. **Dinamani, P.,** Pacific oyster may pose threat to rock oyster, *Catch '74, Minist. Agric. Fish. N. Z.*, 1, 5, 1974.
9. **Dinamani, P.,** Occurrence of the Japanese oyster, *Crassostrea gigas* (Thunberg) in Northland, New Zealand, *N. Z. J. Mar. Freshwater Res.*, 5, 352, 1971.
10. **Dinamani, P.,** Wary researchers watching alien's rapid rise, *Catch '78, Minist. Agric. Fish. N. Z.*, 5, 10, 1978.
11. **Buisson, D. H., Fletcher, G. C., and Begg, C. W.,** Bacterial depuration of the Pacific oyster (*Crassostrea gigas*) in New Zealand, *N. Z. J. Sci.*, 24, 253, 1981.
12. **Chanley, P. and Dinamani, P.,** Comparative descriptions of some oyster larvae from New Zealand and Chile, and a description of a new genus of oyster, *Tiostrea, N. Z. J. Mar. Freshwater Res.*, 14, 103, 1980.
13. **Dinamani, P.,** The morphology of the larval shell of *Saccostrea gomerata* (Gould, 1950) and a comparative study of the larval shell in the genus *Crassostrea,* Sacco, 1897 (Ostreidae), *J. Moll. Stud.*, 42, 95, 1976.
14. **Curtin, L.,** Oyster Hatchery Pilot Scheme, Fish. Tech. Rep., No. 155, Ministry of Agriculture and Fisheries, 1979, 16.
15. **Smith, P. J.,** personal communication, 1985.

INDEX

D

E

F

G

H

I

J

K

L

M

N

O

P

Q

R

S

T

U

V

W

XYZ